Lecture Notes in Chemistry

Volume 107

The series Lecture Notes in Chemistry (LNC), reports new developments in chemistry and molecular science - quickly and informally, but with a high quality and the explicit aim to summarize and communicate current knowledge for teaching and training purposes. Books published in this series are conceived as bridging material between advanced graduate textbooks and the forefront of research. They will serve the following purposes:

- provide an accessible introduction to the field to postgraduate students and nonspecialist researchers from related areas,
- provide a source of advanced teaching material for specialized seminars, courses and schools, and
- be readily accessible in print and online.

The series covers all established fields of chemistry such as analytical chemistry, organic chemistry, inorganic chemistry, physical chemistry including electrochemistry, theoretical and computational chemistry, industrial chemistry, and catalysis. It is also a particularly suitable forum for volumes addressing the interfaces of chemistry with other disciplines, such as biology, medicine, physics, engineering, materials science including polymer and nanoscience, or earth and environmental science.

Both authored and edited volumes will be considered for publication. Edited volumes should however consist of a very limited number of contributions only. Proceedings will not be considered for LNC.

The year 2010 marks the relaunch of LNC.

More information about this series at http://www.springer.com/series/632

Eva Perlt
Editor

Basis Sets in Computational Chemistry

Springer

Editor
Eva Perlt
Department of Chemistry
University of California
Irvine, CA, USA

ISSN 0342-4901 ISSN 2192-6603 (electronic)
Lecture Notes in Chemistry
ISBN 978-3-030-67261-4 ISBN 978-3-030-67262-1 (eBook)
https://doi.org/10.1007/978-3-030-67262-1

This Springer imprint is published by the registered company Springer Nature Switzerland AG
The registered company address is: Gewerbestrasse 11, 6330 Cham, Switzerland

Preface

The expansion of the electronic wavefunction in finite basis sets has been used ever since the beginning of quantum chemistry in 1929 (J. C. Slater *Phys. Rev.* **36**, 57 (1930)). The approach to use linear combinations of Slater, Gaussian, or plane wave functions depending on the specific problem has never changed. Despite the fundamentality and generality of this concept, basis sets are still being developed to this date and are a matter of active research.

Finding the right basis set can be compared to choosing the right solvent in synthesis. There is an ever-growing range of basis sets to choose from and while there is not a unique correct one, in most cases there are better and poorer choices. Which particular basis sets are appropriate depends on a number of factors such as the system investigated, the quantum chemical method used, the properties to be analyzed, and—despite advancing developments in computational hardware—the available computational resources.

Just as for the choice of the solvent in synthesis, the selection of a suitable basis set is usually based on the careful considerations of all these factors. To students who are just entering the field, however, this might at first appear to be based on intuition, experience, or even habit. The same applies to issues like basis set incompleteness or basis set superposition, which students correctly take into consideration but sometimes without being aware of their theoretical origin or their implications in the calculations.

With the incredible development of theoretical chemistry in the past decades, many advanced methods that go well beyond the Hartree–Fock theory have found their way into classrooms, and practical courses in computational chemistry have become a part of most chemistry studies. Today's graduates are equipped with a very good overview of computational methods at hand so that most experimental groups run DFT calculations on their own. At the same time, the knowledge about basis sets is passed rather sparingly, although they contribute just as much to the quality of a quantum chemical calculation—or can even render it useless if chosen badly.

The motivation to compile this contribution was to create a textbook which puts the basis sets first and addresses readers at different levels in their scientific career. All chapters are designed in such a way that they can be read and understood on their own, and knowledge from a previous chapter is not required to understand the contents of the current part. Each chapter is written in a didactic fashion, and illustrative examples from state-of-the-art research are given. Even though an overlap to some extent is unavoidable, each chapter covers a different aspect of basis sets in computational chemistry and they complement each other.

Addressing the target audience of this book, I would like to encourage and motivate students: Keep questioning, keep asking! I believe that the topic of basis sets demonstrates that research—no matter how old-fashioned and established it may appear—can hardly ever be considered complete. It is you who are going to take the next steps, make new discoveries, or modify existing theories.

Finally, I am supremely grateful to all authors who contributed to this volume. Each of them dedicated lots of time and effort to their chapter and I couldn't be more satisfied with the product all of you have created.

Irvine, USA
June 2020

Eva Perlt

Contents

Chapter 1
An Introduction and Overview of Basis Sets for Molecular and Solid-State Calculations

Jeppe Olsen

1.1 Introduction

One-electron wave functions for electrons, in the form of molecular orbitals for molecules and bands for solids, form the basis for all the generally applicable computational methods for describing the electronic structure of molecules and solids. For density functional methods (DFTs), the single-electron functions are used to obtain the density, which then determines the energy and all properties. For the other major group of electronic structure methods, wave function methods, the one-electron functions have a dual purpose, in that they are used to describe the dominating mean-field Hartree–Fock (HF) wave function as well as the correlation between the electrons.

The molecular orbitals are in general written as a linear expansion of a set of analytic functions, the basis set. Basis sets are in general developed and optimized for individual atoms and the basis set for a molecule comprises then the basis sets of the atoms constituting the molecule. The basis functions are therefore atomic orbitals, and the standard separation of atomic orbitals into a radial and a spherical harmonic part allows us to write a basis function as

$$\psi_{n\ell m A}(r_A, \theta_A, \phi_A) = f_n(r_A) Y_\ell^m(\theta_A, \phi_A), \tag{1.1}$$

where n, ℓ, and m are the standard atomic quantum numbers and A indicates a given atom. The coordinates (r_A, θ_A, ϕ_A) are the spherical coordinates in a coordinate system with nucleus A at the origin.

Inspired by the form of the bound state hydrogenic wave functions, the initially used radial basis functions were exponential functions of the form of Slater-Type Orbitals [1] (STOs),

J. Olsen (✉)
Department of Chemistry, Aarhus University, Langelandsgade 140, DK-8000 Aarhus C, Denmark
e-mail: jeppe@chem.au.dk

E. Perlt (ed.), *Basis Sets in Computational Chemistry*, Lecture Notes in Chemistry 107,
https://doi.org/10.1007/978-3-030-67262-1_1

$$f_n^{\rm STO}(r_A) = r_A^{n-1}\, e^{-\zeta r_A}. \tag{1.2}$$

The value of the screening constant ζ differs in general for the various basis functions, but this dependence has for notational convenience been suppressed in the above equation. Although such basis functions were used in the initial calculations of the electronic structure of diatomic molecules and still are much used for atoms, it was early realized that it was very difficult to extend the use of these basis functions to general molecules, in particular, it is very cumbersome to calculate the three- and four-center integrals needed to describe the expectation value of Coulomb repulsion between electrons.

It was early realized [2] that the calculations of integrals of quantum mechanical operators between basis functions were much simpler, when Gaussian- Type Orbitals (GTOs) are used instead of STOs,

$$f_\ell^{\rm GTO}(r_A) = r_A^{\ell}\, e^{-\alpha r_A^2}, \tag{1.3}$$

where the coefficient α again usually differs for the various basis functions. Notice that in addition to the use of a Gaussian form $e^{-\alpha r^2}$ rather than an exponential form, $e^{-\zeta r}$, the GTO uses the monomial r^ℓ rather than the general form of the STO, r^{n-1}. However, single Gaussian functions are very poor approximations to the atomic HF orbitals. To compensate for this deficiency, it was immediately realized that this deficiency of the GTOs could be greatly reduced for most practical purposes [3], if the basis function was a contracted GTO (CGTO), where the basis function is a linear combination of several GTOs

$$f_\ell^{\rm CGTO}(r_A) = r_A^{\ell} \sum_i d_i e^{-\alpha_i r_A^2}. \tag{1.4}$$

To distinguish between the contracted GTO and the individual GTOs of the basis set expansion, the latter are often called primitive GTOs (PGTOs), especially as it is standard practice to call the CGTO of Eq. (1.4) a GTO.

The first part of defining a basis set for a given atom involves the definition of the largest value of ℓ for which basis functions are included and the number of basis functions for each value of ℓ. For a STO basis set, an energy or another target function is then minimized with respect to the screening coefficients. For GTO basis sets, the optimization includes both the number of primitives included in each basis function, as well as the coefficients d_i, α_i making up a given basis function. Often, the screening coefficients of the STOs and GTOs are not optimized individually, but are chosen to constitute a simple, often geometric series, and the few parameters defining this series is then optimized instead of the individual screening constants.

Since the middle of the sixties, GTO basis sets have become the *de facto* standard for calculations on polyatomic molecules, and numerous program packages using GTO basis sets have been developed and are used by quantum chemists as well as scientists not specialized in computational chemistry. An important contribution to this development has been the development and distribution of GTO basis sets. The distributed codes include thus GTO basis sets for all the major types of electronic structure codes, target accuracy, and properties to be calculated. The most important of these developments are outlined in Sect. 1.4.

The development of methods to calculate integrals over STOs has not been completely abandoned, and a group of very skilled and competent researchers have developed improved methods to calculate the various forms of integrals, and significant progress is still being reported. However, due to the limited number of researchers devoted to the development of STO-based methods and the inherent complexity of the task of evaluating the integrals of these basis functions to high precision, the status for STO basis sets is much less advanced than for GTOs. Thus, although recent research on the calculation of multi-center two-electron integrals holds significant promise, the research is still focused on the calculation of individual integrals at a given geometry, rather than on the development of general-purpose basis sets.

For some types of calculations, it is possible to completely abandon the calculation of two-electron integrals *per se*. In particular, for DFT calculations without exact exchange, the two-electron integrals occur only in the form of the Coulomb term, which may be evaluated directly. This was from the onset incorporated in the widely used ADF program package [4], so the use of STOs for the calculation of polyatomic molecules have in this sense been with us from the early days of computational chemistry. We will return to the discussion of this approach as well as the current status for the use of STO basis sets in Sect. 1.5.

For solids and materials that from an atom's point of view are of infinite size, the standard choice of basis functions are plane waves. A plane wave function has the form

$$\phi(\mathbf{r}) = e^{i\mathbf{rk}}, \tag{1.5}$$

where $\mathbf{k}$ is the reciprocal lattice vector and is required to conform to the translational symmetry of the unit cell of the solid. If all possible vectors $\mathbf{k}$ are included, then the basis is complete, but in practice the basis set is truncated according to a kinetic energy criterion, so all $\mathbf{k}$ that transform according to the symmetry of the unit cell and have $|\mathbf{k}| \leq E_{\text{trunc}}$ are included. The plane wave basis functions are orthogonal to each other and a number of integrals of quantum mechanical operators involving these basis functions may be reduced to simple analytic expressions. In particular, the two-electron integrals are trivial to calculate. Although the plane wave basis without truncation is complete, a truncated basis set reproduces highly localized orbitals only with low accuracy These localized orbitals, typically core or inner valence orbitals, are therefore in most solid-state calculations replaced by effective potentials of various forms. We will not discuss such potentials or basis functions here in greater detail, but refer to the Chapter by M. Doll in this book for a more detailed discussion.

The GTOs, STOs, and plane wave basis functions are all global basis functions in the sense that they extend over the complete space. An alternative approach is to use basis functions with only local support, that is functions that are only non-vanishing in a limited region of space. The most prominent of such approaches are the Finite Element (FEM), wavelet, and spline approaches, which are related to the finite difference (FD) approaches of standard numerical analysis. These approaches are briefly discussed in Sect. 1.6.

In the following sections, the most important concepts and developments will be discussed for the GTO, STO, and local support basis sets. I will divert from the historical approach and discuss the developments and uses of GTO basis sets before the STO basis sets. The main reason for this departure is that it is the current status of GTO basis sets that the developers of the STO approach must match and perhaps bypass in order to give the STO basis set a prominent role within electronic structure modeling. I will naturally be brief, and refer to the various chapters of this book for a more detailed and in-depth discussions. In particular, a number of important forms of basis sets have been left out, but the purpose of the present chapter is to provide a general overview, rather than an exhaustive review. Furthermore, the development of basis sets is a very extensive effort, resulting in multiple papers. I will, for brevity again, only reference the paper that introduces a given form of basis set, rather than the complete series of papers.

1.2 Nomenclature for Basis Sets

There is a common nomenclature for GTO and STO basis sets, which gives the number of basis functions of the various angular momenta. This nomenclature is discussed in detail in the standard textbooks of quantum chemistry, including Refs. [5–7], but for completeness this will be briefly recapitulated here. In a minimal or single zeta (SZ) basis set, there is one set of basis functions for each sub-shell occupied in the ground state of the atom. A minimal basis set for the carbon atom contains hence two s-functions and one set of p-functions, shortened as 2s1p. At the next level, double zeta (DZ) basis sets, there are two sets of basis functions for each occupied sub-shell, so such a basis contains 4s2p orbitals for the carbon atom. Triple, quadruple, quintuple, and hextuple zeta basis sets are defined in a similar way, and are in abbreviated forms denoted by TZ, QZ, 5Z, and 6Z. Although the electrons in the core-orbitals give large contributions to the total energy, these orbitals and energies are in general rather constant, so they are often described using fewer basis functions, leading to, for example, valence double zeta (VDZ) basis sets, which for the carbon atom has the constitution 3s2p.

While s- and p-functions are sufficient for Hartree–Fock calculations on the ground states of first-row atoms, molecular calculations require polarization functions, i.e. basis functions of higher angular momenta than needed for the atomic Hartree–Fock calculations, to describe the polarization of the density from the field of the other atoms. As only the electrons in the valence orbitals are polarized to a significant degree, polarization functions are included only for these electrons. At the entry level of basis functions, the VDZ level, a single set of d-functions is added to first-row atoms like C, resulting in VDZP basis sets, which then have the form 3s2p1d. Basis sets of TZ or QZ quality are usually combined with additional polarization functions, f-functions for TZ basis sets and f- and g-sets for QZ basis sets. However, the notation for polarization functions is not in general common across different basis sets so care must be exercised here. For example, the P used to denote the presence of polarization basis functions is used to indicate that the basis set contains a single set of polarization functions, independently of whether the basis set is a double or higher zeta basis set, as well as to label basis sets where the number and type of the polarization functions are aligned with the number of s- and p-functions. A VTZP basis for C may therefore have the form 4s3p2d1f as well as 4s3p1d. For basis sets, where the P denotes a single d-set of polarization functions for first-row atom, the presence of 2 d- and 1 f-sub-shells is denoted 2P. Most basis sets have additional and specialized notations, but this will not be covered here.

1.3 The GTO Versus STO Discussion

Although GTO basis sets have become the *de facto* standard, in particular for correlated wave function calculations, research efforts for obtaining improved methods for calculating integrals over STO functions have continued. This research is motivated by STOs allowing significantly smaller expansions of the occupied orbitals and their advantages in regions around the nuclei and far away from the molecule. When the nuclei are approximated as point charges, the exact wave function Ψ fulfills the nuclear cusp-condition [8] when the coordinates $\mathbf{r}_i$ of electron i equal the coordinates of a nuclei A with charge Z_A,

$$\left.\frac{\partial \Psi_{ex}}{\partial r_i}\right|_{\mathbf{r}_i=\mathbf{r}_A} = -Z_A \Psi_{ex}(\mathbf{r}_i = \mathbf{r}_A). \tag{1.6}$$

From Eq. (1.3), it is seen that the only GTO functions that are non-vanishing at the origin are s-functions, and as these functions here have a vanishing derivative with respect to r, it is obvious that GTO functions in contrast to the STO functions cannot fulfill this cusp-condition. Any GTO basis set is therefore incapable of fulfilling the nuclear cusp condition at the center of this basis set.

For the region far away from the nuclei of the molecule, it can be shown [9] that the exact density $\rho_{ex}(\mathbf{r})$ falls exponentially off,

$$\rho_{ex}(\mathbf{r}) \propto e^{-2^{3/2}\sqrt{I}r} \text{ for large } r, \tag{1.7}$$

where I is the lowest ionization potential of the molecule. For any finite set of GTOs, the decay at long distances will inevitably have a Gaussian form, so Eq. (1.7) cannot be fulfilled in any finite GTO basis set.

The late Jan Almlöf made the counterargument [10] that STOs have two deficiencies: they are incorrect around a nuclei of finite size as the wave function inside such a charge distribution is known to be Gaussian, and they are not accurate at large distances as the above asymptotic cannot be fulfilled whenever too diffuse STOs are present in the basis set. Although the arguments of Almlöf are correct, they are perhaps not very relevant for practical use.

From a formal point of view, both the GTOs and STOs are valid basis set expansions. It has thus been proven that the exponents (α in the GTOs and ζ in the STOs) may be chosen so both types of basis sets are complete both in the standard Hilbert space of square integrable wave functions [11] and in the first and second Sobolev spaces [12]. By systematically extending either of these basis sets, it is thus possible to systematically approach the exact solution to the Schrödinger equation and the exact energy. However, it is important to be precise about what this completeness implies. On the positive side, the completeness implies that we may formally devise a sequence of wave functions, $\Psi_i, i = 1, 2, \ldots$, expanded in increasing basis sets that converge toward the exact wave function Ψ_{ex} in the norm-sense that is $\|\Psi_i - \Psi_{ex}\| \rightarrow 0$ for $i \rightarrow \infty$ and the energy E_i of Ψ_i goes toward the exact energy, E_{ex} for $i \rightarrow \infty$. On the other hand, the completeness of a basis set does not ensure point-wise convergence toward any function in the Hilbert space. Thus, although a complete radial basis set may be made up of s-type GTOs, $e^{-\alpha_i r^2}, i = 1, 2, \ldots$, it is not possible to obtain a sequence of wave functions that have a non-vanishing derivative that fulfills the nuclear cusp-condition, Eq. (1.6), as this equation is the property of the function in a point, rather than in a interval.

1.4 Gaussian-Type Orbitals

After the introduction of GTOs as basis functions by Boys [2] and Mcweeny [3] for general use, it was not until the late sixties that the computer hardware and quantum chemical software were ready to perform *ab initio* calculations on polyatomic molecules. This development was accompanied by the development of the standard basis sets, in particular by Pople and coworkers. In the first general available and used GTO basis sets, the STO-NG basis sets [13], the GTO basis sets were designed as least square fits to STOs. However, it was quickly realized that it was more useful to design the GTOs so that they minimize the Hartree–Fock energy and this lead to the second generation of basis sets, of which the VDZ basis sets 3-21G [14] and 6-31G [15] are still generally used, especially when combined with the polarization function [16]. A number of other basis sets were developed along the same lines, and minor additions and modifications were introduced to allow the calculation of correlation contributions. These basis sets prevailed for more than two decades, not only for Hartree–Fock calculations, but also for correlation calculations.

Until the late eighties, calculations including correlation in the form of configuration interaction (CI), coupled cluster (CC), and Møller-Plesset perturbation theory (MPPT) calculations were thus performed using basis sets that fundamentally were designed for uncorrelated calculations. The state of affair is exemplified by calculations published in 1985 by Ahlrichs, Scharf, and Jankowski [17] on the ground states of several diatomic molecules. Using what at that time was a very large basis,

7s5p3d3f1g CGTOs, and the CPF correlation method, they obtained an electronic dissociation energy for N_2 of 214 kcal/mol which is 14 kcal/mol away from the experimental dissociation energy of 228 kcal/mol. In some sense, this calculation exemplified the midlife crisis of quantum chemistry: using the best available basis set and correlation method, the calculations yielded results that were far away from chemical accuracy of 4 kcal/mol.

However, quantum chemistry did outlive its midlife crisis and regained confidence and purpose. This was due to developments along three lines: improved correlation methods, in particular coupled cluster methods with perturbation estimates of triple corrections (CCSD(T)), basis sets designed for correlation calculations, and systematic basis set extrapolation toward the complete basis set limit (CBS). Of these three developments, the latter two are relevant in the present context. The development of basis sets better suited to describe correlation started with the ANO basis sets of Almlöf and Taylor [18]. These basis sets provided several new insights and strategies that radically changed the accuracy that may be achieved by quantum chemical calculations. Firstly, these basis sets reflect the understanding that higher angular momenta than previously employed were necessary to obtain accurate correlation energies. Secondly, the basis functions were the linear combinations (contractions) of primitive Gaussians with the contraction coefficients being chosen as those of the various natural orbitals of atomic correlated calculations. Thirdly, basis functions with similar contributions to correlation energy were added together. Finally, by providing a hierarchy of basis sets, it was now possible to systematically improve the accuracy of the calculation.

The ANO basis sets provided a significant advance, but they were in general considered as too expensive as they employ a large number of primitive as well as contracted basis functions. More economical basis sets for correlated calculation were introduced in the form of the correlation consistent basis sets of Dunning and coworkers [19], and the ANO basis set from the Lund group [20]. Correlation orbitals have in general the same spatial extent as the corresponding Hartree–Fock orbital, but have additional nodes, so the correlation orbitals for core and valence electrons differ. Basis sets for the correlation of the valence electrons and basis sets for the correlation of both core and valence have been devised.

As groups of basis functions with similar contributions to the correlation energy are added together in the correlation consistent basis sets, the various basis sets have a very simple structure in terms of the number of contracted basis functions per angular momentum. Consider as an example basis sets for valence correlation of a first-row atom. At the Hartree–Fock level, the basis set includes 2s and 1p contracted sub-shells. The first set of correlation orbitals for the valence electrons includes 1s, 1p, and 1d sub-shells, so the total basis set includes 3s, 2p, and 1d sets of orbitals. The second correlation shell adds 1s, 1p, 1d, and 1f sets of functions, so the second set of correlation consistent basis functions includes 4s, 3p, 2d, and 1f sets of orbitals. The basis set may therefore be defined in terms of a cardinal number X, which describes the highest ℓ value included in the basis set, so the basis sets is denoted as cc-pVXZ and includes X+1 s-functions, X set of p-functions,..., and 1 set of functions with

$\ell = X$. Note that the number and distribution over angular momenta of the added functions for correlation are equal to that introduced above for polarization functions.

With the new basis sets, systematic calculations of geometries of other properties were made, see for example Ref. [21], and it was amply demonstrated that the combination of the CCSD(T) method and large basis sets provides geometries with an accuracy that often matches experimentally determined geometries. Energetics are also significantly improved so the goal of chemical accuracy of about 4 kcal/mol is achievable for molecules that are small enough to allow the use of the CCSD(T) or multireference CI methods with large basis sets. The electronic dissociation energy of the nitrogen molecule could thus be calculated to chemical accuracy [22]. The combined use of the CCSD(T) correlation method with large correlation consistent basis functions provided the first gold standard of correlation methods, and the field of quantum chemistry emerged as a mature field with gravitas and confidence—some would claim over-confidence.

Properties depending on energy differences, such as reaction and atomization energies, exhibit a very slow convergence toward the basis set limit, and this reduces the accuracy that may be achieved even using large basis sets. Thus, although the accuracy of 4 kcal/mol was achieved, it seemed impossible or at least very difficult to reduce this error by a factor of 10 or more, which would allow theoretical data to be used for accurate modeling.

The slow convergence of the energies with respect to the size of the basis set is connected with the non-differentiable behavior of the wave function when two electrons have the same spatial coordinates. When two singlet-coupled electrons i and j have the same coordinates, the exact wave function exhibits the cusp

$$\lim_{r_{ij}\to 0}\left(\frac{\partial\Psi}{\partial r_{ij}}\right)_{\text{ave}} = \frac{1}{2}\Psi(r_{ij}=0), \tag{1.8}$$

where the average is over a small sphere surrounding the point with coalescing electrons.

These non-differentiable points in wave function space are obviously only very slow convergent when the wave function as here is expanded as anti-symmetrized products of differentiable wave function. However, it may be shown that this basis set error converges toward the basis set limit as a simple function of the cardinal number, $\left(X+\frac{1}{2}\right)^{-3}$, so the energy $E(X)$ obtained in the basis set with cardinal number X is related to the basis set limit E_∞ as

$$E_X = E_\infty + A\left(X+\frac{1}{2}\right)^{-3}, \tag{1.9}$$

where A is a constant to be determined. This equation may be used to extrapolate toward the basis set limit by combining results from two or more calculations. The simplest and most succesful of these extrapolation schemes [23] use linear interpolation of the energies obtained with cardinal numbers X and $X+1$ to obtain the basis set limit as

$$E_\infty \approx E_{X+1} - \frac{(E_X - E_{X+1})\left(X+\frac{3}{2}\right)^{-3}}{\left(X+\frac{1}{2}\right)^{-3} - \left(X+\frac{3}{2}\right)^{-3}}. \tag{1.10}$$

In general, this simple extrapolation scheme reduces the error of $E(X+1)$ by an order of magnitude.

The convergence factor $\left(X+\frac{1}{2}\right)^{-3}$ is for singlet- coupled electron pairs, whereas the error connected with triplet- coupled electron pairs is significantly smaller and converges faster, with a factor proportional to $\left(X+\frac{1}{2}\right)^{-5}$. A more elaborate extrapolation scheme is thus based on the form

$$E_X = E_\infty + A\left(X+\frac{1}{2}\right)^{-3} + B\left(X+\frac{1}{2}\right)^{-5}, \tag{1.11}$$

with the unknown factors A, B. However, for most basis sets, extrapolation schemes based on Eq. (1.11) have not proven to be more accurate than Eq. (1.10). The use of the simple linear extrapolation scheme of Eq. (1.10) and the coupled cluster wave functions with higher than triple excitations has led to calculations [24] and protocols [25] for quantum chemistry which give atomization and reaction energies with errors less than 1 kJ/mol.

It is important to note that the success of the various GTO basis sets and extrapolation schemes is an indication that these basis functions give energy contributions that are close to the expansion of the atomic natural orbitals in a complete basis. The convergence of the correlation energy toward the complete basis set limit will therefore not be fundamentally changed by another choice of basis functions, such as STOs. However, this does not rule out that even more accurate extrapolation results may be obtained using GTO basis sets with a larger number of primitives or basis sets based on STOs.

The success of the correlation consistent basis sets and the possibilities for basis set extrapolation have led to renewed interest in the construction of basis sets that systematically approach the basis set limit for the Hartree–Fock and Density Functional Theory calculations. Jensen has developed the polarization consistent basis sets [26], which in agreement with the correlation consistent basis sets are defined in terms of the basis function with the highest angular momentum, but where the number basis functions with low ℓ, in particular s- and p-basis functions, are higher than for the comparable correlation consistent basis set. Basis sets for Hartree–Fock as well as DFT calculations have been devised. In agreement with an older analysis of Klopper and Kutzelnigg [27], it was observed [28] that the error compared to the complete basis set limit exhibited exponential convergence. In particular, the forms

$$E = E_\infty + Ae^{-BX}, \tag{1.12}$$

$$E = E_\infty + A(X+1)e^{-B\sqrt{N_s}} \tag{1.13}$$

were found optimal for extrapolation of the Hartree–Fock energy for correlation consistent and polarization consistent basis sets, respectively. In Eqs. (1.12) and (1.13), A, B are extrapolation parameters and N_s is the number of primitive s-functions in the basis set. Jensen argued [28] that the polarization consistent basis sets provided a more accurate and reliable extrapolation than the correlation consistent basis sets, but Karton and Martin [29] showed that the single parameter form

$$E = E_\infty + A(X+1)e^{-9\sqrt{X}} \tag{1.14}$$

provided accurate extrapolation also for the correlation consistent basis sets. Note that the extrapolation formulas in Eqs. (1.12) and (1.13) involve two parameters and require therefore energies from three basis sets, whereas the form suggested by Karton and Martin only requires the energy in two basis sets. The Hartree–Fock energy is converging significantly faster than the correlation energy, so the Hartree–Fock extrapolation schemes reduce typically the error of the energy of the largest basis set by a factor of 2, whereas the extrapolation of the correlation energy typically reduced the error by a factor of 10.

The current status of the GTO basis sets and the use of GTOs for molecular calculations may be summarized as follows:

1. Very efficient algorithms have been developed to determine wave functions of small and large molecules. Often the calculation of individual integrals are completely bypassed or are re-expressed using Cholesky factorization or resolution of the identity. Derivative methods allowing the optimization of the geometry and, for example, vibrational frequencies and intensities have been developed.
2. Basis sets for systematically approaching the complete basis set limit have been developed for DFT, the HF and correlated wave functions for ground as well as excited states. Basis sets targeted for specific properties, for example, magnetic properties have also been developed.
3. The basis sets are so accurate and systematic that extrapolation procedures may be applied to estimate the full basis set values for the correlation energy, as well as for HF and DFT energies.
4. Relativistic basis functions aimed at describing the Dirac equation either at the two- or the four-component level have been developed, allowing systematic procedures also for the relativistic wave functions.

The developed extrapolation schemes do not only give rise to improved energies and other properties, but do also provide estimates of the errors of a given basis set compared to the basis set limit. Although improvements of GTO basis sets, in particular for calculations with very high accuracy, still are needed, the basis sets developed in the last three decades have been one of the important ingredients for developing quantum mechanical calculations into a mature scientific field, where it is possible to provide estimates of the errors associated with a given calculation.

1.5 Slater-Type Orbitals

The development of basis sets over STOs and other exponential-type orbitals (ETOs) has been less extensive than the development of Gaussian basis sets. This is mainly due to the lack of general and accurate procedures for calculating three- and four-center two-electron integrals over these basis functions. Although this chapter and book is devoted to the discussion of basis sets, it is pertinent to first review the current status and progress in the calculation of such integrals as well as other integrals over STOs, thereby summarizing the significant technical and algorithmic challenges that have been overcome in the last few years.

For calculations on atoms, the problem with multi-center integrals is obviously avoided and the STOs has been extensively used for high-accuracy calculations for many decades. A recent addition is the development of the integrals relevant for general calculations involving correlation factors in the form of Hylleraas correlation functions [30].

For diatomic molecules, the recent developments of Lusiak and Moszinski [31–34] have led to algorithms and implementations that allow the evaluation of integrals for STOs with arbitrary angular momentum. In addition to the integrals required for evaluation matrix elements of the non-relativistic Hamiltonian, they have developed and programmed algorithms allowing the calculation of integrals for one-body relativistic operators as well as for effective core potential operators. The integrals are obtained with very high accuracy, 12 digits or more, thereby allowing calculations using basis sets having large overlaps, as typically is the case for basis sets for the accurate calculation of correlation contributions.

For the much-sought generalization to the calculation of three- and four-center integrals, the expansion introduced by Gill [35] of the Coulomb operator in terms of products of functions of one-electron functions has led to significant progress. In particular, Hoggan and coworkers [36, 37] have used the expansion of Gill to devise and program algorithms for all types of integrals needed for general polyatomic molecules and have demonstrated the usefulness of the approach by performing calculations on the dimer of H_2. The method has also been combined with semi-empirical methods and applied to study chemical shifts of benzothiazoles [37]. Unfortunately, to the best of the present author's knowledge, this development has not been extended to correlated calculations of general polyatomic molecules.

An obvious way of bypassing the complications with the development of analytic evaluation of multi-center integrals over STOs is to calculate the integrals, or the contributions from the integrals, using numerical integration. This procedure is typically implemented by first expanding the product of two STOs as a sum over STOs in an auxiliary basis [4], which leads to the two-electron integrals being reduced to integrals between three STOs and these are then evaluated using numerical techniques. This approach has been used for more than four decades in the extensively used ADF program developed of Baerends and coworkers for DFT calculations without exact exchange [4] and later used by Watson, Handy, and Cohen [38] to calculations using Hartree–Fock and DFT with exchange.

In broad terms, there are therefore currently two types of molecular calculations, where STOs are in general use: high-accuracy calculations on diatomic molecules and Hartree–Fock and DFT calculations on general molecules. Starting with the latter approach, new basis sets that systematically approach the full basis set limit have been developed by Van Lenthe and Baerends [39]. The developed basis sets range from double zeta up to quadruple zeta basis sets including polarization functions. For the heavier elements, the basis sets are optimized for a relativistic Hamiltonian. The basis sets have been extensively tested [39] and the results of the smaller basis sets converge toward those of the largest basis sets. The convergence has not been compared to the exponential convergence of Hartree–Fock and DFT energies that were discussed in the previous section. The triple zeta basis have subsequently been slightly extended by Watson, Handy, and Cohen [38] and were shown to be comparable in accuracy to standard triple zeta GTO basis sets. For Hartree–Fock and DFT calculations, there are thus good and well-tested STO basis sets for all atoms, but these have not been demonstrated to provide better energies than GTO basis sets with similar numbers of contracted GTOs. Whether this lack of advantage in the use of STOs is fundamental or is due to other approximations of the approach, in particular the reexpansion of products of STOs into single STOs and the numerical quadrature is not presently known.

For the other current molecular use of STO basis sets, high-accuracy diatomic calculations, basis sets have been developed for selected atoms in connection with calculations. Lesiuk and Moszynski have developed hierarchies of STO basis sets for Be [33], Ca, Sr, and Ba [34]. As there are no general benchmarks available for high-accuracy calculations on diatomic molecules using STO basis sets, the discussion of the performance must necessarily be restricted to these case studies, and we will here focus on the calculation on the Be dimer. For the Be atom, a hierarchy of basis sets was developed for cardinal numbers up to 6 using the ideas of the correlation- consistent GTO basis sets. For numerical stability in the basis set optimization, it was found most convenient to use STO basis with $n = \ell + 1$ in Eq. (1.2) in agreement with the choice for GTO basis sets. The exponents of the primitive functions were not individually optimized, but were defined in three or four parameters and these parameters were then optimized. The use of these basis sets for the Be dimer in connection with basis set extrapolation in the form of Eq. (1.11) was found to provide excellent results for the complete basis set limit. The use of STO basis sets could thus exploit the more involved three-parameter extrapolation, and in particular the reliability of the extrapolation of the contributions from the correlation of the 1s-orbital were found to give better results than for standard GTO correlation consistent basis sets. The binding energy of Be dimer was theoretically determined as 929.0 cm^{-1} with an estimated accuracy of 1.9 cm^{-1}, which constitutes the best computational estimate of this energy, both in terms of the agreement with the experimental value of $929.7 \pm 2.0\ \text{cm}^{-1}$ and in terms of the estimated error of the calculation.

Taking the Be dimer as a prototypical example of the accuracy that may be obtained for small diatomic molecules using STO basis sets, it is concluded that basis sets based on STOs provides the highest accuracy at the moment. Although the evaluation of the integrals over the STO is about an order of magnitude larger than that for GTO basis sets of similar size, the integral evaluation time constitutes only a small fraction of the total time. It is possible that GTO basis sets with similar accuracy may be developed, but these are presently not around.

Many developments are still required before STO basis sets may become a choice on par or even superseding GTO basis for general calculations on polyatomic molecules. Many of the developments that have allowed quantum chemistry to become a general useful tool for describing the structure, energy, and spectra are still missing. For example, the determination of equilibrium structures for molecules requires the calculation of derivatives of integrals with respect to the position of nuclear centers. While the development of such integrals for GTO basis sets was completed about 30 years ago, the corresponding development for STO basis sets has not yet been initiated. Without a fundamental change in the way that integrals over STOs are programmed and calculated, it is not likely that the use of STO will take over the role of GTO basis sets as a general tool for quantum chemical calculations, but they may become a complement for high-accuracy calculations of energetics at given geometries.

1.6 Basis Sets With Local Support, Finite Difference, and Spline Approaches

The various numerical methods that exploit basis functions with local support or spatial grids not only have many common features, but have also a number of important differences. Let us for simplicity restrict our discussion to a single dimension, where these approaches start by introducing a set of points, r_i, spanning the interval of interest. In the finite element [40] and wavelet [41] approaches, the points are collected into groups of k points with the final point of one cell also included as the first point of the following cell. Polynomials of order $k-1$ are then introduced in each of these cells, and it is these polynomials that form the basis functions of the calculation. The Legendre polynomials [42] are simple and useful examples of such local polynomials. In quantum chemistry codes, polynomials of order up to 8 are often used. The polynomials are chosen so that the expansion of the basis functions is continuous and it is also possible to ensure continuity of the first or higher derivatives. For the B-spline approach [43], the points are not grouped into cells, and the defined local polynomials span the interval from a given point r_i and a few of the following points with the polynomials chosen so at least the function and the first derivatives are continuous.

Expansions of orbitals in terms of the basis functions of local polynomials are in principle completely equivalent to the expansion of orbitals in the standard basis functions, but there are two aspects that give their quantum chemical implementation a different flavour: the very large number of basis functions, which may run into hundreds of thousands or millions, and the locality of the basis functions, which greatly simplifies the calculation of integrals between these.

Restricting the discussion to general polyatomic molecules, the extensively used program Octupus [44] is based on the FEM method. The wavelets were introduced in quantum chemistry by Harrison and coworkers [45], and have recently been applied to general molecules at the HF and DFT levels by Frediani and coworkers [46], where they have been used to determine the full basis set limit of energies and magnetic properties.

Predating the FEM and wavelet approaches is the standard finite difference (FD) approach [47] of numerical analysis, which was used to calculate approximate (Hartree) wave functions already in 1927 [48], one year after the papers introducing and defining quantum mechanics were published. The FD approach for atoms has been extensively developed by Froese Fischer [49] and coworkers. There have also been several developments to extend the use of FD methods to molecules and solids. These developments were started by the two-dimensional FD code for diatomic molecules developed by Laaksonen, Sundholm, and Pykko [50], and have most recently been used in the box and boble approach of Sundholm and coworkers [51], where ideas of FEM and FD approaches are combined. A number of real space methods for solids have also been developed in the last decades and are in many contexts giving more predictable accuracy than offered by the plane wave expansions. One of the most prominent and successful developments along these lines is the GPAW code of Thygesen et al. [52].

There is a subtle difference between the way the FEM and FD equations are obtained. Consider, for example, the derivation of the Hartree–Fock equations. In the FD approach, one first obtains a general expression for the Hartree–Fock equation using a formally complete expansion of the orbitals, and one subsequently discretizes this equation on a grid. In the FEM method, one starts by introducing the expansion of the orbitals in the finite elements in the energy expansion, and one then derives the variational condition. Thus, in the FD method, one performs the discretization after having derived the variational condition, whereas in the FEM method, the variational condition is derived for a discretized wave function. This makes the FEM expression somewhat more complicated to derive and program, but ensures that the procedure is variational. In a FEM Hartree–Fock calculation, the final optimized energy is thus the lowest energy that may be obtained with the given set of basis functions, whereas this does not hold for the FD approach. Whether this variational property of the FEM approach is sufficient to justify the added complexity compared to the FD approach depends on the complexity of the problem. For the standard HF wave functions, where the variational problem in general is rather simple, the added complexity of the FEM approach is in general not warranted, but for the much more complex optimization of multi-configurational Hartree–Fock wave functions, the variational aspects of the FEM approach certainly have their advantages.

The above discussion was restricted to the calculation of bound electronic states. An important additional use of quantum mechanical methods is the calculation of cross-sections for scattering, electron detachment, and ionization. Such calculations are important for problems ranging from the reentrance of space vehicles in the atmosphere to the behavior of atoms and molecules in intense laser fields. The wave function of an electron being ionized from a molecule cannot be accurately described in terms of GTOs or STOs, so basis sets with local support, FEM polynomial or splines, are used to describe these. The points of expansion, the grid points, are typical chosen so the matrix elements of a potential operator become diagonal in this basis. This is the fundamental idea of the discrete variable representation (DVR) of FEM [53]. In practice, a molecule is divided into two parts, an inner part where the standard (GTO) basis set is used to describe the bound electrons, and an outer part where the outgoing electron is described using the FEM basis functions or splines.

1.7 Conclusion and Outlook

Although significant developments are being made within each of the abovementioned forms of basis sets, the most innovative developments within recent years have arguably been the combined use of several different forms of basis sets. One may thereby use the various forms of basis set expansions, where they have their strengths. One of the currently most prominent of such approaches is the combined use of a GTO and a plane wave basis to describe solids. The GTO basis is used to describe the core and inner valence electrons, as their highly localized form makes them difficult to describe using plane waves, whereas the plane waves are used to describe the outer delocalized and much slower varying part of the wave function. The abovementioned use of combined GTO and FEM expansions for scattering is another example of such a combination. It is the firm opinion of the present author that much progress is still to be obtained by such combinations of different forms of basis sets. It is the hope that the present book may inspire and facilitate such advances.

References

1. Slater JC (1930) Phys Rev 36:57
2. Boys SF (1950) Proc R. Soc A 200:542
3. McWeeny R (1950) Nature No 4209:21
4. Baerends EJ, Ellis DE, Ros P (1973) Chem Phys 2:41
5. Cremer CJ (2013) Essentials of computational chemistry: theories and models, 2nd edn. Wiley, New York
6. Jensen F (2017) Introduction to computational chemistry, 3rd edn. Wiley, New York
7. Helgaker T, Jørgensen P, Olsen J (2000) Methods of electronic structure theory. Wiley, New York
8. Kato T (1957) Commun Pure Appl Math 10:151
9. Katriel J, Davidson ER (1980) Proc Natl Acad Sci USA 77:4403
10. Almlöf J (1994) Lecture notes in quantum chemistry, II, p 1, Roos, BO (ed) Lecture notes on chemistry, vol 64. Springer, Berlin
11. Klahn B, Bingel W (1977) Theo Chim Act 44:27
12. Hill RN (1998) Int J Quant Chem 68:357
13. Hehre WJ, Stewart RF, Pople JA (1969) J Chem Phys 51:2657
14. Binkley JS, Pople JA, Hehre WJ (1980) J Am Chem Soc 102:939

15. Hehre WJ, Ditchfield R, Pople JA (1972) J Chem Phys 56:2257
16. Hariharan PC, Pople JA (1973) Theo chim Act 28:213
17. Ahlrichs R, Scharf P, Jankowski K (1985) Chem Phys 98:381
18. Almlöf J, Taylor PR (1987) J Chem Phys 86:4070
19. Dunning TH Jr (1989) J Chem Phys 90:1007
20. Widmark P-O, Åke Malmqvist P, Roos BO (1990) Theo Chim Acta 77:291
21. Helgaker T, Gauss J, Jørgensen P, Olsen J (1997) J Chem Phys 106:6430
22. Langhoff SR, Bauschlicher CW, Taylor PR (1987) Chem Phys Lett 135:543
23. Halkier A, Helgaker T, Jørgensen P, Klopper W, Olsen J, Wilson AK (1998) Chem Phys Lett 286:243
24. Ruden TA, Helgaker T, Jørgensen P, Olsen J (2003) Chem Phys Lett 371:62
25. Tajti A, Szalay PG, Csaszar AG, Kallay M, Gauss J, Valeev EF, Flowers BA, Vazquez J, Stanton JF (2004) J Chem Phys 121:11599
26. Jensen F (2001) J Chem Phys 115:9113
27. Klopper W, Kutzelnigg W (1986) J Mol Struct (Theochem) 135:339
28. Jensen F (2005) Theo Chem Acc 113:267
29. Karton A, Martin JML (2006) Theo Chem Acc 115:330
30. Ruiz MB (2011) Advances in the theory of quantum systems in chemistry and physics In: Hoggan PE et al (ed) Progress in theoretical chemistry and physics, vol 22, pp 105–119
31. Lesiuk M, Moszynski R (2014) Phys Rev E 90:063318
32. Lesiuk M, Moszynski R (2014) Phys Rev E 90:063319
33. Lesiuk M, Przybytek M, Musial M, Jeziorski B, Moszynski R (2015) Phys Rev A 91:012510
34. Lesiuk M, Tucholska AM, Moszynski R (2017) Phys Rev A 85:052504
35. Gill PMW, Gilbert ATB (2009) Chem Phys 86:356
36. Hoggan PE (2010) Int J Quant Chem 110:98
37. Hoggan PE, Ruiz MB, Özdogan T (2010) Molecular integrals over slater-type orbitals. From pioneers to recent progress. In: Putz MV (ed) Quantum frontiers of atoms and molecules. Nova Publishing Inc., New York
38. Watson MA, Handy NC, Cohen AJ (2003) J Chem Phys 119:6475
39. Van Lenthe E, Baerends EJ (2003) J Comput Chem 24:1142
40. Reddy JN (2006) An introduction to the finite element method, 3rd edn. McGraw-Hill; Chui CK (1992) An introduction to wavelets. Academic Press
41. Walnut DF (2004) An Introduction to Wavelet Analysis, Birkhaüser Boston
42. Stoer J (2002) Introduction to numerical analysis (Texts in Applied Mathematics), 3rd edn. Springer, Berlin
43. de Boor C (2003) A practical guide to splines, Revised edn. Springer, Berlin
44. Castro A, Marques MAL, Appel H, Oliveira M, Rozzi CA, Andrade X, Lorenzen F, Gross EKU, Rubio A (2006) Physica Status Solidi 243:2465
45. Harrison RJ, Fann GI, Yanai T, Gan Z (2004) J Chem Phys 121:11587
46. Frediani L, Fossgaard E, Flå T, Ruud K (2013) Mol Phys 112:1143
47. Iserlas A (2008) A first course in the numerical analysis of differential equations. Cambridge University Press, Cambridge
48. Hartree DR (1927–1928) Proc Cambr Phil Soc. 24:89
49. Frose-Fischer C, Brage T, Jönsson P (1997) Computational atomic structure: an MCHF approach. IOP Publishing Ltd., Bristol, UK
50. Laaksonen L, Pyykkö P, Sundholm D (1986) Comput Phys Rep 4:313
51. Losilla SA, Sundholm D (2012) J Chem Phys 136:214104
52. Enkovaara J et al (2010) J Phys: Condens Matter 22:253202
53. Rescigno TN, Mccurdy CW (2000) Phys Rev A 62:032706

Chapter 2
Slater-Type Orbitals

Devashis Majumdar, Pabitra Narayan Samanta, Szczepan Roszak, and Jerzy Leszczynski

Abbreviations

ADF	Amsterdam density functional code
a_0	Bohr radius
ALDA	Adiabatic local density approximation
CCSD	Coupled cluster singles and doubles
CCSD(T)	CCSD including triple excitations
CGTO	Contracted GTO
CT	Charge transfer
DFT	Density functional theory
EOM	Equation of motion
$\boldsymbol{F}$	Fock matrix operator
GTO	Gaussian-type orbital
$\boldsymbol{h}$	One-electron operator
HOMO	Highest occupied molecular orbital
$\boldsymbol{J}$	Coulomb operator
$\boldsymbol{K}$	Exchange operator

D. Majumdar (✉) · P. N. Samanta · J. Leszczynski (✉)
Interdisciplinary Center for Nanotoxicity, Department of Chemistry, Physics and Atmospheric Sciences, Jackson State University, Jackson, MS 39217, USA
e-mail: devashis@icnanotox.org

J. Leszczynski
e-mail: jerzy@icnanotox.org

P. N. Samanta
e-mail: pabitra.samanta@icnanotox.org

S. Roszak (✉)
Advanced Materials Engineering and Modelling Group, Faculty of Chemistry, Wroclaw University of Science and Technology, Wyb. Wyspianskiego 27, 50-370 Wroclaw, Poland
e-mail: szczepan.roszak@pwr.wroc.pl

E. Perlt (ed.), *Basis Sets in Computational Chemistry*, Lecture Notes in Chemistry 107,
https://doi.org/10.1007/978-3-030-67262-1_2

LDA	Local density approximation
LCAO	Linear combination of atomic orbital
LUMO	Lowest unoccupied molecular orbital
M-L	Metal–ligand
MO	Molecular orbital
NL	Nonlocal
PA	Polyacetylene
RI	Raman intensity
SCF	Self-consistent field
SERS	Surface-enhanced Raman spectroscopy
SOC	Spin–orbit coupling
STO	Slater-type orbital
TDDFT	Time-dependent density functional theory
VK	Vignale–Kohn
VWN	Vosko–Wilk–Nausair
XC	Exchange–correlation
ZORA	Zeroth-order regular approximation
ζ	Slater exponent

2.1 Introduction

Molecular orbitals in quantum chemistry arise from the solution of Hartree–Fock (HF)-type equations and they are used to compute various electronic properties related to the molecular structure [1]. A few of such properties are dipole moment, charge density on atoms, Bader characteristics, excited-state properties (which include transition moment, oscillator strength of the excited states, etc.), vibrational and Raman spectra of molecules, and so on. These molecular property calculations are not simply restricted to simple HF equations. They also include density functional theory (DFT) calculations [2], coupled cluster (CCSD, CCSD(T), etc.) [3–7], and many other single and multireference techniques at various levels of sophistication [1, 8–10]. For the present purpose, it is easier to consider just HF equations to introduce the concept of molecular orbitals. If we consider an *n*-electron closed-shell molecule, the ground-state wave function Φ could be written as a single Slater determinant form as:

$$\Phi = |\psi_1 \psi_2 \psi_3 \ldots \psi_n| \tag{2.1}$$

The molecular orbitals ψ_i of the ith eigenstate ε_i arise as a solution to the HF equation $\boldsymbol{F}\psi_i = \varepsilon_i \psi_i$ ($\boldsymbol{F}$: Fock matrix operator) which is an integro-differential equation. The $\boldsymbol{F}$ is usually written as a combination of one-electron operator $\boldsymbol{h}$ and two-electron operators $\boldsymbol{J}$ and $\boldsymbol{K}$ as [8],

$$\boldsymbol{F}\psi_i = \boldsymbol{h}\psi_i + (\boldsymbol{J} - \boldsymbol{K})\psi_i \tag{2.2}$$

where

$$\boldsymbol{h} = -\frac{1}{2}\nabla_1^2 + \sum \frac{Z_A}{r_{1A}} \tag{2.3}$$

$$\boldsymbol{J}\psi_i = \left[\int \frac{\psi_i^*(r_2)\psi_i(r_2)}{r_{12}} \mathrm{d}r_2\right]\psi_i \tag{2.4}$$

$$\boldsymbol{K}\psi_i = \left[\int \frac{\psi_j^*(r_2)\psi_i(r_2)}{r_{12}} \mathrm{d}r_2\right]\psi_i \tag{2.5}$$

The coulomb operator $\boldsymbol{J}$ (Eq. 2.4) represents the average local potential at r_1 arising from an electron in ψ_j(at r_2), while the exchange operator K in (2.5) is defined by its effect when operating on spin orbital $\psi_i(\mathrm{r}_1)$. The exchange integrals computed through (2.5) as $\boldsymbol{K}\psi_i(\mathrm{r}_1)$ should vanish unless the spin function of $\psi_j(r_2)$ is the same as $\psi_i(\mathrm{r}_1)$, while the coulomb integrals $\boldsymbol{J}\psi_i(\mathrm{r}_1)$ are non-vanishing. The integro-differential nature of the HF equation is also revealed through operators $\boldsymbol{h}$ (the kinetic energy part revealing the differential component), $\boldsymbol{J}$, and $\boldsymbol{K}$ (providing an integral part). The solutions of such equations are usually achieved *variationally* with a starting guess for ψ_i. The solutions of HF equations are only feasible for atoms and linear molecules, where the angular part of ψ_i could be precisely separated from the radial part due to the presence of axial and full rotational group symmetry. On the other hand, in the case of a nonlinear molecule, the HF equations cannot be solved by following a similar mathematical approach owing to their three-dimensional landscape of ψ_i as well as the potential energy terms that appeared in $\boldsymbol{F}$.

The most commonly used technique is to expand $\psi_i$in terms of basis function φ_μ according to the LCAO-MO (LCAO: linear combination of atomic orbital; MO: molecular orbital) procedure:

$$\psi_i = \sum_\mu c_{\mu i}\varphi_\mu \tag{2.6}$$

This reduces HF equation to the form:

$$\sum_\nu \boldsymbol{F}_{\mu\nu}c_{\nu i} = \varepsilon_i \sum_\nu \boldsymbol{S}_{\mu\nu}c_{\nu i} \tag{2.7}$$

Equation (2.7) is generally called the Roothaan–Hall equation. It introduces the concept of atomic basis functions (or basis sets) to solve the HF equations *variationally*. The $\boldsymbol{S}_{\mu\nu}$ term is the overlap matrix element between the atomic orbitals (AOs) μ and v, $\boldsymbol{F}_{\mu\nu}$ is the Fock matrix element and $c_{\nu i}$ is the LCAO-MO coefficient. It should be noted that the matrix $\boldsymbol{F}_{\mu\nu}$ has a dimension N equal to the number of atomic basis orbitals used in the LCAO-MO expansion, and has N eigenvalues ε_i and N eigenvectors whose elements are $c_{\nu i}$. Thus, there are occupied and virtual molecular orbitals (MOs), which are described by $c_{\nu i}$ coefficients through the solution of Eq. (2.7).

2.2 Concept of Basis Sets

As a consequence of atomic symmetry, atomic orbitals are always of the form [1, 8]:

$$\varphi(r, \theta, \varphi) = R(r)Y_{lm}(\theta, \varphi) \tag{2.8}$$

where $R(r)$ is the radial part and $Y_{lm}(\theta,\varphi)$ represents spherical harmonics. Due to the singularity of the potential at a point nucleus with a charge of Z, the wave function must have a cusp at the nucleus. More specifically, it is required that

$$\left(\frac{\mathrm{d}R}{\mathrm{d}r}\right)_{r=0} = -Z \tag{2.9}$$

At the other end of the range, an electron far away from any molecule would see the remainder of the molecule as a positive charge without any particular structure. Therefore, the wavefunction, like in any single-electron atom (*cf*. the case of H-atom), would decay exponentially. The consideration of such ideas led to the development of atomic basis sets. The basis orbitals (or basis sets) commonly used in LCAO-MO-SCF (SCF: self-consistent field) calculations fall into two classes, viz. Slater-type orbitals (STOs) and Gaussian-type orbitals (GTOs). The STOs could be mathematically represented as:

$$\varphi_{nlm}(r, \theta, \varphi) = N_{nlm\zeta} Y_{lm}(\theta, \varphi) r^{n-1} e^{-\zeta r} \tag{2.10}$$

These STOs are characterized by the quantum numbers n, l, and m (principal, azimuthal, and magnetic, respectively) and the exponent ζ (characterizes the radial *size* of the basis function).

GTOs are usually written in terms of quantum numbers a, b, and c and Cartesian coordinates x, y, and z as:

$$\chi_{abc}(r, \theta, \varphi) = N_{abc\alpha} x^a y^b z^c e^{-\alpha r^2} \tag{2.11}$$

In the Cartesian representation of GTOs in Eq. (2.11), the quantum numbers *a, b, and c* represents different atomic orbitals. For example, orbitals with *a, b, c* values of 1,0,0, or 0,1,0 or 0,0,1 are p_x, p_y, and p_z orbitals; the combinations 2,0,0, or 0,2,0, or 0,0,2 and 1,1,0, or 0,1,1, or 1,0,1 are for five *d* orbital and one *s* orbital (the sum of the 2,0,0, 0,2,0, and 0,0,2 orbitals is an s orbital because $x^2 + y^2 + z^2 = r^2$ is independent of θ and φ). Thus, the quantum numbers *a, b,* and*c* describe the angular shape and direction of the orbital, while the exponent α governs the radial size of the basis function. For both STOs and GTOs, the values of r, θ, and φ signify the localization of electron with respect to an array of axes assigned to the center on which the basis orbital is sited [8].

The 2*p*, 3*d*, etc., in the STOs and GTOs are generalizations of Eqs. (2.10) and (2.11) that have a polynomial in the components r (x, y, etc.) multiplying the same exponential $\left(e^{-\zeta r}\right)$ or Gaussian $\left(e^{-\alpha r^2}\right)$ fall-off. The orbital exponents, which are positive numbers larger than zero, determine the diffuseness or *size* of the basis function. A large exponent implies a small dense function and a small exponent implies a large diffuse function. The major differences between the two functions $e^{-\zeta r}$ and $e^{-\alpha r^2}$ occur at $r = 0$ and at large r. At $r = 0$, the Slater function has a finite slope and the Gaussian function has a zero slope [1],

$$\left[\frac{d}{dr}\left(e^{-\zeta r}\right)\right]_{r=0} \neq 0 \tag{2.12}$$

$$\left[\frac{d}{dr}\left(e^{-\alpha r^2}\right)\right]_{r=0} = 0 \tag{2.13}$$

At large values of r, $e^{-\alpha r^2}$ decays much more rapidly than the Slater function $e^{-\zeta r}$.

All STOs more correctly describe the quantitative features of the molecular orbital ψ_i than do GTOs, and fewer STOs would be needed in basis function expansion of ψ_i for comparable results. As an example, it is possible to show that at larger distances, molecular orbital decays as $\psi_i \rightarrow exp\ (-a_i r)$, which is of the Slater form. For the sake of simplicity, the molecular orbitals of hydrogen atom have been taken into consideration. The *1s*, *2s*, and *2p*$_x$ orbitals could be represented as [8]:

$$1s : \frac{1}{a_0^{3/2}}\pi^{-1/2}\exp(-r/a_0) \tag{2.14}$$

$$2s : \frac{1}{a_0^{3/2}}(2\pi)^{-1/2}\left(\frac{1-\frac{r}{2a_0}}{2}\right)\exp\left(\frac{-r}{2a_0}\right) \tag{2.15}$$

$$2p_x : \frac{1}{a_0^{3/2}}(2\pi)^{-1/2}\frac{r}{4a_0}\cos\theta\exp\left(\frac{-r}{2a_0}\right) \tag{2.16}$$

(r: radial coordinate of the electron, a_0: Bohr radius; nuclear charge = 1) Considering such properties, STOs could be judged as preferable than GTOs in electronic wavefunction calculations. The stumbling block for using STOs appears during the calculations of multicenter multi-electron ***J*** and ***K*** integrals defined in Eqs. (2.4) and (2.5). These integrals could be routinely calculated in GTOs, as they have the advantage of expressing these multicenter integrals in terms of the same center. For example, the two-center products in GTOs could be expressed as [1, 8]:

$$\begin{aligned}&\exp\left[-\alpha_a(r-R_a)^2\right]\exp\left[-\alpha_b(r-R_b)^2\right]\\&=\exp\left[-(\alpha_a+\alpha_b)(r-R_X)^2\right]\exp\left[-\alpha_X(R_a-R_b)^2\right]\end{aligned} \tag{2.17}$$

where

$$R_X=\frac{\alpha_a R_a+\alpha_b R_b}{\alpha_a\alpha_b} \tag{2.18}$$

$$\alpha_X=\frac{\alpha_a\alpha_b}{(\alpha_a+\alpha_b)} \tag{2.19}$$

Subsequently, a four-center two-electron integral over GTOs can be formulated optimally, like a two-center two-electron integral. Such diminution in centers is adequate to calculate multicenter multi-electron integrals effectively. Similar reduction for STO-integrals is not possible in such a straightforward way.

The major drawback of using GTOs over the STOs is prompted by their erratic comportment close to the nuclear center, as revealed by Eqs. 2.12 and 2.13. The nuisance could be resolved by coalescing two, three, or additional GTOs with mixing coefficients, that are not regarded to LCAO-MO parameters, into a different function defined as contracted GTOs (CGTOs). It could be generally written as [1, 11, 12]:

$$\varphi_\mu^{CGTO}(r-R_A)=\sum_{p=1}^{L} d_{p\mu}\varphi_p^{GTO}\left(\alpha_{p\mu},r-R_A\right) \tag{2.20}$$

where L is the length of the contraction and $d_{p\mu}$ is the contraction coefficient. The pth normalized Gaussian φ_p^{GTO} has a functional dependence on the Gaussian orbital exponent $\alpha_{p\mu}$. By the proper choice of L, $d_{p\mu}$, and $\alpha_{p\mu}$, the contracted Gaussian function can be made to assume any functional form consistent with the primitive functions used. Thus, by proper choice of contraction parameters, one can use basis functions that are approximate atomic Hartree–Fock functions, Slater-type functions, etc., while evaluating integrals only with primitive Gaussians. The most widely used procedure, in this respect, is to fit an STO to a linear combination of L = 1, 2, 3…, N primitive Gaussians functions. These are commonly known as STO-NG procedure. Figure 2.1 represents a comparison of Slater- and Gaussian-type functions related to their long-range behavior.

The widely used Gaussian-type basis sets are well documented in regular textbooks [1, 8] and in literature [13]. These types of basis sets are also used in most of the available computer codes for calculating the electronic properties of molecules using quantum chemical techniques. The STOs were used primarily in semi-empirical calculations, and the procedures at different levels of intricacies are well documented [8]. Much efforts have been paid to deploy STOs to directly solve HF-type equations by overcoming the problems of computing molecular integrals, and these techniques in recent times are widely used in density functional theory (DFT) approaches to solve molecular electronic properties in gas-phase, solid-state, and solvated conditions.

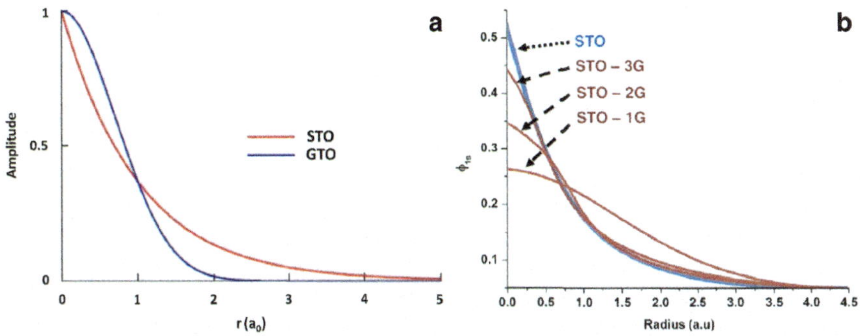

Fig. 2.1 **a** Comparison of Slater- and Gaussian-type orbitals is shown schematically. The orbital exponent for both the orbitals is 1.0. **b** Schematic comparison of the quality of the least-squares fit of a 1s Slater function ($\zeta = 1.0$) obtained at the STO-1G, STO-2G, and STO-3G levels

Such advancements were achievable by continued efforts to generate appropriate Slater-type basis sets and solution techniques for molecular integrals.

2.3 STOs in Quantum Chemistry

The development of the use of STOs in quantum-chemical calculations started in the late 1960s. The main problem was to calculate the multicenter integrals, and a procedure was developed by Harris and Michels [14]. Later, several authors have developed efficient mathematical algorithms for solving multicenter multi-electron integrals in this direction [15–19]. The HF-STOs technique was developed by Ziegler and Rauk [20] to compute the bonding energies of molecules. The use of STOs in DFT was developed later in the 1980s [21, 22] and the techniques are routinely used, in recent times, to predict molecular electronic properties. In the present section, we will briefly outline the ideas of solving the multicenter integrals in DFT, and then introduce the different slater-type basis sets that are commonly used in such calculations.

2.3.1 *Multicenter Integrals*

The Kohn–Sham DFT [23], which adapts a one-electron formalism, represents a viable approach to decode a system of N interacting electrons by accounting for an apposite local potential $V_{XC}(r)$, in addition to any external potentials $V_{\text{ext}}(r)$ and the Coulomb potential of the electron cloud $V_C(r)$ Eq. (2.21).

$$\left[-\frac{1}{2}\nabla^2 + V_{ext}(r) + V_C(r) + V_{XC}(r)\right]\varphi_i(r) = \varepsilon_i\varphi_i(r) \qquad (2.21)$$

The potential V_{XC} is derived from the functional derivative of the exchange–correlation energy $[E_{XC}(\rho)]$ with reference to the electron density ρ. The one-electron molecular orbitals (MOs) φ_i (with corresponding orbital energies ε_i) define the *exact* electronic charge density and give, in principle, access to all properties since they could be expressed as functional of the density (in particular the energy). They also provide an intuitive view of the system as being built from independent-electron orbitals with all ensuing interpretations.

The precise form of the energy density functional $E_{XC}(\rho)$ including the reckoning of characteristics of inherent exchange and correlation effects is obscured. Several sophisticated mathematical algorithms have been proposed previously to quantify $E_{XC}(\rho)$ based on general principles. Their application to real systems has been found to be quite impressive.

The DFT implementations using STOs commonly employ some sort of numerical integration scheme, since the mathematical expression for any relevant density functional makes an analytical integration of matrix elements of the XC potential impossible regardless of the nature of basis functions. The Gaussian-type quadrature method [24] based on the partitioning of space in atomic cells is commonly used in the most popular ADF code [24, 25]. The individual atomic cell is described by a core-region sphere, where the integrands in conjunction with cusps and singularities at the nucleus are manipulated decently in spherical coordinates:

$$\int_{\text{cell}} f(r)\mathrm{d}r = \int_{\text{sphere}} f(r)\mathrm{d}r + \int_{\text{cell - withoutsphere}} f(r)\mathrm{d}r \qquad (2.22)$$

$$\int_{\text{sphere}} f(r)\mathrm{d}r = \int_0^R r^2\mathrm{d}r \int_0^{4\pi} f(r,\Omega)\mathrm{d}\Omega \qquad (2.23)$$

where r^2 *is* the Jacobian factor that eliminates the Coulomb singularity of the integrand at the nuclear position. The cell-without sphere region is described by a Gaussian product theorem in the Cartesian-like coordinates following proper coordinate transformations. The atomic cells fill all space “inside” the molecule. All integrands decline exponentially with the distance in the residual “outer” area, distant from the atoms. Such an integration scheme is literally implemented in the ADF code. In such a numerical grid approach, the grid can be automatically tuned by varying the number of points in the different regions (atomic spheres, cells, and the outer region), while monitoring a series of test integrals such that these are evaluated with an (input-adjustable) precision. The detailed procedure regarding the geometric manipulations for the grid base integration scheme is available in references [16] and [17] and need not be discussed here in detail.

2.3.2 Coulomb potential and fit function

All terms in the Kohn–Sham operator of Eq. (2.21) are local in the sense that they can be evaluated readily as functions of r, except for the Coulomb potential $V_C(r)$. Slater-type basis functions yield entangled multicenter integrals in the evaluation of the Coulomb potential. This problem is resolved by using an auxiliary set of *fit* functions. The fit functions can be expressed in terms of prototype of Slater exponential functions placed on the atoms in the same way as basis functions. The actual density furnished by a sum of products of basis functions is subsequently estimated by a linear combination of the fit functions. The combination coefficients (c_i) are called the fit coefficients [24].

$$\rho(r) = \sum_i c_i f_i(r) \tag{2.24}$$

Approximate Coulomb potential is obtained as an expansion in fit *potential* functions $\left(f_i^c(r)\right)$ through the solution of the Poisson equation for the fit functions (Eqs. 2.25 and 2.26).

$$f_i^c(r) = \int \frac{f_i(r')}{|r - r'|} dr' \tag{2.25}$$

$$V_C^{fit}(r) \approx \sum_i c_i f_i^c(r) \tag{2.26}$$

The generation of $V_C^{fit}(r)$ involves the least-squared solution of Eq. (2.25) under the constraint that the number of electrons in the fit representation is preserved. For an STO-type fit function $f(r) = Y_{lm}(\Omega) r^n e^{-\zeta r}$, the corresponding Coulomb potential function $f^c(r)$ [24] is obtained through Eq. (2.25) as given below:

$$f_i^c(r) = \frac{4\pi}{2l+1} \varphi_{lm}(\Omega) \left[\frac{\int_0^r (r')^{n+l+2} \exp(-\zeta r' dr')}{r^{l+1}} + r^l \int_0^\infty (r')^{n+l+1} \exp(-\zeta r') dr' \right] \tag{2.27}$$

For the higher values of r, the multipolar term of long-range interactions becomes a crucial factor, and it disintegrates as q/r^{l+1}, where q is the electronic charge and l represents the angular momentum quantum number for a given STO-type fit function. q could be considered as the sum of charges enclosed in the fit function on account of the $l = 0$ monopole function. Thus, it could be determined by performing the integration over all space of the fit function itself. On the other hand, the long-range multipole term is altered by an exponentially decaying small-range term due to the smaller values of r, and the function reaches zero when $r = 0$. For practical application, the repository of Slater basis function is generated for each atom by including the details of basis functions, the frozen core orbitals, and the fit functions centrally located on that specific atom to facilitate the computation.

2.3.3 Slater basis sets

Various types of Slater basis sets are usually available from the computational chemistry code like ADF [24, 25]. ADF exploits Slater-type functions (STOs) $f(r) = Y_{lm}(\Omega)r^n e^{-\zeta r}$ (Ω contains the information of θ and φ of Eq. 2.10) as basis set elements, in terms of which the molecular orbitals (MOs) are expanded. A large number of basis sets for all elements of the periodic table, ranging from minimum single-ζ (SZ) to doubly polarized triple-ζ (TZ2P) basis sets, are available (2.25). It incorporates special extended sets with very diffuse functions for polarizability and hyperpolarizability calculations [26–30] and others with very contracted deep-core function sets for relativistic calculations within the ZORA formalism [31, 32]. The implemented method is based upon even-tempered STO-basis sets [33] developed by Chong et al. [34]. The implementations were made with a few modifications to generate proper molecular properties. The details of the basis sets are available in the ADF manual [25]. The information required to describe the basis set in ADF code for a given element includes the Slater-type basis functions, core expansion functions, core description, and fit functions.

The information about quantum numbers viz. principal and angular together with the exponential decay factor for a given set of Slater-type basis functions is acquired by the details of the basis record. The core data looks very much like the basis functions: a list of Slater-type function descriptions. The parenthetical record of the core contains four integers viz. *ns*, *np*, *nd*, and *nf*. They are the numbers, respectively, of *s*-, *p*-, *d*-, and *f*- frozen core shells in the atom. For the sake of illustration, let us examine the numbers of frozen core shells belonging to the Ruthenium atom under the wing of the frozen core prior to the 4p shell. Therefore, for the Ru atom, the core comprises four *s*-frozen core shells (1*s*, 2*s*, 3*s*, 4*s*), three *p*-frozen core shells (2*p*, 3*p*, 4*p*), one *d*-frozen core shell (3*d*), and no frozen *f*-shells. The idea behind the frozen core is to lower the computational time by reducing the size of the variational basis set. The crucial objective for such reduction is to ensure the least involvement of the deep-core atomic orbitals upon bond formation and thus save time without sacrificing the quality. The valence basis set used in the SCF equations is explicitly orthogonalized to the frozen core to ensure orthogonality between the frozen core orbitals and the valence basis functions. Each valence function $\left(\chi_{\upsilon}^{valence}\right)$ is replaced by a linear combination, as shown in Eq. (2.28) [24]:

$$\chi_{\upsilon}^{\text{valence}} \Rightarrow \chi_{\upsilon}^{\text{valence}} + \sum_{\mu} c_{\mu\upsilon}\chi_{\upsilon}^{\text{core}} \tag{2.28}$$

The condition that each such modified valence function be orthogonal to each frozen core orbital determines exactly all the coefficients $c_{\mu\nu}$ (these values are actually in the core data as 1s 24.40..., etc.). The frozen core orbitals are deduced by performing meticulous single-atom computations in conjunction with substantial STO-basis sets.

The core description data describes the frozen core shells as linear combinations of the core expansion functions. All $n\,l$ frozen shells corresponding to the individual angular momentum quantum numbers ($l = s, p, d, f$) are accounted by expansion coefficients. To every function set associated with the angular momentum quantum number l value, there correspond a coefficient in the list of expansion functions. There cannot be distinct coefficients for all m values since all m values are identical in a spherically symmetric model atom. The last dataset is based on the fit functions, and the basis of such data has already been discussed in section 2.3.2.

2.4 Examples on the Use of STOs in Chemistry

The STOs are nowadays used routinely in quantum chemistry as a tool for molecular electronic property calculations. The DFT-STO techniques are used in almost all spheres of quantum chemical analysis and a huge number of publications, in this respect, have appeared in various journals. To name a few of such applications are electronic structure analysis of molecules, spectroscopic properties, reaction path analysis, spin–orbit effect on molecular properties, NMR analysis of molecules, solid-state structures and properties. We will mention about a few specific applications to complete this present discussion.

2.4.1 Benchmark molecular structure calculations

The initial benchmark calculations using DFT-Slater calculations were directed to understand the metal–ligand (M-L) bonds [22]. It is, in general, accepted that electron correlation effects are important in *M-L* bond description, and the degree of electron correlation in such bonding varies considerably with the ligand. The synergic ligands for instance CO, cyclopropane (CP), and olefins thereby necessitate a substantial amount of correlation because of their intricate bonding characteristic mediated by electron donation and back donation. Simple two-electron donor ligands such as halides or phosphines, on the other hand, demand a more modest degree of electron correlation. As a consequence, it is not uncommon that approximate *ab-initio* calculations predict a distorted structure in which bonds involving Cl or PR_3 (R: H or any alkyl group) are close to experimental observations, whereas bonds involving synergic ligands are misrepresented. Even chemically distinct bonds involving the same ligand might have different demands on electron correlation. An important example is $Fe(CO)_5$, in which Hartree–Fock calculations provide a distorted structure where the Fe-C_{ax} bonds are too long by 0.24 Å., whereas the Fe-C_{eq} bonds are more in line with the experiment.

In this regard, efforts are being made to optimize molecular geometries of metal–ligand systems such as MnO_4^-, $Cr(CO)_6$, $Mo(CO)_6$, $Fe(CO)_5$, $Ni(CO)_4$, $HMn(CO)_5$, $HCo(CO)_4$, $Fe(C_5H_5)_2$, and C_5H_5NiNO employing Local Density Approximation (LDA)-based method together with its nonlocal (NL) extension including Perdew's [36] correction to correlation and Becke's [37] correction to exchange. Uncontracted triple-ς quality STO-basis set, augmented by three (n + 1)*p* STOs, were used for the transition metals. The main group elements were described by a double-ς STO-basis augmented by one *2p* function for hydrogen and a single *3d* polarization function for the other elements. A group of auxiliary s, *p, d, f,* and *g* STOs placed on each nucleus were deployed to fit the atomistic density. The computed metal–ligand bond distances using LDA-based methods are underestimated by about 0.05 Å compared to the experimental results, while such deviations minimize to 0.01 Å upon nonlocal corrections [22].

Several other organic molecules containing more than a single heavy atom were also investigated in this context. They included H_2CO (C_{2v}), C_2H_2 ($D_{\infty h}$), C_2H_4 (D_{2h}), CO_2 ($D_{\infty h}$), CH_3OH (C_1), and oxiran (ethylene oxide, C_{2v}). The results showed no significant difference between LDA and LDA/NL for bond lengths involving hydrogen atoms. Single, double, and triple C–C bond lengths in oxiran, ethylene, and acetylene, respectively, were well reproduced by either local or nonlocal methods and the deviations from experiments were found to be less than 0.02 Å [22]. The three types of C-O bonds present in CO_2 ($D_{\infty h}$), CH_3OH (C_1), and oxiran (ethylene oxide, C_{2v}), were somewhat better represented by LDA/NL. The C-O bond in oxiran, for example, was 1.416 Å according to LDA. It is 0.018 Å shorter than the experimental value of 1.434 Å. This discrepancy between the LDA/NL value of 1.442 Å and the experiment is only 0.008 Å. It is clear from the results that nonlocal corrections to bond lengths between main group elements are modest, adding little to the already good agreement between experiment and local results. This is in contrast to metal–ligand bond lengths where nonlocal corrections are substantial and important for a quantitative estimate [22]. These benchmark results obtained by the nonlocal corrections to LDA (LDA/NL) for geometries of transition-metal compounds using STOs were quite encouraging, especially for metal–ligand bond lengths, and paved the way for the further development of DFT-STO techniques in quantum chemistry.

2.4.2 Excitation Energy Calculations

In this section, we present a specific example of excitation energy calculations using Slater orbitals within the framework of time-dependent density functional theory (TDDFT). The present example reports the $\pi \rightarrow \pi^*$ singlet excitations of the π-conjugated oligomers of polyacetylene, polydiacetylene and the lowest singlet excitations of the hydrogen chain. TDDFT calculations within the Vignale–Kohn (VK) [37] and adiabatic local density approximations (ALDA) were used for such studies [38]. The lowest energy structures of these oligomers on the potential energy hypersurface were determined by exploiting generalized gradient approximation (GGA)-based BP functional comprising exchange potential proposed by Becke [35] and correlation counterpart developed by Perdew [36]. The oligomer geometries were forced to be planar (C_s). A triple-ζ Slater-type basis set augmented with two polarization functions was used in such calculations, and the cores were kept frozen for carbon up to 1*s* and for sulfur up to 2*p*.

The excitation energies for polyacetylene and polythiophene molecules are shown in Fig. 2.2a, b, respectively. The ground states of the studied molecules were computed with the local density approximation (LDA) functional in the VWN (Vosko, Wilk, and Nusair) parametrization [39]. The response calculations were carried out with the ALDA derived from the ground-state LDA expression and the VK functional [37].

Figure 2.2a shows the results for the dipole allowed ^1B_u excitation energies and corresponding oscillator strengths of polyacetylene (PA) against the number of oligomer (C_2H_2) units. The results are compared with experiment [41] and CCSD-EOM findings [40]. The predicted 1^1B_u excitation energies for the shorter chain segments (3–4 units) employing ALDA-based method are near to the CCSD-EOM-derived results; however, there are certain discrepancies for the longer chains. For instance, the calculated excitation energy using ALDA for the longest oligomer comprising of 15 units is underestimated by 1.39 eV with respect to the CCSD result.

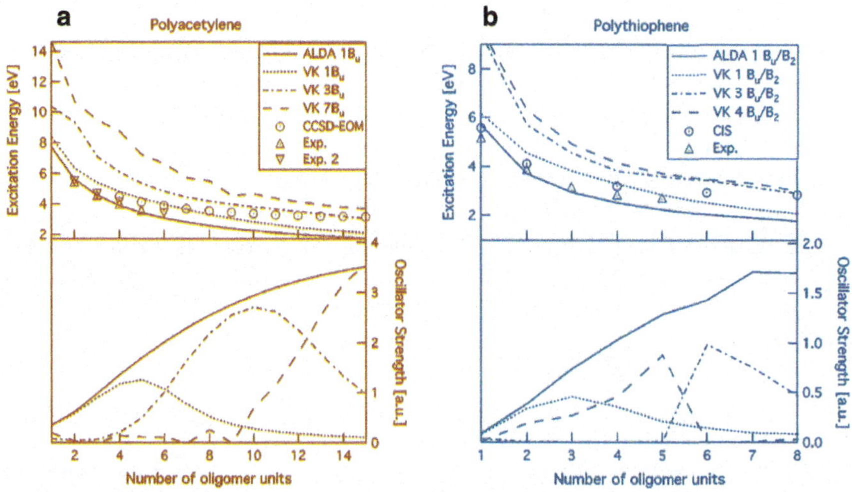

Fig. 2.2 **a** Excitation energies and oscillator strengths of polyacetylene oligomers compared with CCSD-EOM and experimental results [38]. Experiments 1 and 2 are both absorption spectra from ref. [41]. **b** Excitation energies and oscillator strengths of polythiophene oligomers compared with CIS [42] and experimental results (monomer and dimer: ref. [43]; tetramer and seximer: ref. [44]; octamer: ref. [45]). Adapted with permission from Ref. 38 Copyright (2004) (AIP Publishing)

It is known that ALDA overestimates the polarizability, and thus the reduction of the static polarizability is yielded by using the VK functional. On the other hand, it diminishes static polarizability assisted by the diminution of the oscillator strength of the lowest transitions or surge of the excitation energies, or both. The calculated excitation energies using VK functional are higher than the ALDA results. The VK results also coincide with the CCSD results for the small oligomers, although the computed excitation energy for the longer polymers demonstrates a deviation from the CCSD results.

Figure 2.2b shows the variation of excitation energies and oscillator strengths of polythiophene (PT) for the 1^1B_u (even number of thiophene rings) and 1^1B_2 (odd number of thiophene rings) transitions as a function of SC_4H_2 units. The predicted trends of excitation energies with ALDA and VK functional show the same behavior as observed for PA. The figure also compares the results with CIS and experimental results. It is evident that the excitation energies compare quite well with experimental and CIS calculations. There are many other observations related to these investigations, which need not be discussed here. The primary objective here is to show that TDDFT calculations with Slater orbitals can produce information regarding excited states, which are comparable to the other standard techniques. Interested readers can consult the original manuscript for further detailed information [38].

2.4.3 Relativistic studies on UO_2^{2+} spectrum

UO_2^{2+} was detected in the gas phase by mass spectrometry [46] and pulsed laser reaction of evaporated uranium and oxygen [47]. Experimental spectroscopic data, of course, were not available for this ion. A number of molecular orbital calculations at different levels of theory (including DFT with several functionals and ab-initio methods at different correlated levels) have reported [48–53] the molecule to be linear with $^1\Sigma_g^+$ as the ground state. The computed U–O distances range from 1.65 to 1.78 Å, and symmetric stretching frequencies lie between 740 and 1220 cm^{-1}. High correlation coupled cluster (CCSD(T)) [51] and CASPT2 [53] computed the distance to be 1.715 Å and symmetric stretching frequency as 974 cm^{-1}.

The excited-state spectrum of UO_2^{2+} was also computed at the CASPT2 level [53], and the results were later used to benchmark DFT-STO calculations [54]. The DFT results for the excited states of UO_2^{2+}, presented here, were computed using SAOP functional [55]. This functional was especially designed for the calculation of response properties. The ALDA has been exploited to evaluate the exchange-correlation kernel. Scalar relativistic effects were included via the ZORA (zeroth-order regular approximation) formalism [32] due to the presence of heavy U atom (spin–orbit free (SOF) calculation). Two-component ZORA-TDDFT formalism was further used for the inclusion of spin–orbit coupling (SOC) effects. The optimal QZ4Pae basis sets (quadruple-ζ all electron basis sets including four polarization functions) of Slater-type orbitals have been employed within the framework ZORA formalism [54].

SAOP vertical excitation energies for UO_2^{2+} including SOC for all states indicated that except the highest $c\Delta_g$ state, the corresponding SOF-state contributions added up to 90% or more, and thus pointing to the absence of considerable mixing with higher-lying SOF states. This is illustrated in Fig. 2.3. The lowest excitations out of the highest π_u-type bonding, in fact, already appear at slightly lower energies than the highest $\sigma_u^+ \rightarrow \delta_u, \varphi_u$ state $c\Delta_g$, thus explaining the larger multiconfigurational character found for this state. There is an overall agreement between the DFT and CASPT2 [53, 54] results, and all DFT results are somewhat higher than the CASPT2 values, with differences ranging between 339 and 3680 cm^{-1}. The largest differences are found for the states with predominant $^3\Delta_g$ character $a\Pi_g$ (3680 cm^{-1}), $b\Delta_g$ (3110 cm^{-1}), and $b\Pi_g$ (3030 cm^{-1}), close to the difference of 3408 cm^{-1} found for the parent $^3\Delta_g$ state.

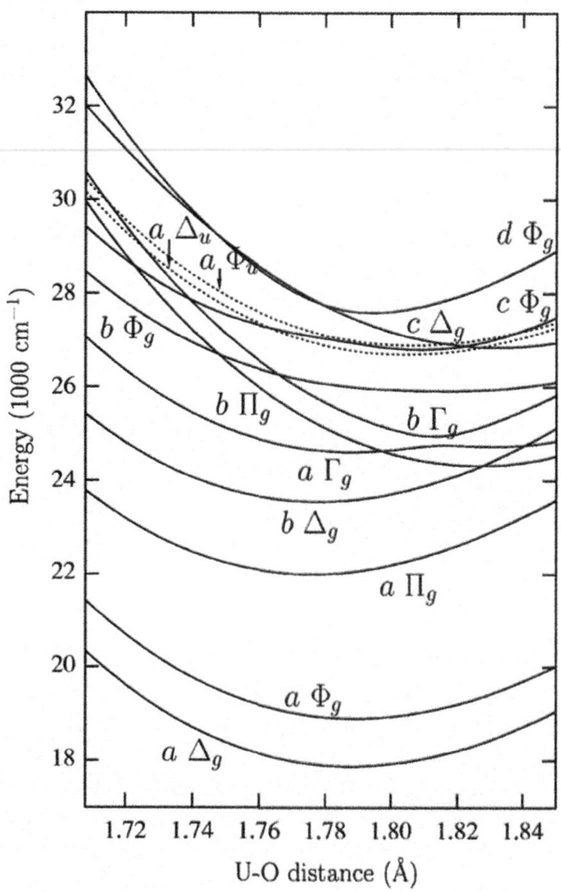

Fig. 2.3 Excited-state curves of UO_2^{2+} along the U–O symmetric stretching path obtained from SAOP-TDDFT calculations. Reprinted with permission from Ref. [54] Copyright (2007) (AIP Publishing)

The differences are smaller for the $a\Delta_g$, $b\Phi_g$, and $a\Gamma_g$ states (757–1166 cm^{-1}). This conforms to the difference of 1225 cm^{-1} obtained for their parent $^3\Phi_g$ state. For the two highest states $c\Phi_g$ and $c\Delta_g$, with predominant singlet character, the DFT-STO (SAOP functional) and CASPT2 results are very close (within 1000 cm^{-1}), as was the case for the corresponding singlet states. The DFT-derived energy gaps between the singlet and triplet manifolds of the SOF states are found to decrease compared to the CASPT2 results. There is an extensive mixing of singlet and triplet characters in the DFT results. The results confirm the experimental interpretation, and the outcomes of the previous theoretical calculations manifest that this part of the spectrum is built from excitations out of the HOMO σ_u^+ orbital into the nonbonding δ_u or φ_u uranyl orbitals. The detailed account of the research investigation need not be discussed here. It is available in the main manuscript [54]. The present discussion is only a part of the main paper and demonstrates the effective use of DFT-STO formalism to explain the complicated excited-state spectrum of molecules containing heavy atoms.

2.4.4 Applications of DFT-STO formalism to Resonance Raman Spectrum calculations

Resonance Raman (RR) and surface-enhanced resonance Raman spectroscopy (SERS) are nowadays used routinely in the detection of nanoparticles and many complexes of nanoparticles (with specific ligands) having Raman-active modes. Quantum chemistry offers a technique to compute the Raman spectra of molecules adsorbed on metals and many other surfaces. The enhanced Raman-active modes of the associated molecules could be used as a finger-print technique for their detection. DFT-STO technique has been used quite successfully for such Raman mode analysis. The Raman intensity (RI) is usually computed as the differential Raman scattering cross section using the expression [56]:

$$\frac{d\sigma}{d\Omega} = \frac{\pi^2}{\varepsilon_0^2}\left(\widetilde{\nu}_{in} - \widetilde{\nu}_p\right)^4 \frac{h}{8\pi^2 c\widetilde{\nu}_p}\left[45\alpha'^2_p + 7\gamma'^2_p\right]\frac{1}{45\left(1 - \exp\left(-hc\widetilde{\nu}_p/k_B T\right)\right)} \qquad (2.29)$$

where $\widetilde{\nu}_{in}$ and $\widetilde{\nu}_p$ are the frequency of the incident light and of the pth vibrational mode, respectively. $[45\alpha'^2_p + 7\gamma'^2_p]$ correspond to the scattering factor. The terms α'_p and γ'_p denote the isotropic and anisotropic polarizability derivatives with respect to the vibrational mode p, respectively. The polarizability derivatives are computed by exploiting numerical three-point differentiation with respect to the normal mode displacements [57]. The other technique uses the following expression (Eq. 2.30), which is valid within a short time approximation [58] to compute relative RI.

$$\frac{I_j}{I_k} = \frac{\nu_j^2 \Delta_j^2}{\nu_k^2 \Delta_k^2} \quad (2.30)$$

The left-hand side of Eq. (2.30) represents the intensity ratio of the two vibrations j and k. ν_j and ν_k are the frequencies of the vibrations j and k and Δ_j and Δ_k represent displacements of the excited-state minimum along these modes in terms of dimensionless normal coordinates. In this section, we will present examples related to both Eqs. (2.29) and (2.30) to demonstrate the use of DFT-STO calculations in this context.

The first example is related to Eq. (2.29). It unveils the SERS of molecular junction involving pyrazine molecule sandwiched between two Ag_{20} nanoclusters (Fig. 2.4a). A comprehensive theoretical investigation has been conducted by examining Raman spectra of the molecular junction [59] followed by simulating absorption spectra within the framework of TDDFT. All the calculations related to geometry optimization, vibrational frequency, excitation energy, and polarizability on and off resonance have been performed using the BP86 exchange–correlation functional [35, 36] in conjunction with triple-ζ polarized Slater-type (TZP) basis set [25] considering the 1s-4p frozen core shells for Ag. The optimized geometry for the full system, including both the silver clusters and the pyrazine molecule, was used for the SERS analysis.

Excitation energies were calculated for $\hbar\omega$ up to 4 eV for the *Junction*, and the simulated absorption spectra are displayed in Fig.2.4b. To elucidate the impact of coupling between the organic moiety and the Ag_{20} cluster, the absorption spectra of the pristine Ag_{20} cluster and the dimer of the Ag_{20} cluster (symbolized as *Dimer*) are further compared in Fig. 2.4b. The obtained results clearly demonstrate that the *Dimer* exhibits the lowest energy absorption band at 370 nm which is absent in the case of the pristine Ag_{20} cluster. The higher energy peak centered at 359 nm is slightly perturbed in comparison with the tetrahedral Ag_{20} cluster. As is immediately manifest from Fig. 2.4b, the lowest-lying electronic excitations of the *Junction* centered at 361 nm

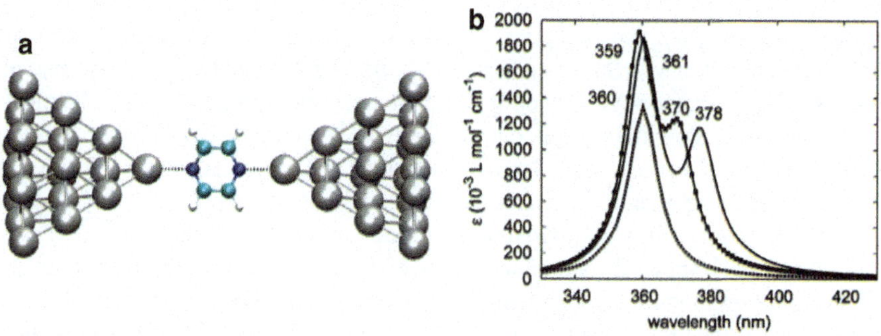

Fig. 2.4 **a** Configuration of the Junction. **b** Simulated absorption spectra (solid line) of the Junction with a molar absorption coefficient in units of 10^{-3} L mol^{-1} cm^{-1} and wavelength in nm. For comparison, absorption spectra for the isolated Ag_{20} tetrahedron (dashed line) and Dimer (solid line with squares) are also plotted [59]. Reprinted with permission from Ref. [59] Copyright (2006) American Chemical Society

and 378 nm are further red shifted by 2 nm and 8 nm, respectively, when compared with the characteristic absorption peaks of the *Dimer*.

The robust coupling between the two Ag-clusters is evident by the presence of two typical electronic absorptions of the *Dimer* as well as the *Junction* material. An analysis of orbitals involved in the excitations showed that both the absorption peaks correspond to excitations involving several orbitals from the two Ag_{20} clusters ("collective" excitations). Albeit the deviation of energies of the low-lying electronic excitations is small, the excitation properties of the *Dimer* bear a resemblance to the dimers of larger nanoparticles [30, 31, 60]. In such nanoparticles, the shift of plasmon wavelength toward the higher wavelength takes place due to the electronic excitations which are parallel to the interparticle axis, whereas the excitations which are perpendicular to the interparticle axis yield a blue-shift of the plasmon wavelength. These excitations are further characterized by the metal–molecule charge transfer (MMCT) transitions even though the "collective" excitations contribute immensely to the Ag_{20}-pyrazine-Ag_{20} junction. The band centered at 360 nm (oscillator strength, $f = 0.425$) with a shoulder at 379 nm (oscillator strength, $f = 0.622$) can be mainly attributed to the MMCT from orbitals beneath the highest occupied molecular orbital (HOMO) of the silver nanocage to the lowest unoccupied molecular orbital (LUMO) of pyrazine [59].

Figure 2.5a, b delineates the calculated Raman spectra (RS) of the pristine pyrazine and the *Junction*, respectively, using the BP86 exchange–correlation functional in combination with triple-ζ polarized Slater-type (TZP) basis set. The polarizability derivatives for Eq. (2.29) were computed at zero frequency. The Raman differential cross section $\left(d\sigma/d\Omega\right)$, however, was computed at the wavelength of 514.5 nm. The enhancement due to the chemical interactions between the molecule and the silver cluster(s) is only seen here (Chemical (CHEM) enhancement), as the silver clusters are not excited.

The augmentation of Raman intensities (RIs) for the *Junction* is estimated to be ~10^5 in general. This amplification is a bit of surprise since CHEM augmentation of only a factor of 10 was predicted in the earlier investigation on the pyridine-Ag_{20} vertex complex [61]. This happens although the pyridine moiety forms a stronger bonding interaction with the Ag_{20} cluster (a shorter Ag–N bond length) in comparison with the Ag_{20}-pyrazine-Ag_{20} junction. As evident by the computed polarizabilities, a substantial alteration in the polarizability along the bonding (x) axis for the *Junction* system (α_{xx}: 2921.6 a.u. [59]) leads to a higher CHEM enhancement. The onset of Raman scattering caused by chemical interactions is thus better captured by the changes in polarizability compared to the modification of bond lengths.

Figure 2.6a, b shows the simulated resonance-enhanced Raman scattering spectra of the *Junction* at incident wavelengths of both 361 and 378 nm, respectively, that come close to the two absorption maxima of the system. Since this is reminiscent of SERS experiments where the plasmon resonance of the metal is excited, it is usually designated as SERS spectra. Nonetheless, the model system is radically different from the experiment because the size of the metal clusters is not enough to possess tangible plasmon resonance.

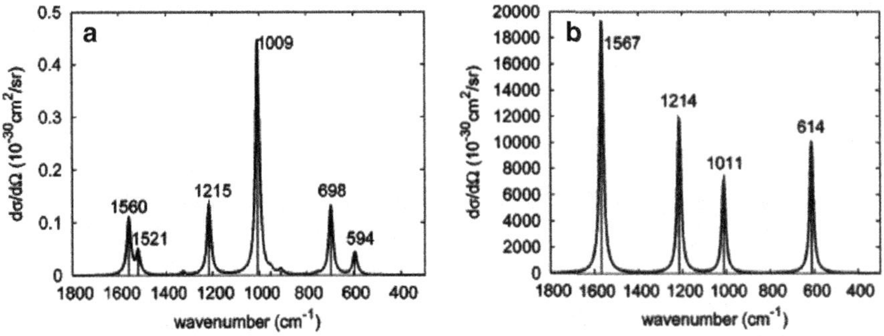

Fig. 2.5 Simulated RS of the isolated pyrazine (**a**), and the Junction (**b**) at an incident wavelength of 514.5 nm based on static polarizability derivatives. Raman differential cross section is in 10^{-30} cm^2/sr and wavenumber in cm^{-1} [59]. Reprinted with permission from Ref. [59] Copyright (2006) American Chemical Society

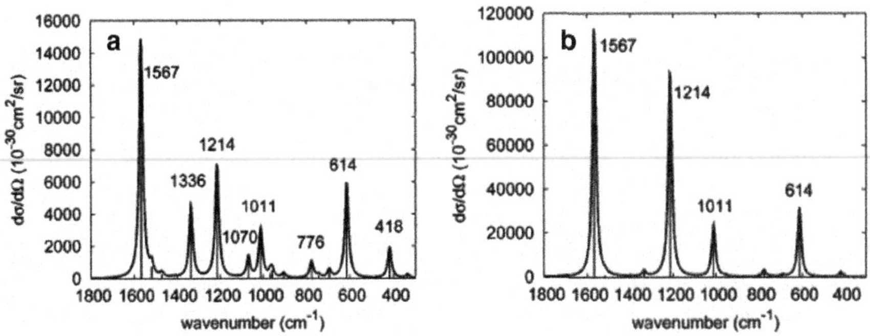

Fig. 2.6 Simulated SERS spectra of the Junction at an incident wavelength of 361 nm (**a**), and 378 nm (**b**), respectively. Raman differential cross section is in 10^{-30} cm^2/sr and wavenumbers in cm^{-1} [59]. Reprinted with permission from Ref. [59] Copyright (2006) American Chemical Society

There are a few observations related to the SERS spectra in Fig. 2.6. The total SERS augmentations for the Ag_{20}-pyrazine-Ag_{20} junction are of the order of 10^6. This is commensurate with the order of enhancement of 10^5 obtained for isolated silver nanoparticles. A CHEM enhancement is observed for the Junction, and it is large as 10^5. This is further significant than earlier understood specifically for nanoparticle aggregates.

The next example is related to the use of the DFT-STO technique to compute the relative resonance Raman intensity using Eq. (2.30). The system chosen is related to the enhancement of $P = O$ stretching mode of the nerve agent (NA) DFP (diisopropy-lfluorophosphate) on the Au_{20} surface. The NA (DFP) is physisorbed on small MgO ($Mg_{16}O_{16}$) and CaO ($Ca_{16}O_{16}$) clusters through the oxygen of the $P = O$ bond. The Raman intensity of the $P = O$ stretching mode ($RI_{P=O}$) due to the surface adsorption is found to be augmented compared to the isolated DFP [62]. This intensity enhancement is interpreted due to the chemical effect, as oxides are not the materials to cause

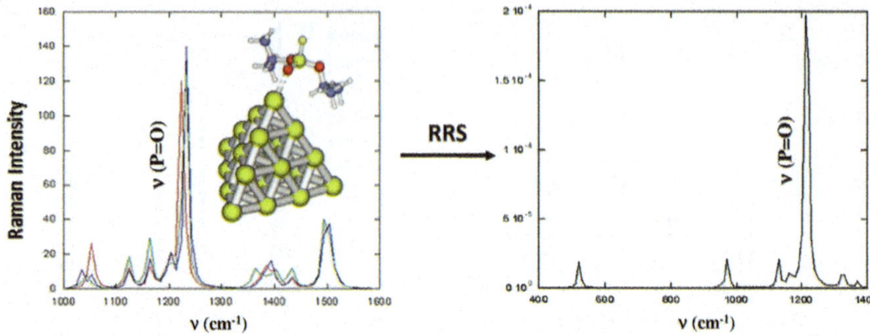

Fig. 2.7 Schematic representation of the Raman spectra and RRS of DFP adsorbed on the Au-cluster [66]. The RRS spectra shows how the ν(P = O) of DFP could be assigned as a characteristic of the NA (DFP) (The detailed studies are available in Ref. [63]. Adapted with permission from Ref. [66] (Advances in In Silico Research on Nerve Agents, Devashis Majumdar, Szczepan Roszak, Jing Wang et al.), Copyright (2014) Springer

plasmonic resonance. It was further shown through quantum-chemical calculations that Au and metal oxide ($Mg_{16}O_{16}$, $Ca_{16}O_{16}$, and their single oxygen defect clusters)-supported Au clusters can trap DFP, and the $RI_{P=O}$ is higher than those of the isolated metal oxide clusters. This enhancement of intensity due to the chemical effect is the key factor to select materials to generate effective $RI_{P=O}$ in RRS. When this molecule is adsorbed on such clusters, the Raman signal regarding the intensity of the P = O stretching mode could be really enhanced many times with respect to the isolated DFP molecule. This is usually done by generating Raman spectra using the excitation wavelength around the specific λ_{max} of gold clusters (or the oxide-supported gold clusters), and such frequency-dependent Raman signal (usually called RRS) is a typical characteristic of these NAs and could be used for their detection [63]. In this discussion, we are citing the example highlighting the RRS enhancement of DFP adsorbed on the Au_{20} cluster.

The BP86 XC potential has been employed to perform relativistic calculations. The calculations utilized triple-ζ Slater-type basis sets for all the atoms with a [$1s^2$-$4d^{10}$] frozen core for Au, [$1s^2$] frozen core for the first-row elements, and a [$1s^2$-$2p^6$] frozen core for P [25]. The simulation of the RRS spectra of Au_{20}...DFP cluster needed the knowledge of the absorption spectra of the system. The experimental and previously estimated λ_{max} for the Au_{20} cluster is 516 nm [64]. The estimated surface plasmon absorption spectra of gold (radius < 25 nm) amounts to 520 nm [65]. The calculated λ_{max} values for Au_{20} (504 – 668 nm using various DFT functionals orbitals DFT/B3LYP/GTO, DFT/BP86/GTO, and DFT/BP86/STO) were found to be quite reasonable (but a bit overestimated) in this respect [63]. The simulated absorption spectra further showed that the λ_{max} in Au_{20}...DFP cluster is red shifted with respect to the pure gold clusters. Moreover, all of the calculated spectra using various strategies as discussed above had broad overlapping zones [63] and thus excitation wavelengths chosen within this range were safely used for the RRS calculations using DFT/BP86/STO technique.

The simulated Raman spectra of Au_{20}...DFP are shown schematically in Fig. 2.7. The field-free Raman spectra are shown on the left panel. The blue-colored spectrum is for the Au_{20}...DFP complex. It has also the Raman spectra of other DFP bound systems and they are not discussed here. The right panel represents the RRS at the 740 nm. The picture clearly shows how the relative Raman intensity could be used to identify the enhanced $P = O$ stretching mode. It is to be mentioned that it is a non-ZORA calculation. ZORA-calculations were also used in such analysis and the results did not show any qualitative change (supplementary section of Ref. [63]).

2.5 Concluding Remarks

Slater-type orbitals (STOs) are nowadays routinely used in quantum-chemical calculations to solve various problems in chemistry. In this article, we have discussed, in detail, about the use of Slater-type orbitals. At the outset, a general discussion is being made to introduce the concept and importance of GTOs and STOs in quantum-chemical calculations (in the framework of Hartree–Fock equation) and compared the advantages and disadvantages of such atomic orbital-based calculations, excluding plane-wave orbitals (mostly used to solve condensed-phase problems), in solving chemistry-related problems of molecules in gas-phase and solvated condition. The concept of STOs is then discussed in detail omitting, of course, the details of mathematical technicalities. These discussions are based on the DFT calculations, and the concept of various Slater atomic basis sets are being introduced. This concept of Slater atomic basis sets is an important part in carrying out DFT-STO-type calculations. In the final part of the article, a few specific examples related to the application of DFT-STOs for different chemical problems are discussed. These included basic benchmark calculations on simple molecular structures, excitation energy calculations, excitation energy spectrum of UO_2^{2+}, and examples of SERS spectrum analysis. We have thus covered with such examples the use of DFT-STOs in a few important fields of quantum chemistry. The applications of DFT-STOs are quite numerous nowadays, and our examples covered only a small part of it to introduce the importance of such an approach in quantum chemistry. Interested researchers can easily find many of such applications in various high-level journals.

Acknowledgements This work has been supported by NSF-CREST (Award No. 154774) and EPSCOR R-II (Award No. OIA - 1632899). One of the authors (S.R.) acknowledges the financial support by a statutory activity subsidy from Polish Ministry of Science and Technology of Higher Education for the Faculty of Chemistry of Wroclaw University of Technology.

References

1. Szabo A, Ostlund NS (1996) Modern quantum chemistry. Introduction to advanced electronic structure theory. Dover Publications Inc., New York
2. Parr RG, Yang W (1989) Density-functional theory of atoms and molecules. Oxford University Press, Oxford
3. Purvis GD III, Bartlett RJ (1982) A full coupled-cluster singles and doubles model – the inclusion of disconnected triples. J Chem Phys 76:1910–1918
4. Scuseria GE, Janssen CL, Schaefer HF III (1988) An efficient reformulation of the closed-shell coupled cluster single and double excitation (CCSD) equations. J Chem Phys 89:7382–7387
5. Scuseria GE, Schaefer III, HF (1999) Is coupled cluster singles and doubles (CCSD) more computationally intensive than quadratic configuration-interaction (QCISD)? J Chem Phys 90:3700–3703
6. Watts JD, Gauss J, Bartlett RJ (1993) Coupled-cluster methods with noniterative triple excitations for restricted open-shell Hartree-Fock and other general single determinant reference functions. Energies and analytical gradients. J Chem Phys 98:8718–8733
7. Pople JA, Head-Gordon M, Raghavachari K (1987) Quadratic configuration interaction—a general technique for determining electron correlation energies. J Chem Phys 87:5968–5975
8. Simons J, Nichols J (1999) Quantum mechanics in chemistry. Oxford University Press, New York (1997); Jensen F (1988) Introduction to computational chemistry. Wiley, New York
9. Roos BO (ed) (1994) Lecture notes in quantum chemistry II. Springer, Heidelberg
10. Roos BO (ed) (1992) Lecture notes in quantum chemistry VII. Springer, Heidelberg
11. Hehre WJ, Stewart RF, Pople JA (1969) Self-Consistent molecular orbital methods. I. Use of Gaussian expansions of Slater-type atomic orbitals. J Chem Phys 51:2657−2664
12. Magalhães AL (2014) Gaussian-type orbitals versus Slater-type orbitals: a comparison. J Chem Edu 91:2124–2127
13. EMSL basis set exchange, https://bse.pnl.gov/bse/portal
14. Harris FE, Michels HH (1967) The evaluation of molecular integrals for Slater-type orbitals. In: Prigogine I (ed) Advances in chemical physics vol XIII, pp 205– 265. Wiley, New York (1967)
15. Tai H (1979) On the evaluation of molecular multicentre integrals for Slater-type orbitals. J Phys B 12:177–185
16. Boerrigter PM, Te Velde G, Baerends E (1988) Three-dimensional numerical integration for electronic structure calculations. Int J Quantum Chem 33:87–113
17. Te Velde G, Baerends E (1992) Numerical integration for polyatomic systems. J Comput Phys 99:84–98
18. Barnett MP (2000) Symbolic calculations of auxiliary functions for molecular integrals over Slater orbitals. Int J Quantum Chem 76:464–472
19. Talman JD (2003) Numerical methods for multicenter integrals for numerically defined basis functions applied in molecular calculations. Int J Quantum Chem 93:72–90
20. Ziegler T, Rauk A (1977) On the calculation of bonding energies by the Hartree-Fock Slater method. Theoret. Chim Acta (Berl.) 46:1–10
21. Versluis L, Ziegler T (1987) The determination of molecular Structures by density functional theory. The evaluation of analytical energy gradients by numerical integration. J Chem Phys 88:322–328
22. Fan L, Ziegler T (1991) Optimization of molecular structures by self-consistent and nonlocal density functional theory. J Chem Phys 95:7401–7408
23. Kohn W, Sham L (1965) Self-Consistent equations including exchange and correlation effects Phys Rev A 140:1133–1138
24. Velde GT, Bickelhaupt FM, Baerends EJ, Fonseca Guerra C, Van Gisbergen SJA, Snijders JG, Ziegler T (2001) Chemistry with ADF. J Comput Chem 22:931–967
25. ADF2017, SCM, Theoretical Chemistry, Vrije Universiteit, Amsterdam, The Netherlands, https://www.scm.com

26. van Gisbergen SJA, Snijders JG, Baerends EJ (1995) A density functional theory study of frequency-dependent polarizabilities and Van der Waals dispersion coefficients for polyatomic molecules. J Chem Phys 103:9347–9354
27. van Gisbergen SJA, Osinga VP, Gritsenko OV, van Leeuwen R, Snijders JG, Baerends EJ (1996) Improved density functional theory results for frequency-dependent polarizabilities, by the use of an exchange-correlation potential with correct asymptotic behavior. J Chem Phys 105:3142–3151
28. van Gisbergen SJA, Snijders JG, Baerends EJ (1996) Application of time-dependent density functional response theory to Raman scattering. Chem Phys Lett 259:599–604
29. van Gisbergen SJA, Snijders JG, Baerends EJ (1998) Calculating frequency-dependent hyperpolarizabilities using time-dependent density functional theory. J Chem Phys 109:10644–10656
30. van Gisbergen SJA, Snijders JG, Baerends EJ (1998) Accurate density functional calculations on frequency-dependent hyperpolarizabilities of small molecules. J Chem Phys 109:10657–10668
31. van Lenthe E, Snijders JG, Baerends EJ (1996) The zero-order regular approximation for relativistic effects: The effect of spin–orbit coupling in closed shell molecules. J Chem Phys 105:6505–6516
32. van Lenthe E, Ehlers AE, Baerends EJ (1999) Geometry optimizations in the zero order regular approximation for relativistic effects. J Chem Phys 110:8943–8953
33. Raffenetti RC (1973) Even-tempered atomic orbitals. II. Atomic SCF wavefunctions in terms of even-tempered exponential bases. J Chem Phys 59:5936–5949
34. Chong DP, Van Lenthe E, Van Gisbergen S, Jan Baerends E (2004) Even-tempered Slater-type orbitals revisited: from hydrogen to krypton. J Comput Chem 25:1030–1036
35. Becke AD (1988) Density-functional exchange-energy approximation with correct asymptotic behavior. Phys Rev A 88:3098–3100
36. Perdew JP (1986) Density-functional approximation for the correlation energy of the inhomogeneous electron gas. Phys Rev B 33:8822–8824
37. van Faassen M, de Boeij PL (2004) Excitation energies for a benchmark set of molecules obtained within time-dependent current-density functional theory using the Vignale-Kohn functional. J Chem Phys 120:8353–8363
38. van Faassen M, de Boeij PL (2004) Excitation energies of π conjugated oligomers within time-dependent current-density-functional theory. J Chem Phys 121:10707–10714
39. Vosko SH, Wilk L, Nusair M (1980) Accurate spin-dependent electron liquid correlation energies for local spin density calculations: a critical analysis. Can J Phys 58:1200–1211
40. Shuai Z, Brédas JL (2000) Coupled-cluster approach for studying the electronic and nonlinear optical properties of conjugated molecules. Phys Rev B 62:15452–15460
41. Granville MF, Kohler BE, Snow JB (1981) Franck-Condon analysis of the $1^1A_g \rightarrow 1^1B_u$ absorption in linear polyenes with two through six double bonds. J Chem Phys 75:3765–3769
42. Chakraborty D, Lagowski JB (2001) Configuration interaction study of singlet excited state of thiophene and its cyano derivative oligomers. J Chem Phys 115:184–194
43. Chadwick JE, Kohler BE (1994) Optical spectra of isolated s-cis- and s-trans-bithiophene: torsional potential in the ground and excited states. J Phys Chem 98:3631–3637
44. Colditz R, Grebner D, Helbig M, Rentsch S (1995) Theoretical studies and spectroscopic investigations of ground and excited electronic states of thiophene oligomers. Chem Phys 201:309–320
45. Guay J, Kasai P, Diaz A, Wu R, Tour JM, Dao LH (1992) Chain-length dependence of electrochemical and electronic properties of neutral and oxidized soluble .alpha.,.alpha.-coupled thiophene oligomers. Chem Mater 4:1097–1105
46. Cornehl HH, Heinemann C, Marqalo J, de Mates AP, Schwarz H (1996) The "bare" uranyl (2^+) ion, $UO_2{}^{2+}$. Angew. Chem. Int Ed Eng 35:891–894
47. Hunt RD, Andrews L (1993) Reactions of pulsed-laser evaporated uranium atoms with molecular oxygen: Infrared spectra of UO, UO_2, UO_3, $UO_2{}^+$, $UO_2{}^{2+}$, and UO_3-O_2 in solid argon. J Chem Phys. 98:3690–3696

48. Zhang Z, Pitzer RM (1999) Application of relativistic quantum chemistry to the electronic energy levels of the uranyl ion. J Phys Chem A 103:6880–6886
49. Craw JS, Vincent MA, Hillier IH, Wallwork AL (1995) Ab initio quantum chemical calculations on uranyl UO_2^{2+}, plutonyl PuO_2^{2+}, and their nitrates and sulfates. J Phys Chem 99:10181–10185
50. Ismail N, Heully J-L, Saue T, Daudey J-P, Marsden CJ (1999) Theoretical studies of the actinides: method calibration for the UO_2^{2+} and PuO_2^{2+} ions. Chem Phys Lett 300:296–302
51. de Jong WA, Visscher L, Nieuwpoort WC (1999) On the bonding and the electric field gradient of the uranyl ion. J Mol Struct: Theochem 458:41 – 52
52. Majumdar D, Balasubramanian K, Nitsche H (2002) A comparative theoretical study of bonding in UO_2^{++}, UO_2^{+}, UO_2, OUCO, $O_2U(CO)_2$ and UO_2CO_3. Chem Phys Lett 361:143–151
53. Pierloot K, van Besien E (2005) Electronic structure and spectrum of UO_2^{2+} and $UO_2Cl_4^{2-}$. J Chem Phys 123:204309–204310
54. Pierloota K, van Besien E, van Lenthe E, Baerends EJ (2007) Electronic spectrum of UO_2^{2+} and $[UO_2Cl_4]^{2-}$ calculated with time-dependent density functional theory. J Chem Phys 126:194311–194318
55. Schipper PRT, Gritsenko OV, van Gisbergen SJA., Baerends EJ (2000) Molecular calculations of excitation energies and „hyper…polarizabilities with a statistical average of orbital model exchange-correlation potentials. J Chem Phys 112:1344–1352
56. Neugebauer J, Reiher M, Kind C, Hess BA (2002) Quantum chemical calculation of vibrational spectra of large molecules—Raman and IR spectra for Buckminsterfullerene. J Comput Chem 23:895–910
57. Reiher M, Neugebauer J, Hess BA (2003) Quantum chemical calculation of Raman intensities for large molecules: the photoisomerization of $[\{Fe'S_4'(PR_3)\}_2(N_2H_2)]$ ($'S_4'^{2-}$ = 1,2-bis(2-mercaptophenylthio)-ethane(2−)). Z Phys Chem 217:91–103
58. Neugebouer J, Baerends EJ, Efremov EV, Ariese F, Gooijer C (2005) Combined theoretical and experimental deep-uv resonance raman studies of substituted pyrenes. J Phys Chem A 109:2100–2106
59. Zhao LL, Jensen L, Schatz GC (2006) surface-enhanced raman scattering of pyrazine at the junction between two Ag_{20} nanoclusters. Nano Lett 6:1229–1234
60. Zhao LL, Kelly KL, Schatz GC (2003) The extinction spectra of silver nanoparticle arrays: influence of array structure on plasmon resonance wavelength and width. J Phys Chem B 107:7343–7350
61. Zhao LL, Jensen L, Schatz GC (2006) Pyridine−Ag_{20} cluster: a model system for studying surface-enhanced raman scattering. J Am Chem Soc 128:2911–2919
62. Kolodziejczyk W, Majumdar D, Roszak S, Leszczynski J (2007) Probing the role of P=O stretching mode enhancement in nerve-agent sensors: Simulation of the adsorption of diisopropylfluorophosphate on the model MgO and CaO surfaces. Chem Phys Lett 450:138–143
63. Majumdar D, Roszak S, Leszczynski J (2010) Density functional theory based studies on the nature of Raman and resonance Raman scattering of nerve agent bound to gold and oxide-supported gold clusters: a plausible way of detection. J Phys Chem A 114:4340–4353
64. Cao YC, Jin R, Mirkin CA (2002) Nanoparticles with Raman spectroscopic fingerprints for DNA and RNA detection. Science 297:15636–15640
65. Link S, Wang ZL, El-Sayed MA (1999) Alloy formation of gold−silver nanoparticles and the dependence of the plasmon absorption on their composition. J Phys Chem B 103:3529–3533
66. Majumdar D, Roszak S, Wang J, Dinadayalane TC, Rasulev B, Pinto H, Leszczynski J (2014) Advances in In Silico Research on Nerve Agents. In: Leszczynski J, Shukla MK (eds) Practical aspects of computational chemistry III. Springer, New York, pp 283–320

Chapter 3
Local Orbitals in Quantum Chemistry

Nadia Ben Amor, Stefano Evangelisti, Thierry Leininger, and Dirk Andrae

3.1 Introduction

The formalism of Quantum Mechanics is inseparably connected to the concept of wave function, which is a non-local object by definition. This fact was already clear from the very beginning of this theory, and pointed out in the famous debates [1] that followed the development of the Schrödinger Equation, in 1926 [2].

This aspect leads to difficulties in conciliating Quantum Mechanics with other branches of Physics or different sciences, like Chemistry or Biology. Things are particularly evident in the case of Chemistry. Traditionally, Chemistry is associated with the concept of bond, which is the entity that ties together two neighbor atoms in a molecule. Quantum Chemistry, and in particular the Electronic Structure branch of it, has the ambition of deriving the bonding mechanism in a molecule from the principles of Quantum Mechanics. Therefore, one has to use the wave function of a molecule, a non-local object, in order to define the bonds of the molecule, which are essentially local objects. Despite the tremendous success of Quantum Chemistry in explaining and predicting the behavior of a countless number of molecular systems, it is not always easy to conciliate the two points of view. The difficulties connected to the interpretation of Quantum Mechanics are obviously totally beyond the scope of this work. Nevertheless, we believe that the duality between the non-local nature of the wave function and the locality of our common everyday experience is at the root of the localized/delocalized description of the Molecular Orbitals of a system.

Let us consider, for instance, the case of a cyclic polyene, composed of N units of CH_2 type, placed in a regular way on a circle. The situation is particularly evident

N. Ben Amor · S. Evangelisti (✉) · T. Leininger
Laboratoire de Chimie et Physique Quantiques, Université de Toulouse et CNRS, UT3 - Paul Sabatier 118, Route de Narbonne, 31062 Toulouse, France
e-mail: stefano.evangelisti@irsamc.ups-tlse.fr

D. Andrae
Physikalische und Theoretische Chemie, Institut für Chemie und Biochemie, Freie Universität Berlin, Arnimallee 22, 14195 Berlin, Germany

E. Perlt (ed.), *Basis Sets in Computational Chemistry*, Lecture Notes in Chemistry 107,
https://doi.org/10.1007/978-3-030-67262-1_3

when we are in presence of spatial symmetry, and we assume that the symmetry of the resulting structure is (at least) C_N. At the equilibrium geometry, the usual chemical intuition tells us that the whole structure is held together by a series of C-C simple bonds that are established between every pair of neighboring carbon atoms. However, if we perform a straightforward calculation, for instance at SCF or DFT level, or even a semi-empirical one, the picture that emerges is totally different. In a one-electron picture, the electrons occupy orbitals that belong to irreducible representations of the symmetry group of the system. For this reason, the MOs of the system will be totally delocalized over the entire chain, regardless of its length. Notice that this situation is by no means specific to Chemistry. A similar situation occurs, in fact, in Solid-State Physics, with the two alternative and complementary approaches of Bloch and Wannier wave functions.

This dichotomy between a localized versus a delocalized philosophy goes back to the very beginning of Quantum Chemistry. The concept of a bond resulting from an electron pair shared between two atoms was first proposed by Lewis [3] in 1916, even before the Schrödinger equation. This concept was reformulated by Heitler and London [4] in 1927, who used Schrödinger's wave equation (published just one year earlier!) to show how two hydrogen-atom wave functions combine themselves to form the bond. This was the birth to the Valence-Bond (VB) formalism. In VB, each bond is built by using orbitals that have essentially an atomic character. The resulting orbitals are therefore strictly localized. On the opposite, the Canonical Orbitals that emerge from an SCF or CASSCF calculation are often spread out over several atoms (sometimes the entire system, as was seen in the previous example). Historically, the canonical description became much more popular than the localized one since it is possible to associate the difference between two MO energies with a well-defined measurable quantity, i.e., the frequency of the electronic transition associated with the two orbitals.

In the present context, we are neither interested in the subtle, and still open, problems posed by the nonlocality of the wave function nor in the emergence of locality out of an entangled wave function. From our point of view, it is interesting to study how the different representations of the orbitals, localized versus delocalized, are related, and which advantages and/or disadvantages they both have. In fact, the global many-electron wave function is a tremendously complex object, and the orbitals are usually the building blocks needed for its description. Since the important object is the electronic wave function, it turns out that it is possible to perform a *unitary transformation* on these orbitals without changing the final wave function, and this fact is the crucial point that is behind the freedom of using either a localized or a delocalized description of the system. The type of unitary transformation that can be applied to the orbitals depends on the wave function we are using to describe our system. Broadly speaking, in the vast majority of cases, we have to choose from one of the three following possibilities: Full Configuration Interaction (FCI), Hartree–Fock Self Consistent Field (HF-SCF), and Complete Active Space SCF (CASSCF).

1. FCI: the wave function is invariant under any arbitrary unitary transformation of the orbitals, provided the CI coefficients are accordingly transformed.
2. HF-SCF: the wave function is invariant under any unitary transformation that does not mix the occupied orbitals with the virtual ones. Under this condition, the HF determinant remains a single determinant.
3. CASSCF: the wave function is invariant under three separate unitary transformations, in each of which only orbitals within one of the three classes "doubly occupied orbitals", "active orbitals", and "virtual orbitals" are separately rotated. Notice that the rotation of active orbitals needs a transformation of related CI coefficients as well.

In most cases, the localization procedure is aimed at obtaining a set of local orbitals, that will be subsequently used to express the desired wave function, or to investigate its nature. Broadly speaking, the existing localization procedures can be divided into two large classes:

1. A posteriori: The Localized Molecular Orbitals (LMO) are obtained from the set of "Canonical" Molecular Orbitals (CMO) (HF-SCF, CASSCF, or other) of a previous calculation by using a suitable localization procedure.
2. A priori: By starting from a guess of LMO, the HF-SCF or CASSCF convergence procedure is imposed, by taking into account the fact that the orbitals must remain as local as possible (in practice, they are often required not to be too different from the guess orbitals [5, 6]).

In the recent past, local orbitals enjoyed a renewed interest, since they have been used to implement Linear-Scaling (LS) algorithms. LS approaches consist of a reformulation of already existent Quantum Chemistry methods in order that they scale linearly as a function of the system size. This is possible since the interaction between different fragments goes to zero with increasing distance. In order to take advantage of these vanishing interactions, it is necessary to work with localized orbitals. LS has been achieved so far for Single-Reference Approaches (SCF), and for the treatment of the Dynamical Correlation: Perturbation Theory, Coupled Cluster, Configuration Interaction, and also for Density Functional Theory. Indeed, using localized orbitals allows to reduce the size of the determinant or configuration space by restricting the excitations to neighbor orbitals. However, non-Dynamical Correlation, needed to deal with Quasi-Degenerate systems, is often very important in chemistry. In these cases, localized orbitals may be useful for selecting the molecular region where the physical processes take place. Some examples are as follows:

- Chemical reactions: different determinants can be dominant on the two sides of the Transition State.
- Electronically excited states: they are often of a Multi-Reference (MR) nature.
- Magnetic systems: there is usually a competition among a large number of Quasi-Degenerate determinants.

In the following, we will concentrate on the construction of local orbitals with some widely used localization procedures for typical molecular systems like small linear molecules, aromatic compounds, and metal complexes. In Sect. 3.2, the mathematical formalism used in the different localization procedures is described. Then the different techniques are illustrated in Section 3.3 via application to a particularly straightforward system, two weakly interacting H_2 molecules described by the Hückel approximation. Section 3.4 is dedicated to selected applications: small molecules (Sects. 3.4.1–3.4.3); coordination complexes (Sect. 3.4.4); and Polycyclic Aromatic Hydrocarbons (Sect. 3.4.5). Finally, some Conclusions are drawn is Sect. 3.5. The Reference section is split into two parts. The first includes all references explicitly quoted in the text. A second part from reference 67 to the end contains pertinent articles that are not cited in the work, organized in chronological order.

3.2 Localization Formalism

We start from the usual Born–Oppenheimer (BO) scheme, and we use the non-relativistic "exact" Hamiltonian. It is composed of zero-, one-, and two-body terms ($\mathcal{H}_0$ is actually a *constant*, E_0):

$$\mathcal{H} = \mathcal{H}_0 + \mathcal{H}_1 + \mathcal{H}_2, \tag{3.1}$$

where

$$\mathcal{H}_0 \equiv E_0 = \sum_{A<B} \frac{Z_A Z_B}{\|\mathbf{R}_A - \mathbf{R}_B\|}, \tag{3.2}$$

$$\mathcal{H}_1 \equiv \mathcal{T} + \mathcal{V} = -\frac{1}{2}\sum_a \nabla_a^2 + \sum_{A,a} \frac{-Z_A}{\|\mathbf{R}_A - \mathbf{r}_a\|}, \tag{3.3}$$

$$\mathcal{H}_2 \equiv \mathcal{W} = \sum_{a<b} \frac{1}{\|\mathbf{r}_a - \mathbf{r}_b\|}. \tag{3.4}$$

In the following, atomic units are used; the labels A and B indicate nuclei, while a and b indicate electrons. The "exact" (in the BO approximation) wave function of the system is a solution of the electronic Schrödinger Equation

$$\mathcal{H}\Psi(\mathbf{x}) = E\Psi(\mathbf{x}), \tag{3.5}$$

where $\mathcal{H}$ is the BO Hamiltonian and E the BO energy. The vector $|\Psi\rangle$ belongs to the complete Hilbert space spanned by all Slater determinants having the given number of electrons, m, with spatial and spin coordinates, $\mathbf{x}$. The wave function is supposed to be normalized, i.e., $\langle\Psi|\Psi\rangle = 1$. The total energy E is the sum of the zero-, one-, and two-body energies:

$$E = E_0 + E_1 + E_2. \tag{3.6}$$

We place ourselves in the Linear Combination of Atomic Orbital (LCAO) formalism and expand the Molecular Orbitals (MO) of the system in terms of a set of (in general) non-orthogonal Atomic Orbitals (AO). Notice that, however, in what follows the adoption of a LCAO scheme is not really required. In most ab initio approaches, there is a first step that involves an orbital optimization, in order to obtain the best MO set according to some criterion. These orbitals are often called the "Canonical" Molecular Orbitals (CMO) of the system according to a given method. In most approaches, the CMO are obtained through the diagonalization of a one-electron Hamiltonian, of which the CMO are eigenvectors. If, as it is, in general, the case, this Hamiltonian commutes with the operators associated with the symmetry point group of the system, the CMO will belong to some irreducible representation (irrep) of the symmetry group. Therefore, corresponding AO placed on atoms that are equivalent by symmetry will have coefficients that are dictated by the symmetry irrep to which the CMO belongs. This is, particularly for highly symmetric small molecules, the main reason of the delocalization of the CMO. Notice that this phenomenon is not related to the physical interaction between different regions of the molecule, but is an unavoidable consequence of the structure of Quantum Theory.

In the case of large non-symmetric systems, exact equivalence among different atoms is missing. The CMO, however, are still spread over a large number of different centers. The localization techniques here described are precisely used in order to minimize the number of centers over which the LMO are significantly different from zero.

We will indicate by χ_μ the AO set, by ψ_i the CMO set, usually obtained via some SCF (either HF or DFT) or CASSCF calculations, while the set ϕ_k is composed of the localized orbitals:

$$\psi_i = \sum_\mu C_{\mu i} \chi_\mu \tag{3.7}$$

and

$$\phi_k = \sum_i U_{ik} \psi_i . \tag{3.8}$$

Here, U_{ik} are the elements of a unitary matrix **U**. The elements $C_{\mu i}$, on the other hand, define a matrix **C** that is usually not unitary, since the AO basis set is, with few exceptions, not orthogonal.

3.2.1 A Posteriori Techniques

Most localization formalisms that have been proposed in Quantum Chemistry are a posteriori techniques: a set of CMO is transformed into a set of LMO through the action of a unitary matrix **U**. In order to define **U**, several techniques can be used: in most cases, one searches for a **stationary point** of a suitably chosen functional $\mathcal{L}$ [7].

We consider here in detail three among the most popular localization schemes: Foster–Boys [8], Edmiston–Ruedenberg [9], and Pipek–Mezey [10]. These localization methods are used for localizing occupied orbitals while the virtual orbitals can be obtained from the atomic orbital basis by projecting the atomic basis functions into the virtual subspace as in the case of the Projected Atomic Orbitals (PAOs). These virtual orbitals are local and orthogonal to the occupied ones but not to one another. The computational cost of the localization methods scales differently. The Foster–Boys, Pipek–Mezey, and Cholesky decomposition [11] procedures scale as N^3, while the Edmiston–Ruedenberg localization scales formally as N^5, where N is the number of electrons. However, several numerical techniques have been proposed to reduce the computational effort of these procedures [12, 13].

3.2.1.1 Foster–Boys

In the Foster–Boys scheme [8, 14] (FB), the functional is based on the spatial position of the orbitals. See reference [15] for an application of the FB formalism in a Many-Body Perturbation-Theory context. The FB localization functional is given by

$$\mathcal{L}_{\text{FB}} = \sum_{k=x,y,z} \sum_{i \in \text{occ}} (\langle \phi_i | r_k^2 | \phi_i \rangle - \langle \phi_i | r_k | \phi_i \rangle^2). \tag{3.9}$$

The local orbitals are obtained by imposing the condition that $\mathcal{L}_{\text{FB}}$ is a **minumum** [16, 17]. As noticed by Resta [18], this is the one-electron part of the trace of the TPS tensor [19–22].

Let us consider a MO $|\phi(\theta)\rangle$ expanded in terms of two pointlike AO $|\chi_1\rangle$ and $|\chi_2\rangle$, placed at $(1, 0, 0)$ and $(-1, 0, 0)$, respectively:

$$|\phi(\theta)\rangle = \cos\theta\, |\chi_1\rangle + \sin\theta\, |\chi_2\rangle. \tag{3.10}$$

We have

$$\begin{aligned} \langle x^2 \rangle_c &= \langle \phi(\theta)|x^2|\phi(\theta)\rangle - \langle \phi(\theta)|x|\phi(\theta)\rangle^2 \\ &= \cos^2\theta + \sin^2\theta - (\cos^2\theta - \sin^2\theta)^2 = 1 - \cos^2 2\theta \end{aligned} \tag{3.11}$$

with similar expressions for y and z. The localization effect is due to fact that $\langle x^2 \rangle_c$ takes its minimum for $\theta = 0 + k\pi/2$ where k is an integer number. This corresponds to $|\phi(\theta)\rangle$ being perfectly localized either on $|\chi_1\rangle$ or $|\chi_2\rangle$.

A variant of the FB procedure is obtained by replacing the second moment of the position by the fourth moment [23, 24]. This procedure has been introduced in order to obtain orbitals restricted to small volume in space with a thin tail as required in most of the local correlation methods. We put

$$\mathcal{L}_{4\mathrm{M}} = \sum_{k=x,y,z} \sum_{i \in \mathrm{occ}} (\langle \phi_i | r_k^4 | \phi_i \rangle - \langle \phi_i | r_k | \phi_i \rangle^4). \tag{3.12}$$

Since this is not conceptually very different, we will not describe this approach in detail. We notice that a drawback of this choice is that $\mathcal{L}_{4\mathrm{M}}$, and hence the final result, is not rotationally invariant.

3.2.1.2 Edmiston–Ruedenberg

The Edmiston–Ruedenberg procedure [9, 25] (ER) is based on the minimization of interorbital Coulomb repulsions, or, equivalently, on the maximization of the intraorbital Coulomb repulsion:

$$\mathcal{L}_{\mathrm{ER}} = \sum_{i \in \mathrm{occ}} \langle \phi_i \phi_i | \frac{1}{r_{\mu\nu}} | \phi_i \phi_i \rangle. \tag{3.13}$$

Therefore, this method is not suitable, for instance, for Hückel or Tight-Binding Hamiltonians. For a variant of the ER method, see reference [26].

Let us consider the same orbitals as in the FB case. For pointlike orbitals, the Coulomb interaction has to be regularized, and we set, as is usually done,

$$\frac{1}{r_{\mu\nu}} \to \frac{1}{1 + r_{\mu\nu}}. \tag{3.14}$$

We have

$$\begin{aligned} \langle \phi(\theta)\phi(\theta) | \frac{1}{1 + r_{\mu\nu}} | \phi(\theta)\phi(\theta) \rangle &= \cos^4\theta + \sin^4\theta + \cos^2\theta \sin^2\theta \\ &= (\cos^2\theta + \sin^2\theta)^2 - \cos^2\theta \sin^2\theta \\ &= 1 - \cos^2\theta \sin^2\theta. \end{aligned} \tag{3.15}$$

The requirement that self repulsion is a maximum leads again to $\theta = 0 + k\pi/2$.

3.2.1.3 Pipek–Mezey

The Pipek–Mezey scheme [10] is based purely on the atomic charges, originally Mulliken's charges. Different other charge definitions [27–30] have also been suggested, in particular to overcome the problem introduced by unbalanced basis sets. For a mathematically defined definition of localization, based on Mulliken's population analysis, see reference [31]. The Pipek–Mezey scheme preserves the σ-π separation in linear and planar systems, unlike the schemes introduced by Foster–Boys and Edmiston–Ruedenberg. The distances between atoms in the molecule do not play any direct role, but they act only indirectly by determining the system wave function, and hence the charges.

The functional to be **maximized** is given by the sum of the squares of the partial charges (orbitals and atoms):

$$\mathcal{L}_{\mathrm{PM}} = \sum_{A} \sum_{i \in \mathrm{occ}} q_{Ai}^2, \tag{3.16}$$

where the sum over A runs over all atoms in the molecule. The application to the example used for the previous methods yields the following charges on atoms 1 and 2:

$$q_1 = \cos^2\theta \tag{3.17}$$

and

$$q_2 = \sin^2\theta. \tag{3.18}$$

Hence, the contribution to the functional is

$$\mathcal{L}_{\mathrm{PM}} = \cos^4\theta + \sin^4\theta = 1 - 2\cos^2\theta\sin^2\theta. \tag{3.19}$$

The requirement that the sum of the squares of the partial charges is a maximum leads, once again, to the condition $\theta = 0 + k\pi/2$.

3.2.1.4 Natural Bond Orbitals and Natural Localized Molecular Orbitals

Another widely used method in computational chemistry is the method of the Natural Bond Orbitals (NBO) or its extension, the Natural Localized Molecular Orbitals (NLMO) approach, developed by Reed and Weinhold [32–34]. This type of analysis is very frequently used, as it is a standard tool [35] available in many popular quantum chemistry packages.

As indicated above, usual localization procedures use unitary transformations from the Canonical Molecular Orbitals (CMO) to obtain localized orbitals. In the NBO approach, the first step consists in obtaining orthogonal Natural Atomic Orbitals (NAO). These NAOs are obtained by block diagonalization of the density matrix $\Gamma^{[1]}$ built from the CMOs. The blocks are defined over all angular momenta of each atom, resulting in NAOs with atomic symmetry. Once the NAOs are obtained, they are used to get the Natural Hybrid Orbitals (NHO) by diagonalizing atomic and two-center density matrices in the NAO basis. These NHOs typically look like lone pairs and usual hybrid orbitals. Finally, the bonding and antibonding NBOs, whose occupation numbers are, respectively, close to 2 and to 0, are obtained for each bond by a final diagonalization of the bond density matrix in the basis of the involved NHOs. Once the NBOs are obtained, they may be efficiently transformed to NLMOs which are similar to the Boys or Edmiston–Ruedenberg orbitals (see [34]).

NBOs provide a representation of the many-electron wave function that is very close to the chemists' Lewis structure, the deviation of the occupation numbers to 2 and 0 being an indicator of electronic delocalization. Further decomposition of the NLMOs over NBOs and complementary atomic hybrids, characterizing the delocalization tail, brings some additional information.

3.2.1.5 Cholesky Decomposition

It has recently been suggested to use the Cholesky decomposition of the one-body density matrix $\Gamma^{[1]}$ in the AO basis in order to obtain a set of suitable LMOs [11]. In principle, the Cholesky decomposition does not ensure *per se* the locality of the resulting orbitals. It should be noticed, however, that if the system breaks down into several non-interacting subsystems, the matrix $\Gamma^{[1]}$ is block diagonal. Since the decomposition of a block-diagonal matrix gives a set of non-overlapping Cholesky vectors, the orbitals obtained in this way are localized onto each one of the fragments. In fact, this approach has several appealing properties:

- Once the AO set is defined, the decomposition is unique.
- High numerical efficiency.
- Non-iterative approach.
- Absence of an initial guess as starting orbitals.
- If needed, virtual molecular orbitals can be obtained exactly in the same way as occupied ones.

Being based on the knowledge of the $\Gamma^{[1]}$ matrix only, this procedure is obviously particularly attractive for density matrix based approaches. We notice that the knowledge of the CMOs is not needed.

The method scales as N^3, but is faster than the Foster–Boys and the Edmiston–Ruedenberg schemes due to the non-iterative approach. There are, however, some drawbacks that should be pointed out. Although the decomposition is unique, it depends on the *order* of the orbitals. Moreover, as shown later in this section, equivalent orbitals are treated in a different way for the simple fact of being placed in a different position in the orbital list. Finally, as recognized by the authors themselves, "the main disadvantage is that the Cholesky MOs are less local than the localized orbitals obtained by the conventional procedures. The major culprit seems to be the inability of the Cholesky localization to reproduce two-center MOs, (i.e., the common "chemical bond")" [11].

As a stringent numerical test, we applied the Cholesky decomposition to a cyclic polyene C_nH_n, containing an even number n of carbon atoms, and treated the π-system at the Hückel level. The atomic $2p_\pi$ orbitals are assumed to be arranged in consecutive order, with every center connected to the previous and the following ones. Since the system is cyclic, the last center is connected to the first one. In absence of dimerization (a unique value of the hopping parameter β for all the bonds), the system is gapless and has a metallic character in the limit $n \to \infty$. It is well known

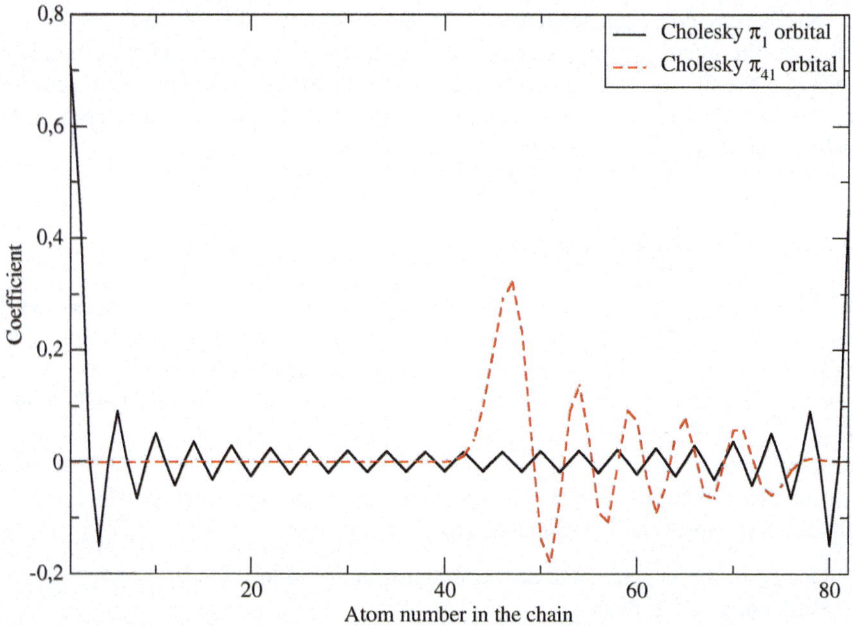

Fig. 3.1 Cholesky orbitals for cyclic polyene $C_{82}H_{82}$

that this is a challenging situation since the localized orbitals in this case can have at most a rational decay [36].

The Cholesky decomposition is applied to the CMOs of π symmetry for the case $n = 82$. Notice that this is a closed-shell system since the Hückel $4k + 2$ rule is satisfied. There are 41 doubly occupied orbitals (a lowest nondegenerate one and twenty doubly degenerate pairs) and 41 empty ones. In Fig. 3.1, the coefficients of the first and the last *occupied* Cholesky orbitals are plotted as a function of the AO number. The first orbital is well localized and symmetrical, with a peak on χ_1, large identical coefficients on χ_2 and χ_{82}, and oscillating coefficients on the remaining atoms that become smaller and smaller as the topological distance form χ_1 is increased. The following orbitals (not shown in the figure) have a less local character. The last doubly occupied Cholesky orbital ϕ_{41}, on the other hand, has strictly zero coefficients on the first 40 atoms (as ensured by the properties of the Cholesky decomposition). It is peaked on atom 46, and has relatively large coefficients on most of the remaining atoms, up to the last AO of the chain. It is by no means symmetrical, and very poorly localized. It is clear that for such a highly symmetric system, the Cholesky decomposition, although perhaps computationally efficient, from a formal point of view is very unsatisfactory.

3.2.2 A Priori Localization

A posteriori techniques are very effective, but are difficult to apply in all those cases where localization is used to identify the physically "interesting" region of the system. For instance, the region where the Active Space should be placed in a CASSCF calculation. In such cases, a priori formalisms can be extremely useful.

A set of local orbitals is built, often based on chemical intuition, and then used as a guess in order to start the iterative SCF or CASSCF procedure. Special care is needed during this procedure, since the diagonalization step usually destroys the locality of the orbitals.

The best option in order to obtain local orbitals is through the action of a localizing unitary operator $\mathbf{U}$ which is defined via an exponential expansion,

$$\mathbf{U} = \exp(\mathbf{A}) \; , \tag{3.20}$$

where $\mathbf{A}$ is an antihermitian operator, $\mathbf{A}^{+} = -\mathbf{A}$ [37]. Notice that in practical calculations $\mathbf{A}$ is usually a real, antisymmetric matrix. We consider now the case where the orbital space is split into several subspaces, and we assume that, for the sake of notational simplicity, the orbitals are ordered in such a way that orbitals belonging to a given subspace are contiguous. The unitary transformations that leave each subspace invariant are block diagonal. Without loss of generality, we can restrict ourselves to the case where only two blocks are present. This is the case, for instance, of HF description, or a CASSCF in the absence of core orbitals. As already discussed, the origin of delocalization in HF formalism is connected to diagonalization. This step is very convenient in order to obtain a well-defined set of orbitals (the canonical orbitals), but in almost any case leads to a set of orbitals strongly delocalized over the whole system. This is the case in particular if the system is composed of two or several equivalent subsystems.

Let us start from a guess of local orbitals. If this set does not satisfy the optimal conditions, the one-body density matrix $\Gamma^{[1]}$ will not be block diagonal. The key point in getting local orbitals is, in the course of the optimization procedure that produces a block-diagonal density matrix, to avoid as much as possible orbital rotations within each block ("in-block rotations"), and to concentrate the rotations between orbitals belonging to different blocks ("out-of-block rotations"). Of course, some in-block rotations are in general unavoidable in order to maintain the orthogonality of the final orbital set, but these operations should be reduced as much as possible.

For this purpose, let us consider the density matrix $\Gamma^{[1]}$ at some point of the iterative procedure. It will be of the form

$$\Gamma^{[1]} = \left\| \begin{matrix} \mathbf{A} & \mathbf{D}^{+} \\ \mathbf{D} & \mathbf{B} \end{matrix} \right\| \, . \tag{3.21}$$

The unitary operator $\mathbf{U}$ is defined in such a way as to block-diagonalize $\Gamma^{[1]}$ by setting the off-diagonal blocks $\mathbf{D}$ and $\mathbf{D}^{+}$ equal to zero, while leaving as much as possible unchanged within the diagonal blocks **A** and **B**. A suitable operator for such an action will be given by

$$\mathbf{U} = \exp(\mathbf{X}) \tag{3.22}$$

with the operator **X** given by

$$\mathbf{X} = \begin{Vmatrix} \mathbf{0} & -\mathbf{Y}^{+} \\ \mathbf{Y} & \mathbf{0} \end{Vmatrix}. \tag{3.23}$$

The **0** diagonal block ensures that, at the lowest order, in-block rotations are absent in **U**. We want to impose the block-diagonal condition

$$\mathbf{U}\,\Gamma^{[1]}\,\mathbf{U}^{+} = \Gamma^{[1]'}, \tag{3.24}$$

which means that $\Gamma^{[1]'}$ has the form

$$\Gamma^{[1]'} = \begin{Vmatrix} \mathbf{A}' & \mathbf{0} \\ \mathbf{0} & \mathbf{B}' \end{Vmatrix}. \tag{3.25}$$

Since we introduce this scheme into an iterative procedure, it is enough to impose Eq. (3.25) *at the first order* in **Y**. At this level, **U** has the form

$$\mathbf{U} = \begin{Vmatrix} \mathbf{1} & -\mathbf{Y}^{+} \\ \mathbf{Y} & \mathbf{1} \end{Vmatrix} \tag{3.26}$$

while $\mathbf{U}^{-1} = \mathbf{U}^{+}$ is given by

$$\mathbf{U}^{-1} = \begin{Vmatrix} \mathbf{1} & \mathbf{Y}^{+} \\ -\mathbf{Y} & \mathbf{1} \end{Vmatrix}. \tag{3.27}$$

Equation (3.25) implies

$$\mathbf{YA} - \mathbf{BY} - \mathbf{YDY}^{+} + \mathbf{D} = \mathbf{0} \tag{3.28}$$

or, to the first order,

$$-\mathbf{YA} + \mathbf{BY} = \mathbf{D}. \tag{3.29}$$

In order to solve this equation and obtain **Y** in a non-iterative way, it is convenient to fully diagonalize the two diagonal blocks **A** and **B** with the help of two unitary operators $\mathbf{V}_A$ and $\mathbf{V}_B$, such that

$$\mathbf{V}_A \mathbf{A} \mathbf{V}_A^{+} = \Lambda_A \tag{3.30}$$

and

$$\mathbf{V}_B \mathbf{B} \mathbf{V}_B^{+} = \Lambda_B, \tag{3.31}$$

where Λ_A and Λ_B are two diagonal matrices with elements λ_a and λ_b, respectively. Now we multiply Eq. (3.29) by $\mathbf{V}_A^+$ from the right and $\mathbf{V}_B$ from the left, and we get

$$-\mathbf{V}_B\mathbf{YAV}_A^+ + \mathbf{V}_B\mathbf{BYV}_A^+ = \mathbf{V}_B\mathbf{DV}_A^+ \quad . \tag{3.32}$$

By taking into account that $\mathbf{V}_A^+\mathbf{V}_A = \mathbf{1}_A$ and $\mathbf{V}_B^+\mathbf{V}_B = \mathbf{1}_B$, the latter can be rewritten as

$$-\mathbf{V}_B\mathbf{YV}_A^+\mathbf{V}_A\mathbf{AV}_A^+ + \mathbf{V}_B\mathbf{BV}_B^+\mathbf{V}_B\mathbf{YV}_A^+ = \mathbf{V}_B\mathbf{DV}_A^+ \quad , \tag{3.33}$$

and finally, with Eqs. (3.30) and (3.31), as

$$-\mathbf{V}_B\mathbf{YV}_A^+\Lambda_A + \Lambda_B\mathbf{V}_B\mathbf{YV}_A^+ = \mathbf{V}_B\mathbf{DV}_A^+ \quad . \tag{3.34}$$

The solution of the equation system (3.34) is now straightforward, because, due to the diagonal form of the matrices Λ_A and Λ_B, the different equations are decoupled. Therefore, we have

$$(\mathbf{V}_B\mathbf{YV}_A^+)_{ab} = \frac{(\mathbf{V}_B\mathbf{DV}_A^+)_{ab}}{\lambda_b - \lambda_a} \quad . \tag{3.35}$$

Finally, the elements of the rectangular matrix $\mathbf{Y}$ can be obtained by transforming the matrix elements of equation (3.35) back into the original basis set, i.e., by multiplying $\mathbf{V}_B\mathbf{YV}_A^+$ by $\mathbf{V}_A$ from the right and $\mathbf{V}_B^+$ from the left.

In practice, in our previous work [38], we implemented the block-diagonalization of the density matrix by fully diagonalizing it through a unitary transformation $\mathbf{U}$, and then avoiding the complete delocalization by multiplying $\mathbf{U}$ from the left by the operator $\mathbf{U}_D^{-1}$, with $\mathbf{U}_D$ given by

$$\mathbf{U}_D = \begin{Vmatrix} \mathbf{V}_A & \mathbf{0} \\ \mathbf{0} & \mathbf{V}_B \end{Vmatrix} \quad . \tag{3.36}$$

Notice that the two procedures coincide for quasi-diagonal density matrices.

Again, we applied this localization scheme to the π system of the cyclic polyene $C_{82}H_{82}$ previously discussed in connection with the Cholesky decomposition. Starting from a guess of localized bonding π orbitals that alternate every two bonds, $\frac{1}{\sqrt{2}}(\chi_1 + \chi_2)$, $\frac{1}{\sqrt{2}}(\chi_3 + \chi_4)$, ..., $\frac{1}{\sqrt{2}}(\chi_{81} + \chi_{82})$, the SCF conditions are iteratively imposed up to convergence. We denote as $\phi_{i,i+1}$ the LMO that derives from the guess orbital $\frac{1}{\sqrt{2}}(\chi_i + \chi_{i+1})$. Two LMOs at convergence are illustrated in Fig. 3.2. The orbitals partly respect the symmetry of the system, since they are equivalent by the D_{41h} subgroup of the full symmetry group of the system, which is D_{82h}. Notice that the symmetry lowering of the LMOs is due to the half-filled character of the energy band, and hence the metallic nature of the wave function. An alternative possibility, showing the same D_{41h} symmetry, would be obtained by starting from the atomic orbitals χ_1, χ_3, ..., χ_{81} (or, alternatively, χ_2, χ_4, ..., χ_{82}), and obtain SCF orbitals that have somehow an "atomic" character. In any case, since there are a total of 41 doubly occupied orbitals, it is impossible to completely respect the full

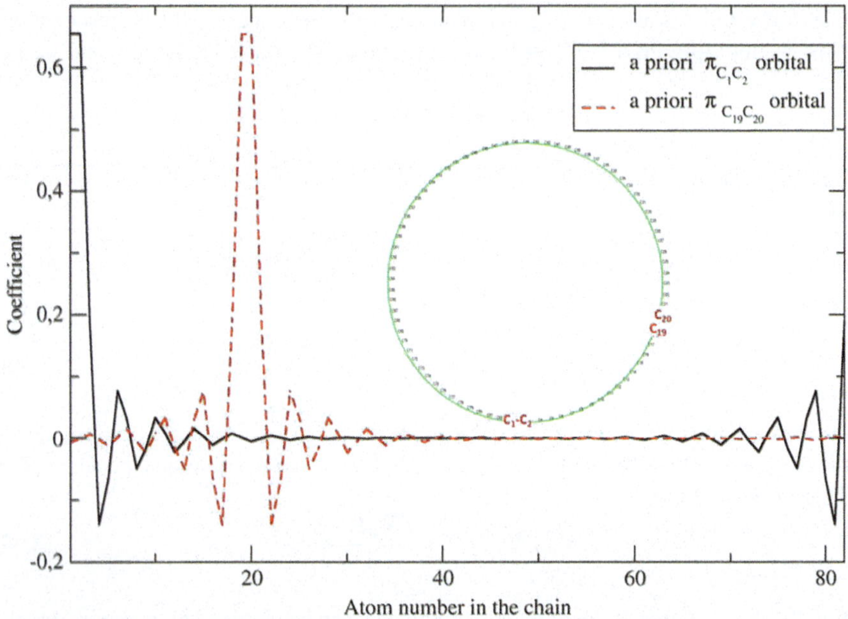

Fig. 3.2 A priori orbitals for cyclic polyene $C_{82}H_{82}$

symmetry of the system. Other, less symmetric, localization would also be possible. Needless to say, *all* these choices of orbitals give the same total wave function, that has the full D_{82h} symmetry of the system.

Finally, one can note that the cost of the a priori localization scales as N^2, N being the number of electrons.

3.3 A Simple Application: The Hückel Hamiltonian

In this section, we examine a simplified model, composed of two weakly interacting Hydrogen molecules, $(H_2)_2$, described by a Hückel-type Hamiltonian. Because of its extreme simplicity, and due to the one-electron nature of interactions of the model, an explicit solution can be found. Moreover, because of the absence of two-electron interactions, the convergence to the HF solution does not require any iterative process.

Each Hydrogen atom of the two dimers is placed at the vertex of a rectangle, as shown in Fig. 3.3. We assume that the intramolecular interaction appearing in the one-electron Hückel-type Hamiltonian is given by the usual Hückel parameter β while the intermolecular interaction is described by a parameter δ. Both β and δ are negative quantities, and we have $|\beta| > |\delta|$.

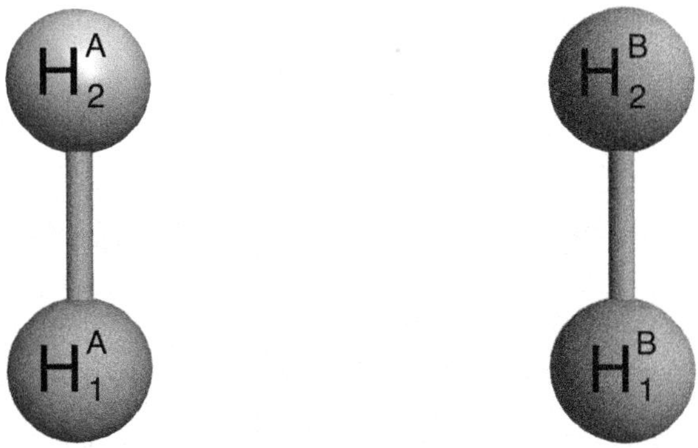

Fig. 3.3 A simple application of the Hückel Hamiltonian: the $(H_2)_2$ dimer

$$\mathbf{H} = \begin{Vmatrix} 0 & \beta & \delta & 0 \\ \beta & 0 & 0 & \delta \\ \delta & 0 & 0 & \beta \\ 0 & \delta & \beta & 0 \end{Vmatrix} . \tag{3.37}$$

Since the Hückel Hamiltonian does not take into account the electron–electron interaction, different-spin electrons do not interact in any way. Electrons having equal spin, on the other hand, interact indirectly via the Pauli exclusion principle. For this reason, it is sufficient in this case to consider electrons of one spin type only, say α.

The two lowest eigenvalues of $\mathbf{H}$ are

$$\varepsilon_+^+ = \beta + \delta \tag{3.38}$$

and

$$\varepsilon_+^- = \beta - \delta \tag{3.39}$$

with the corresponding eigenvectors

$$|\psi_+^+\rangle = \frac{1}{2}(|\chi_1^A\rangle + |\chi_2^A\rangle + |\chi_1^B\rangle + |\chi_2^B\rangle) \tag{3.40}$$

and

$$|\psi_+^-\rangle = \frac{1}{2}(|\chi_1^A\rangle + |\chi_2^A\rangle - |\chi_1^B\rangle - |\chi_2^B\rangle) . \tag{3.41}$$

These two states form the occupied manifold. For completeness, we give also the virtual manifold, which is given by the eigenvalues

$$\varepsilon_-^+ = -\beta + \delta \tag{3.42}$$

and

$$\varepsilon_-^- = -\beta - \delta \tag{3.43}$$

with the corresponding eigenvectors

$$|\psi_-^+\rangle = \frac{1}{2}(|\chi_1^A\rangle - |\chi_2^A\rangle + |\chi_1^B\rangle - |\chi_2^B\rangle) \tag{3.44}$$

and

$$|\psi_-^-\rangle = \frac{1}{2}(|\chi_1^A\rangle - |\chi_2^A\rangle - |\chi_1^B\rangle + |\chi_2^B\rangle)\,. \tag{3.45}$$

We consider in the following subsections the localization of the CMO via a posteriori or a priori techniques. Within the first group, the Edmiston–Ruedenberg approach cannot be used, since the electron–electron repulsion is not taken into account within the Hückel model. The fourth moment procedure, on the other hand, is extremely similar to the Foster–Boys one, so we will concentrate our attention on Foster–Boys and Pipek–Mezey schemes.

3.3.1 A Posteriori Approach

We define the two linear combinations

$$\begin{aligned}|\phi_+^{\text{cs}}(\theta)\rangle &= \cos\theta\,|\psi_+^+\rangle + \sin\theta\,|\psi_+^-\rangle \\ &= \frac{1}{2}[(\cos\theta + \sin\theta)(|\chi_1^A\rangle + |\chi_2^A\rangle) + (\cos\theta - \sin\theta)(|\chi_1^B\rangle + |\chi_2^B\rangle)]\end{aligned} \tag{3.46}$$

and

$$\begin{aligned}|\phi_+^{\text{sc}}(\theta)\rangle &= -\sin\theta|\psi_+^+\rangle + \cos\theta|\psi_+^-\rangle \\ &= \frac{1}{2}[(-\sin\theta + \cos\theta)(|\chi_1^A\rangle + |\chi_2^A\rangle) + (-\sin\theta - \cos\theta)(|\chi_1^B\rangle + |\chi_2^B\rangle)]\,.\end{aligned} \tag{3.47}$$

3.3.1.1 Foster–Boys

In order to localize the occupied orbitals by using the Foster–Boys procedure, we have to minimize the quantity

$$\Lambda(\theta) = \langle x^2(\theta)\rangle_c + \langle y^2(\theta)\rangle_c, \tag{3.48}$$

where

$$\begin{aligned}\langle x^2(\theta)\rangle_c = {} & \langle\phi_+^{\text{cs}}(\theta)|x^2|\phi_+^{\text{cs}}(\theta)\rangle - \langle\phi_+^{\text{cs}}(\theta)|x|\phi_+^{\text{cs}}(\theta)\rangle^2 \\ & + \langle\phi_+^{\text{sc}}(\theta)|x^2|\phi_+^{\text{sc}}(\theta)\rangle - \langle\phi_+^{\text{sc}}(\theta)|x|\phi_+^{\text{sc}}(\theta)\rangle^2\end{aligned} \tag{3.49}$$

and a similar expression holds for $\langle y^2(\theta)\rangle_c$.

Since δ is smaller, in absolute value, than β, orbital localization can happen with respect to the y position. In fact, $\langle x^2(\theta)\rangle_c$ does not depend on θ, and is a constant equal to one for each orbital. The spread in the y direction, on the other hand, is given by two identical contributions for the two orbitals, given by

$$\langle y^2(\theta)\rangle_c = 1 - 4\sin\theta\,\cos\theta\,. \tag{3.50}$$

The total spread, therefore, becomes

$$\Lambda(\theta) = 4 - 2\sin^2 2\theta\,. \tag{3.51}$$

The function $\Lambda(\theta)$ is bounded between 2 and 4, and its plot as a function of θ is shown in Fig. 3.4. For $\theta = 0$ or $\theta = \pi/2$, the function adopts its maximum $\Lambda = 4$. This corresponds to the situation where the new orbitals coincide with the canonical ones, $|\phi_+^{\mathrm{cs}}\rangle = |\psi_+^+\rangle$ and $|\phi_+^{\mathrm{sc}}\rangle = |\psi_+^-\rangle$ (or vice versa). For $\theta = \pi/4$ or $\theta = 3\pi/4$, on the other hand, the functional has its minima, that correspond to $|\phi_+^{\mathrm{cs}}\rangle = |\phi_+^A\rangle$ and $|\phi_+^{\mathrm{sc}}\rangle = |\phi_+^B\rangle$ (or vice versa), where we have defined $|\phi_+^A\rangle = \frac{1}{\sqrt{2}}(|\chi_1^A\rangle + |\chi_2^A\rangle)$ and $|\phi_+^B\rangle = \frac{1}{\sqrt{2}}(|\chi_1^B\rangle + |\chi_2^B\rangle)$, i.e., localized occupied molecular orbitals on each hydrogen molecule.

We notice that, while the Hamiltonian is *diagonal* in its eigenbasis $\{|\psi_\pm^\pm\rangle\}$:

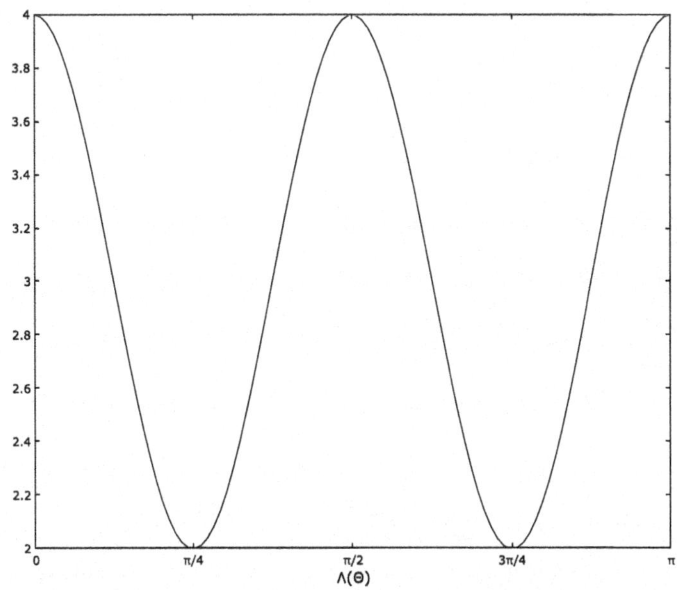

Fig. 3.4 A simple application of the Hückel Hamiltonian: the $(H_2)_2$ dimer—Foster–Boys localization

$$\mathbf{H}_\psi = \begin{Vmatrix} \beta+\delta & 0 & 0 & 0 \\ 0 & \beta-\delta & 0 & 0 \\ 0 & 0 & -\beta+\delta & 0 \\ 0 & 0 & 0 & -\beta-\delta \end{Vmatrix} , \tag{3.52}$$

it is only *block-diagonal* in the basis $\{|\phi_+^A\rangle, |\phi_+^B\rangle, |\phi_-^A\rangle, |\phi_-^B\rangle\}$:

$$\mathbf{H}_\phi = \begin{Vmatrix} \beta & \delta & 0 & 0 \\ \delta & \beta & 0 & 0 \\ 0 & 0 & -\beta & \delta \\ 0 & 0 & \delta & -\beta \end{Vmatrix} . \tag{3.53}$$

However, the Slater determinants that describe the HF ground state in the two cases are the same,

$$|\psi_+^+ \psi_+^-\rangle = |\phi_+^A \phi_+^B\rangle \equiv |\Phi_0\rangle , \tag{3.54}$$

and this shows that the two treatments are equivalent.

3.3.1.2 Pipek–Mezey

We define the Mulliken charges for each occupied orbital and atom ($i = 1, 2$) as the sum of the squares of the coefficients of the orbitals. The Mulliken charges for the $|\phi_+^{\rm cs}\rangle$ and $|\phi_+^{\rm sc}\rangle$ are

$$Q_i^A(\text{cs}) = \frac{1}{2}(\cos\theta + \sin\theta)^2 = \frac{1}{2}(1 + 2\cos\theta\,\sin\theta) , \tag{3.55}$$

$$Q_i^B(\text{cs}) = \frac{1}{2}(\cos\theta - \sin\theta)^2 = \frac{1}{2}(1 - 2\cos\theta\,\sin\theta) , \tag{3.56}$$

$$Q_i^A(\text{sc}) = \frac{1}{2}(\cos\theta - \sin\theta)^2 = \frac{1}{2}(1 - 2\cos\theta\,\sin\theta) , \tag{3.57}$$

$$Q_i^B(\text{sc}) = \frac{1}{2}(\cos\theta + \sin\theta)^2 = \frac{1}{2}(1 + 2\cos\theta\,\sin\theta) . \tag{3.58}$$

The sum of charges over the occupied orbitals and the centers gives obviously 2, since two electrons are in the molecule (we remind that only one type of spin is considered, and hence the charge is to be multiplied by 2).

The Pipek–Mezey localizing functional, $\mathcal{L}_{\rm PM}$, is given by the sum of the squares of the charges,

$$\mathcal{L}_{\rm PM} = 1 + 4\cos^2\theta\,\sin^2\theta . \tag{3.59}$$

The maximum of this functional is reached for $\theta = \pi/4 + k\pi/2$.

3.3.2 A Priori Local Orbitals

We consider now a priori localization. In this case, one selects the physical nature of the occupied and virtual localized orbitals and then imposes the HF procedure while keeping as much as possible of the nature of these localized orbitals. Let us assume we want for our LMOs something located either on fragment A or on fragment B. We start from the LMO guess $|\phi_o^X\rangle = |\chi_1^X\rangle$ and $|\phi_v^X\rangle = |\chi_2^X\rangle$, for $X = \mathrm{A, B}$. This corresponds to a one-body Density Matrix given by

$$\Gamma_X^{[1]} = \begin{Vmatrix} 1 & 0 & 0 & 0 \\ 0 & 0 & 0 & 0 \\ 0 & 0 & 1 & 0 \\ 0 & 0 & 0 & 0 \end{Vmatrix} . \tag{3.60}$$

The unitary transformation that mixes the occupied and virtual orbitals *on each one* of the two fragments, but *not between* them, has the general form

$$\mathbf{U}_X = \begin{Vmatrix} \cos\xi_\mathrm{A} & \sin\xi_\mathrm{A} & 0 & 0 \\ -\sin\xi_\mathrm{A} & \cos\xi_\mathrm{A} & 0 & 0 \\ 0 & 0 & \cos\xi_\mathrm{B} & \sin\xi_\mathrm{B} \\ 0 & 0 & -\sin\xi_\mathrm{B} & \cos\xi_\mathrm{B} \end{Vmatrix} \tag{3.61}$$

with $\xi_\mathrm{A}, \xi_\mathrm{B} \in [0, 2\pi]$.

The transformed density matrix becomes

$$\Gamma^{[1]} = \begin{Vmatrix} \cos^2\xi_\mathrm{A} & \sin\xi_\mathrm{A}\cos\xi_\mathrm{A} & 0 & 0 \\ \sin\xi_\mathrm{A}\cos\xi_\mathrm{A} & \sin^2\xi_\mathrm{A} & 0 & 0 \\ 0 & 0 & \cos^2\xi_\mathrm{B} & \sin\xi_\mathrm{B}\cos\xi_\mathrm{B} \\ 0 & 0 & \sin\xi_\mathrm{B}\cos\xi_\mathrm{B} & \sin^2\xi_\mathrm{B} \end{Vmatrix} . \tag{3.62}$$

The energy associated with this density matrix is

$$E = \mathrm{tr}(\Gamma^{[1]}\mathbf{H}) = 2\beta\,(\sin\xi_\mathrm{A}\cos\xi_\mathrm{A} + \sin\xi_\mathrm{B}\cos\xi_\mathrm{B}) = E(\xi_\mathrm{A}, \xi_\mathrm{B}) .$$

By differentiating this expression with respect to ξ_A and ξ_B and setting the derivatives equal to zero, one is able to find maxima, minima, and saddle points of the energy (see Fig. 3.5). The extrema correspond to $\xi_X = \pi/4, 3\pi/4, 5\pi/4, 7\pi/4$, with $X = \mathrm{A, B}$. The values $\xi_X = \pi/4,\ 5\pi/4$ on both A and B yield minima, while $\xi_X = 3\pi/4,\ 7\pi/4$ correspond to maxima (we remind that β is negative). All other combinations give saddle points.

Notice that the local orbitals obtained via the two procedures are identical. This is due to the extreme simplicity of the present model, and this property does not hold in general.

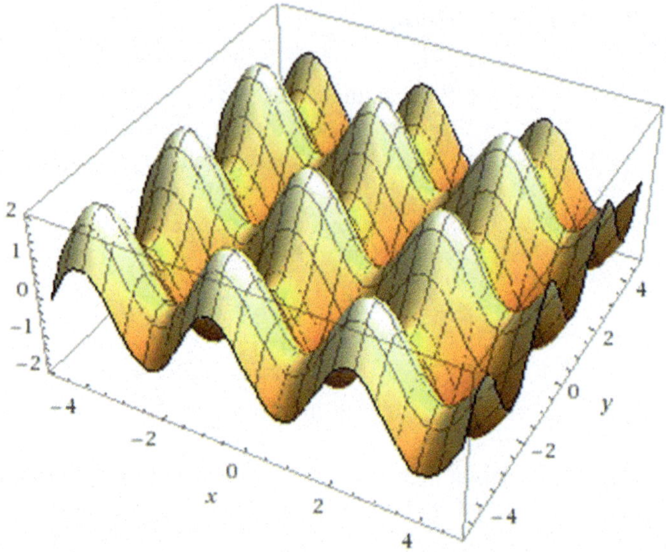

Fig. 3.5 A simple application of the Hückel Hamiltonian: the $(H_2)_2$ dimer—Energy associated with the density matrix in the a priori localization

3.4 Selected Applications of Localized Orbitals

We consider now some examples related to small-size molecules. In particular, we will consider here linear molecules (like C_2H_2, C_2F_2) for which several a posteriori localization schemes are probed and compared to the localized orbitals obtained with an a priori localization scheme (dubbed DoLo). The aspect of the selection of a suitable active space thanks to the localization of the orbitals will also be considered for the bond dissociation of a linear system (C_2HF). Flexibility of the a priori localization is illustrated for the planar benzene molecule and *trans*-stilbene, for both planar and non-planar conformers. Subsequently, the common Boys localization technique is successfully applied to two coordination compounds. Finally, localization in large Polycyclic Aromatic Hydrocarbons (PAH) is investigated with both the Boys and the DoLo methods.

The a priori local orbitals are obtained thanks to the DoLo program. DoLo requires an initial guess of non-orthogonal local orbitals. Usually, two steps follow: (i) the projection of this guess of local orbitals onto the SCF or CASSCF orbitals; the quality of the guess of local orbitals is then improved (`proj_scf` code), and (ii) the Schmidt orthogonalization of the obtained orbitals (separately within the three subspaces of occupied/active/virtual sets of orbitals, with the `schmudort` code). Finally, this multi-step process provides a set of local and orthogonal orbitals, with SCF or CASSCF quality, i.e., it gives the same total energy. In some cases, the projection step is not relevant, as for example when a molecule presents several resonance structures. In that case, a suitable guess set of local orbitals has to be built for each structure. The local orbitals are orthogonalized and optimized by iterative procedures keeping their locality (localized CASSCF [39, 40], Monte–Carlo method [41]). However, most of the time, the local orbitals are used as starting orbitals of a CASSCF calculation or to introduce dynamical correlation, and the three-step (DoLo/`proj_scf`/`schmudort`) procedure is relevant. All the DoLo orbitals presented in this chapter are obtained in this way. Input data for the a priori localization are given in the Appendix 1.

Beyond its role as a tool for analyzing the wave function, orbital localization is also particularly useful to select the active space, in particular for the new-generation Multi-Reference Self Consistent Field (MR-SCF) methods like GASSCF [42, 43], ORMAS [44], FRACCIS [45], etc., or to reduce reference space in the case of Configuration-Interaction calculations. In addition, localized orbitals are a necessary ingredient for the development of linear-scaling methods required to treat dynamical correlation.

The role of symmetry in a localized-orbitals context is particularly important (see also [46]). In fact, the Canonical Molecular Orbitals transform according to irreducible representations of the symmetry point group of the considered molecule. For this reason, the CMOs are intrinsically delocalized, and spread out, in principle, over all the equivalent centers of the molecule. Generally speaking, Local MOs break the symmetry of the system. This fact is unpleasant, since the symmetry of a molecule is extremely useful in order to reduce the computational complexity of the studied system. This is true, in particular, for highly symmetric small molecules, while it is usually less crucial in the case of large systems that often belong to the C_1 symmetry group (in other words, they do not have any symmetry at all). In presence of symmetry, one faces an alternative: either use localized orbitals, and renounce to the symmetry advantages, or keep the Canonical Orbital approach. This fact is well illustrated by the Boys localization, where even the symmetry distinction between σ and π orbitals in the case of conjugated systems is lost. It is possible, however, to use a somehow intermediate approach, and define Symmetry-Adapted LMO, i.e., symmetry combinations of strictly localized orbitals. In this way, all the advantages of localization are essentially kept, while the symmetry information is conserved, in order to reduce the computational cost. This possibility has been implemented, in particular, in the EXSCI [47–49] and DoLo [38] chains. EXSCI is a quasi-linear-scaling MRCI code, which takes into account all the symmetry information at the CI step, in order to reduce the computational complexity of a calculation.

Computational Details

For the small linear systems C_2H_2, C_2F_2, C_2HF, we used the Atomic Natural Orbital (ANO) basis set optimized by Widmark and coworkers [50], taking small contractions: $2s1p$ for H, $3s2p1d$ for the atoms of the second period (C, F). In all cases, we decided not to use higher-angular-momentum orbitals, like f or g orbitals. The same basis sets and contractions are used for the *trans*-stilbene while for the benzene, we used different contractions: $3s2p$ for H, $4s3p2d$ for C. For the polycyclic aromatic hydrocarbons and for the coordination complexes, we used Dunning-type basis sets [51], with overall singly-polarized valence double-zeta quality. The basis set for Ni thus includes f functions.

Different software packages were used, like MOLCAS [52], MOLPRO [53], TURBOMOLE [54], DOLO [55], and the figures were generated with GV [52, 56], Avogadro[57], and MOLDEN [58, 59]. Notice that throughout this article, even if the orbitals were drawn with different orbital visualization software, we tried to keep the same contour value to allow fair comparison.

3.4.1 Small Linear Molecules: C_2H_2, C_2F_2

As a general choice, in an HF context, we will use the notation On as label for the n-th Occupied orbital and, conversely, the notation Vn to label the n-th Virtual orbital (Figs. 3.6, 3.7, 3.8, 3.9, 3.10, 3.11, 3.12, and 3.13). At CASSCF level, this convention, will be augmented by An as label for the n-th Active orbital.

C_2H_2

In Figs. 3.6 and 3.7, the Boys and DoLo localized orbitals of C_2H_2 are shown. The most striking difference between the two sets is the fact that the Boys Orbitals do not reflect the σ–π separation of this system. This is a well-known defect of the Boys orbitals: in order to minimize the Boys functional, combinations of orbitals of different symmetries are generated, so that the resulting orbitals do not belong to any irreducible representation of the symmetry group of the molecule. In fact, the obtained orbitals are hybrid orbitals that transform into a corresponding orbital under the action of some symmetry operation. Notice that this is not related to the non-Abelian nature of the group (in this case, $D_{\infty h}$). As we can see in Fig. 3.6, the C–H σ bonds are well represented by O1 and O3 and do not pose particular problems. The C≡C triple bond, on the other hand, is built with the combination of σ, π_x, and π_y C–C bonds. Three equivalent orbitals are obtained that mutually transform under the action of the C_{3z} symmetry operation. They are O2, O4, and O5 in the figure.

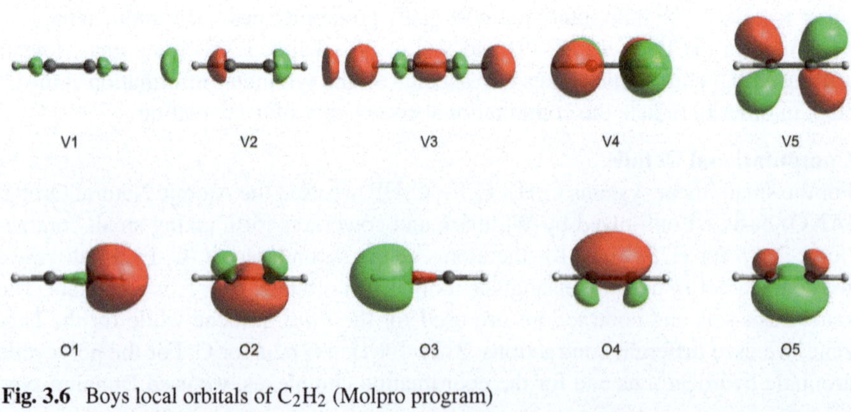

Fig. 3.6 Boys local orbitals of C_2H_2 (Molpro program)

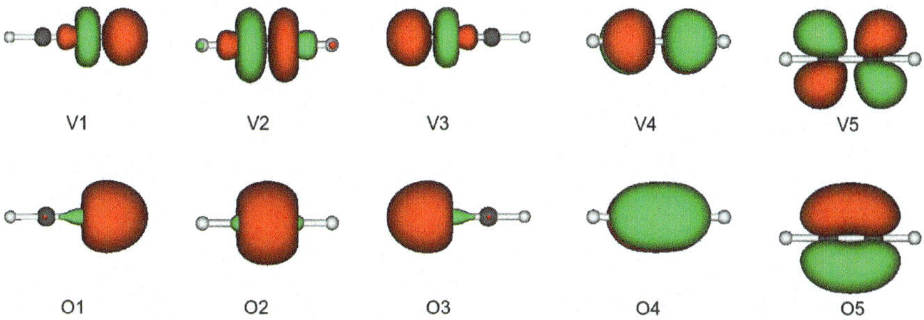

Fig. 3.7 DoLo local orbitals of C_2H_2

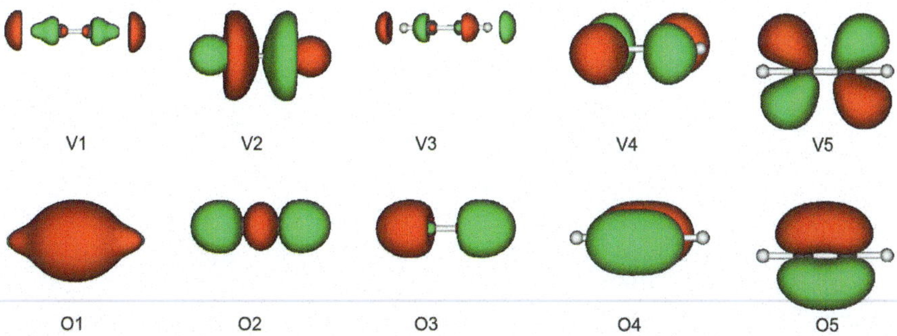

Fig. 3.8 Canonical orbitals of C_2H_2

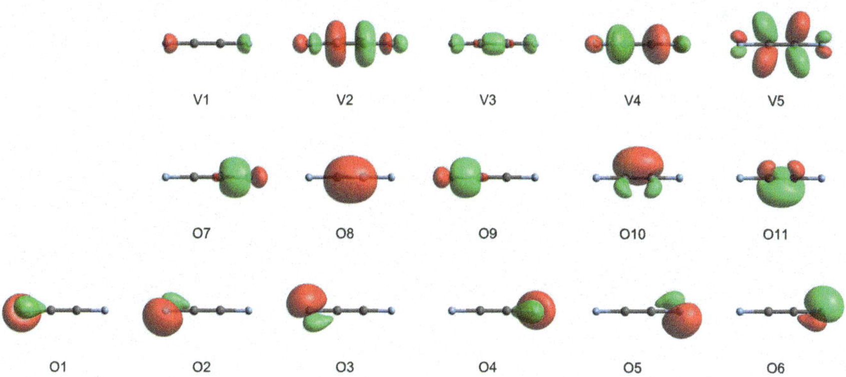

Fig. 3.9 Boys local orbitals of C_2F_2 (Molpro program)

The Boys localization procedure is usually applied only to occupied orbitals, and the virtual ones are left at the canonical level. However, if the procedure is applied to the virtual orbitals obtained with a small basis set, the same type of behavior would be observed.

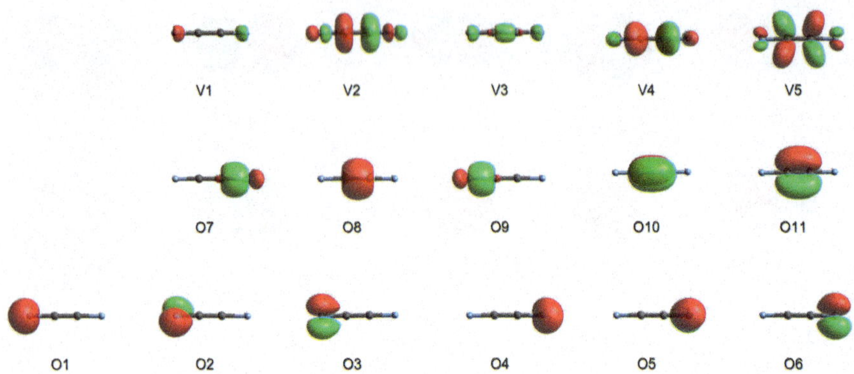

Fig. 3.10 Pipek local orbitals of C_2F_2 (Molpro program)

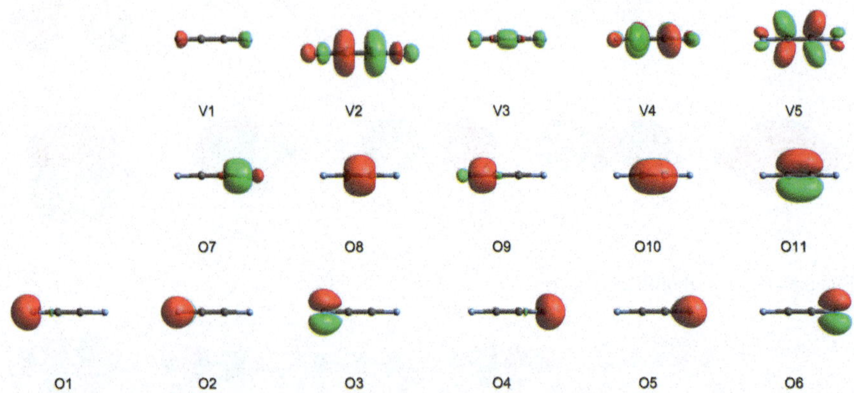

Fig. 3.11 Natural local orbitals of C_2F_2 (Molpro program)

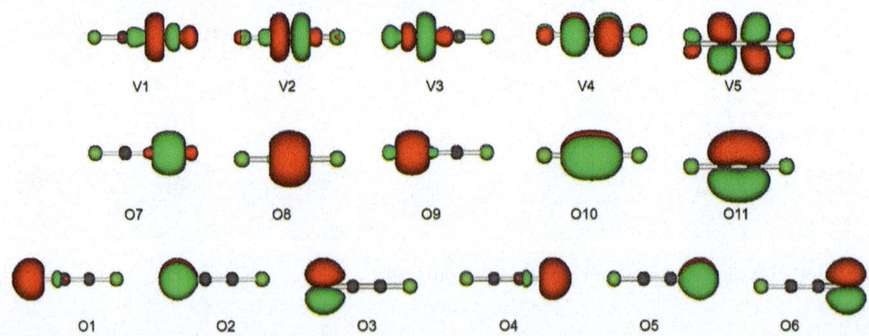

Fig. 3.12 DoLo local orbitals of C_2F_2

In Fig. 3.7, some of the localized orbitals obtained with the DoLo procedure are shown. The σ–π symmetry separation is respected. Notice that this would be true for many of the localization procedures, except for the Boys scheme. Moreover,

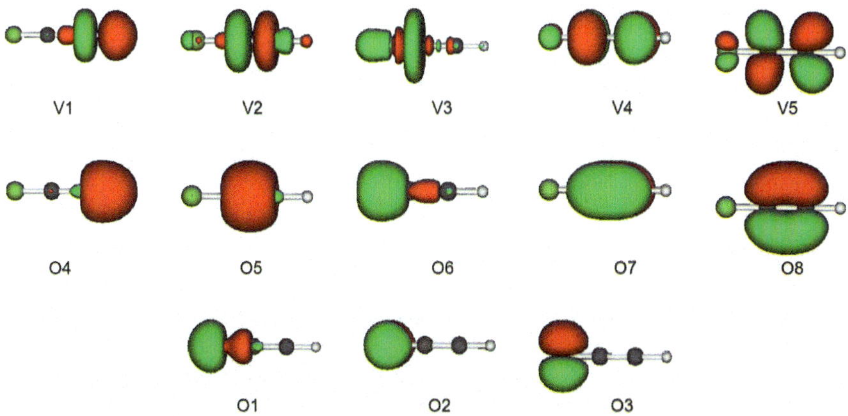

Fig. 3.13 DoLo local orbitals of C_2HF. The suitable active space CAS(2, 2) for the dissociation of the H atom corresponds to the O4 and V1 orbitals and for the dissociation of the F atom to O6 and V3

the virtual orbitals are local. The sigma antibonding orbitals (V1, V2, and V3) are each one on a bond, with delocalization tails somehow more important than for the corresponding occupied orbitals. Because of the symmetry property, the virtual π orbitals are essentially identical to the canonical ones (Fig. 3.8).

C_2F_2
In Figs. 3.9, 3.10, 3.11, and 3.12, the localized orbitals are shown for the C_2F_2 molecule. The situation is very much the same as for the C_2H_2 system, except for the presence of the lone-pair orbitals on the Fluorine atoms. The Boys orbitals, Fig. 3.9, mix the σ and π orbitals. The Pipek orbitals, Fig. 3.10, and Natural orbitals, Fig. 3.11, keep this separation. They look extremely similar. The DoLo orbitals, Fig. 3.12, as in the previous example, give local virtual orbitals too.

3.4.2 Bond Dissociation of C_2HF

As we already pointed out, one of the most interesting aspects of localization is the possibility to focus on the description of a part of the electronic structure in a molecule in that region of space where "the interesting things" happen. We illustrate this fact by considering the two possible dissociation processes of C_2HF, whose structure is F-C-C-H (Fig. 3.13 shows the structure and the DoLo localized orbitals). Two possible fragmentation processes considered here are fragmentation either into F + CCH or into FCC + H. In this particular case, the complete valence space could be chosen as an active space, and this would not pose any problems. However, this choice corresponds to an active space of CAS(16/13) type, which is unreasonably large. Moreover, the addition of just a few more atoms would transform this system from a computationally difficult one to a simply unfeasible one. Indeed, it is clear that the use of such a large active space is completely useless in the present case. In each of the two bond-breaking processes just mentioned, only *two orbitals* and *two electrons* are involved (this is the case of simple single bonds; if the bond is a double or triple bond, the number of orbitals and electrons would become four or six, respectively).

In Figs. 3.14 and 3.15, the energy profiles from short bond distances up to dissociation are reported. In particular, Fig. 3.14 shows the curve for the two different processes, obtained by using a CAS(2, 2) active space of localized and delocalized orbitals. Let us consider first the CAS defined on energetic criteria (the "delocalized" one, obtained by starting from the HOMO-LUMO pair of HF orbitals). The point is that, while for long distances, the active space is always located, for energetic reasons, on the breaking bond, close to equilibrium, it will be located on C-H, since this is the bond that has the largest correlation contribution at CAS(2,2) level, as shown in Fig. 3.15. For this reason, the corresponding energy curve for the C-F dissociation will be discontinuous (Fig. 3.14, bottom), while it is continuous for C-H (Fig. 3.14, top). The curves obtained by localizing the active space on the studied bond, on the other hand, show no sign of discontinuity.

3.4.3 Benzene and Trans-Stilbene

The use of localized orbitals allows the understanding of the electronic structure of a molecule in terms of chemical concepts such as bonds and lone pairs. In the case of an aromatic molecule, "delocalized electrons" are not associated with a bond but with several molecular orbitals. For Edmiston and Ruedenberg [9], *the localized orbitals provide a quantitative basis for the qualitative chemical concepts of "localized electrons" and "delocalized electrons"*. This will be illustrated here with the benzene and *trans*-stilbene molecules.

Consider first the benzene molecule (Fig. 3.16). The molecular orbitals coming from a CAS(6, 6)SCF calculation are localized according to the core of the C atoms, the occupied and virtual σ bond orbitals as well as the occupied and virtual π ones, and the remaining atomic orbitals. There are several possibilities to localize the π bonds. When the π orbitals are localized according to CC bonds, the wave function of the ground state ψ_π^0, expressed hereafter only on the active orbitals of the main determinants (weight $\geqslant 0.02$), becomes multi-referential:

$$\begin{aligned}\psi_\pi^0 = \; & 0.34\,|222000\rangle + 0.05\,|221100\rangle + 0.05\,|221010\rangle + 0.05\,|212100\rangle \\ & + 0.05\,|212001\rangle + 0.05\,|122010\rangle + 0.05\,|122001\rangle + 0.02\,|211011\rangle \\ & + 0.02\,|112110\rangle + 0.02\,|121101\rangle\,,\end{aligned} \tag{3.63}$$

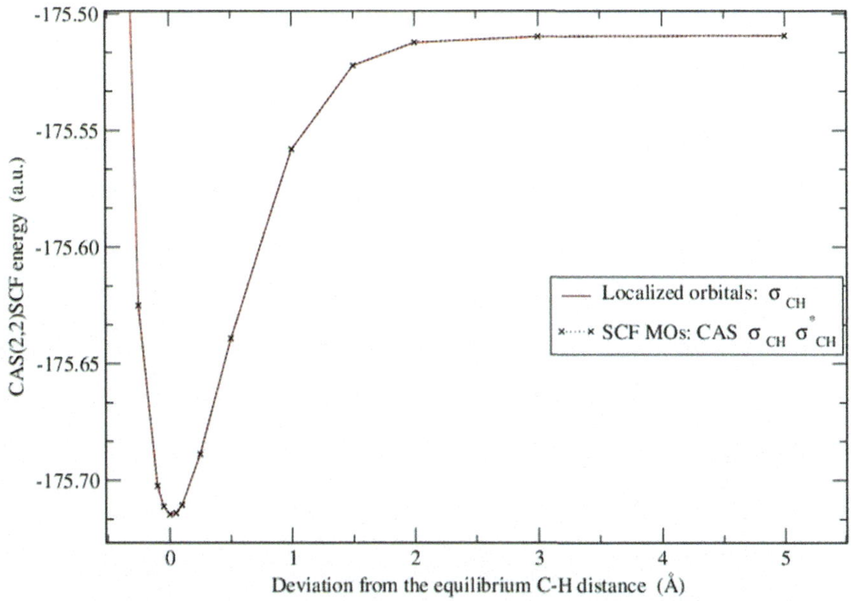

a) Dissociation along the CH distance: CAS(2,2)SCF starting from SCF orbitals

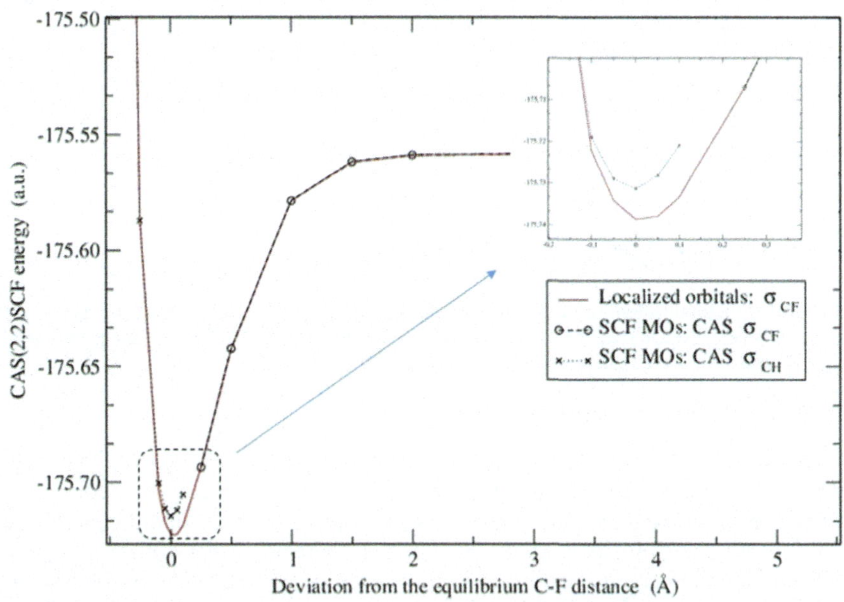

b) Dissociation along the CF distance: CAS(2,2)SCF starting from SCF orbitals

Fig. 3.14 Dissociation of C_2HF along the CH or CF bond. The active space is formed by the HOMO and LUMO SCF orbitals

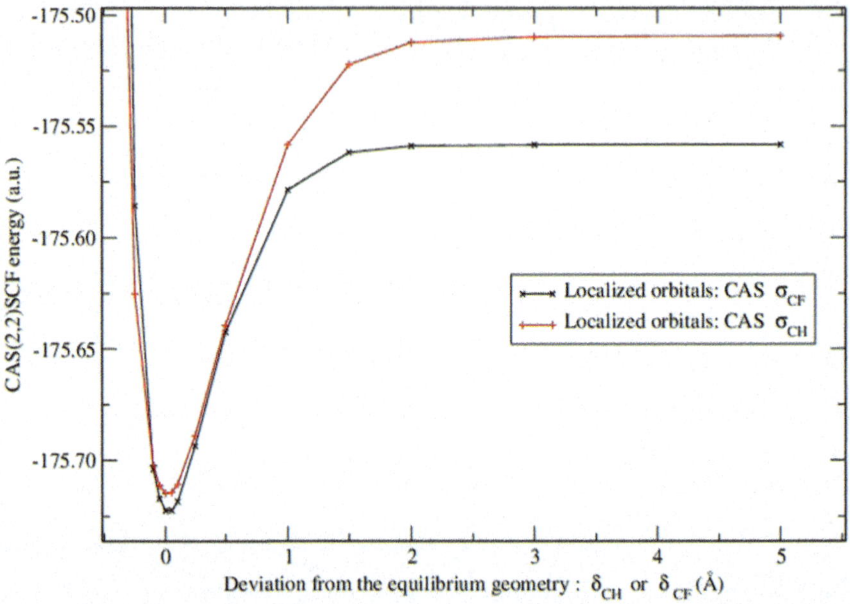

Fig. 3.15 Dissociation of C_2HF along the CH or CF bond at CAS(2, 2). The active space is constituted by the CH or CF localized bond, respectively. At the (common) equilibrium distance, the active space localized on the C-H bond gives the lowest energy

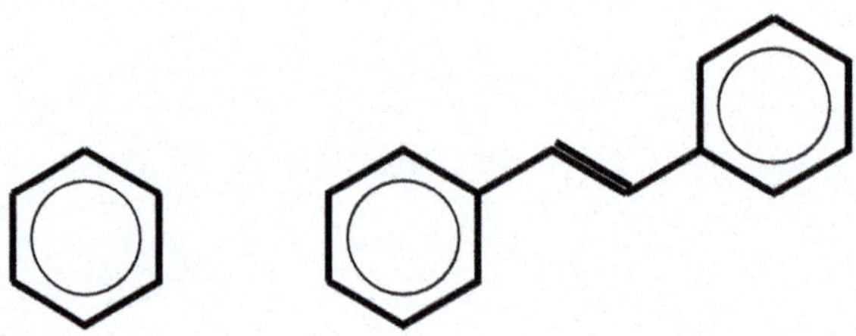

Fig. 3.16 Structural formulas for the molecules of benzene, C_6H_6, and of (E)-1,2-diphenylethene (*trans*-stilbene), H_5C_6-CH=CH-C_6H_5, showing circles representing Robinson–Clar π sextets for the π-systems in the six-membered rings

where the order of the active orbitals corresponds from left to right to the orbitals presented in Fig. 3.17b. With this localization scheme, the $|222000\rangle$ determinant is defined on three occupied π_{CC} bonds and their corresponding antibonding orbitals.

Of course, the multi-referential wave function character is then enhanced in order to reintroduce the conjugation. It is also the case of the excited states of benzene which are no longer described simply by a few mono-excitations on the $|222000\rangle$ determinant as in a delocalized description, but by numerous excitations (mainly mono- and di-excitations) with small weights. Furthermore, the first excited state ψ_{π}^{1} wave function has a very similar weight on the $|222000\rangle$ determinant as the ground state:

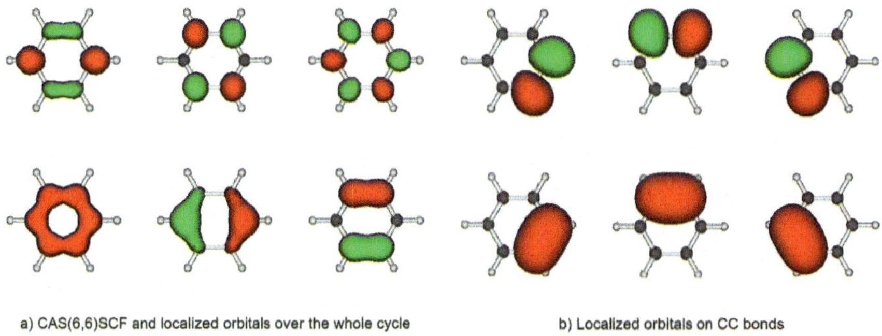

Fig. 3.17 CASSCF and Localized π active orbitals of the benzene molecule

$$\begin{aligned}\psi_{\pi}^{1} &= 0.34\,|222000\rangle + 0.09\,|111111\rangle + 0.05\,|211011\rangle + 0.05\,|121101\rangle \\ &+ 0.05\,|112110\rangle + 0.02\,|220002\rangle + 0.02\,|202020\rangle + 0.02\,|022200\rangle\,.\end{aligned} \tag{3.64}$$

In some cases, the localization can then reduce the understanding of the electronic structure instead of making it easier. Some solutions have been proposed, as the intermediate localization of Zimmerman [60], leading to orbitals preserving the almost diagonal nature of the Fock matrix while maximizing orbital locality.

With the a priori DoLo localization, a different solution is proposed. Indeed, the π orbitals can be localized on the entire cycle while keeping the locality of all the other orbitals. The delocalized and localized π orbitals are shown in Fig. 3.17. The DoLo input file is given in the Appendix, while all the valence localized orbitals are shown in Fig. 3.18. With this mixed solution, the eigenfunctions of the ground and the first excited states (weight $\geqslant 0.02$) are as follows:

$$\psi_{\text{cycle}}^{0} = 0.89\,|222000\rangle + 0.03\,|211110\rangle + 0.02\,|220020\rangle + 0.02\,|202200\rangle \tag{3.65}$$

and

$$\begin{aligned}\psi_{\text{cycle}}^{1} &= 0.40\,|212010\rangle + 0.40\,|221100\rangle + 0.05\,|112110\rangle + 0.04\,|211101\rangle \\ &+ 0.02\,|121020\rangle + 0.02\,|121200\rangle,\end{aligned} \tag{3.66}$$

where the order of the active orbitals corresponds from left to right to the orbitals presented in Fig. 3.17a. The weights of these determinants are now similar to those obtained with a delocalized CASSCF calculation, but the determinants are expressed on localized doubly occupied orbitals, delocalized active orbitals, and localized virtual orbitals (Fig. 3.18).

Concerning the *trans*-stilbene (also on Fig. 3.16), two conformers are studied. In the first one, the two cycles are in the same plane and present a delocalization through the central π_{CC} bond. This conformer is called "in-plane". In the second (highly unstable) conformer, "out-of-plane", the two cycles are in two parallel planes, but there is no more delocalization between the two cycles. In both cases, all the orbitals

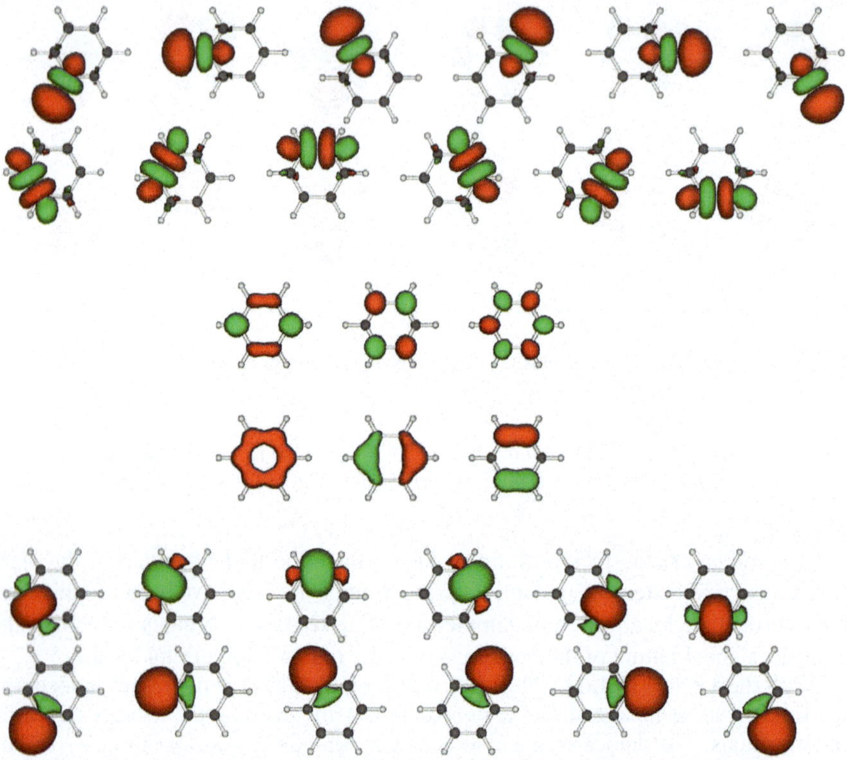

Fig. 3.18 Localized valence molecular orbitals of the benzene molecule. The π orbitals are defined on the whole cycle

are localized on the bonds (or kept atomic for the core and virtual ones) except the π orbitals which are localized on each cycle, and on the central bond (Fig. 3.19). The SCF molecular orbitals of the in-plane conformer are completely delocalized over the two cycles and the double bond connecting them (Occupied π orbitals are presented in Fig. 3.20), whereas the out-of-plane SCF MOs (Fig. 3.21) present a break-down of the delocalization at the level of the central bond. In the latter case, the localized orbitals (Fig. 3.22) on each cycle show a similarity with those found for the benzene molecule, only slightly distorted by the presence of the first neighbor carbon of the central bond. This similarity is lost when the two cycles are completely delocalized (Fig. 3.19).

Similar conclusions are obtained when we analyze the wave functions. The ground state wave functions of the two conformers (weight $\geqslant 0.02$) are as follows:

$$\begin{aligned} \psi_{\text{in}} = \; & 0.65\,|2222222000000000\rangle + 0.03\,|2220222000020000\rangle \\ & + 0.02\,|2212222000010000\rangle + 0.02\,|2222221000010000\rangle \\ & + 0.02\,|2221222000000100\rangle + 0.02\,|2221222100000000\rangle \end{aligned} \tag{3.67}$$

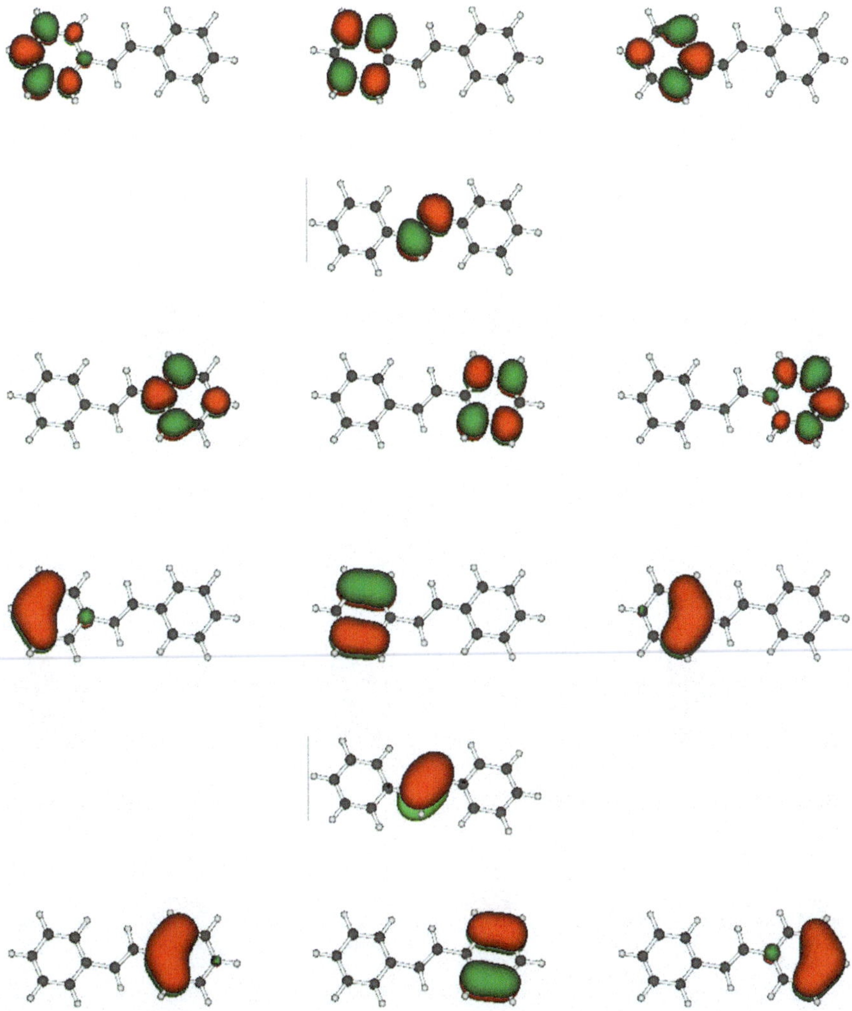

Fig. 3.19 In-plane *trans*-stilbene: localization on the cycles and on the central bond of the SCF MOs

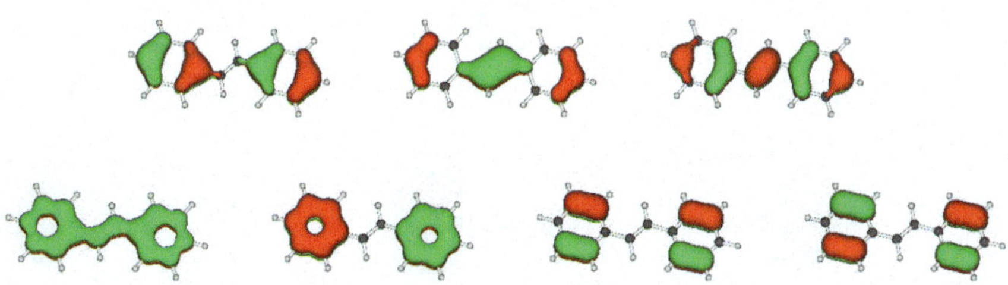

Fig. 3.20 In-plane *trans*-stilbene: Occupied π SCF MOs

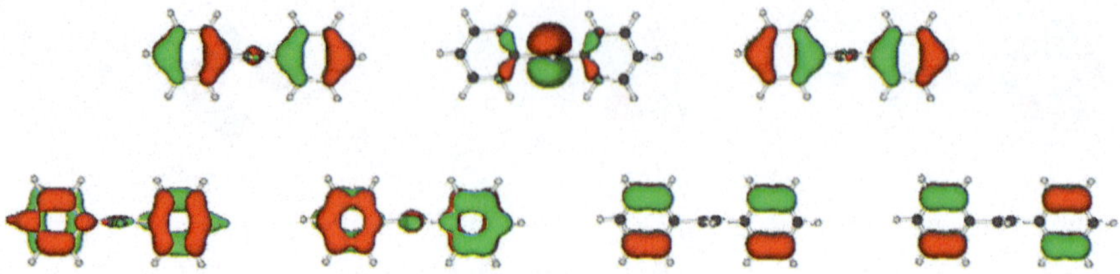

Fig. 3.21 Out-of-plane *trans*-stilbene: Occupied π SCF MOs

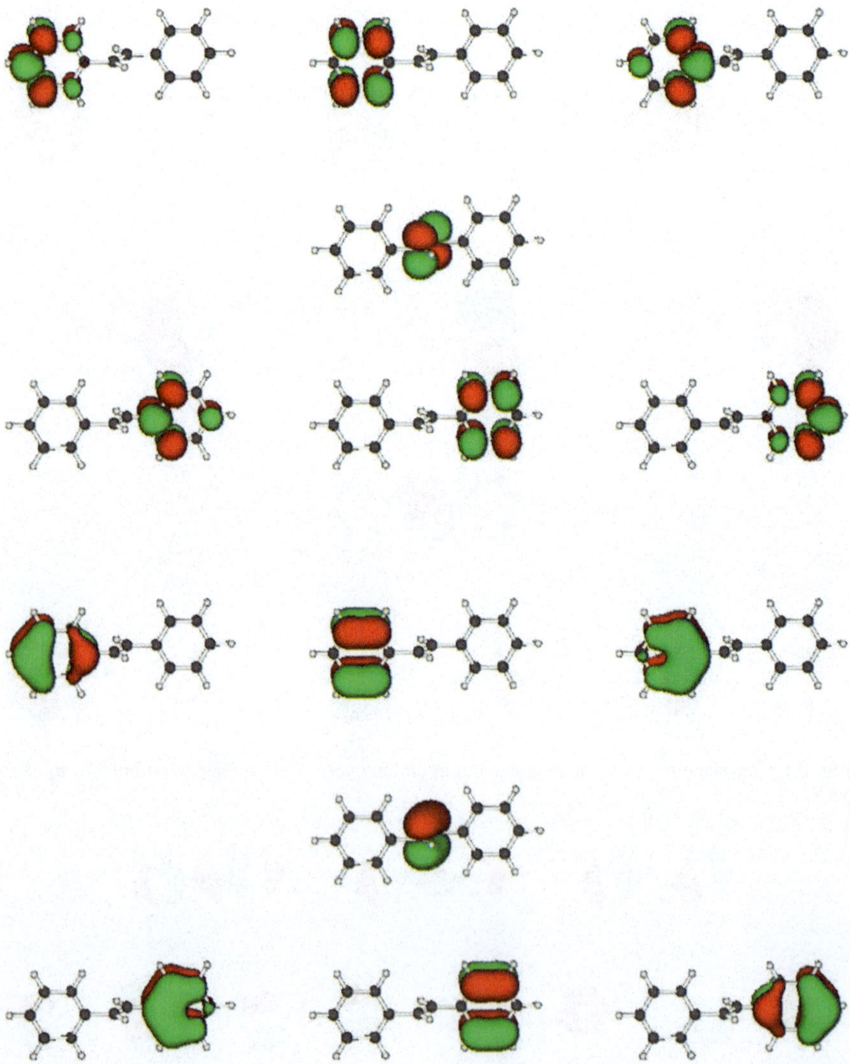

Fig. 3.22 Out-of-plane *trans*-stilbene: localization on the cycles and on the central bond of the SCF MOs

and

$$\begin{aligned}\psi_{\text{out}} = \quad & 0.76|22222220000000\rangle + 0.03|22202220002000\rangle \\ & + 0.02|22222110000110\rangle + 0.02|21122221100000\rangle\end{aligned} \tag{3.68}$$

Numbers in red refer to one cycle, those in cyan to the other one, and numbers in black refer to the central bond. First of all, the weight of the main determinant is larger in the out-of-plane conformer than in the in-plane one, as expected for a lower delocalization of the electrons. In the two conformers, the double excitation on the central bond has the same weight while the excitation from one cycle to the central bond (or from the central bond to one cycle) occurs only in the in-plane conformer and is representative of the delocalization of the electrons between the two cycles. The double excitations localized on one cycle have a larger weight in the out-of-plane conformer than in the in-plane one. This mixed localization, separating the two cycles, illustrates the possibility to obtain wave functions that have a chemical meaning.

The flexibility of the a priori DoLo localization allows the description of "delocalized electrons" keeping at the same time all the other orbitals localized, a necessary feature for the development of linear-scaling methods [47–49].

Orbital localization can be seen as a tool to select different regions of a molecule in order to choose the relevant active space, as in the example of the dissociation of C_2HF, or to study the approach of an atom or a molecule to one side of another one [49]. Another example is the description of excited states by limiting the size of the active space to those orbitals involved in the excitations [61].

3.4.4 Coordination Complexes

Now we lift the restriction to molecules composed exclusively of atoms from light main-group elements, at least partly, and exemplarily consider two diamagnetic molecular compounds of metals: tetracarbonylnickel, $[Ni(CO)_4]$, and tris(acetyl-acetonato)aluminium, $[Al(C_5H_7O_2)_3]$ or $[Al(acac)_3]$ (Hacac is pentane-2,4-dione or acetylacetone). Under standard ambient conditions, tetracarbonylnickel is a colorless transparent liquid, whereas tris(acetylacetonato)aluminium forms a white or pale yellow crystalline powder. These few experimental facts already suffice as strong indications for a closed-shell electronic ground state and for the absence of low-lying excited electronic states. In fact, the electronic ground state of the molecules forming these two compounds is found to be a totally symmetric singlet state, in both cases.

The structure of a tetracarbonylnickel molecule is shown in Fig. 3.23. The molecular point group is T_d. The electron configuration in its electronic ground state can be derived from the superposition (or combination) of a Ni atom with closed-shell electron configuration $3d^{10}$ and of the filled MOs of the four CO molecules in their closed-shell electronic ground states ($|C{\equiv}O|$), see the Appendix 2 for the summarized final result. The electron configuration of $[Ni(CO)_4]$ is built up from seventeen atomic core orbitals (Ni $1s^2$ $2s^2$ $2p^6$ $3s^2$ $3p^6$, four C $1s^2$, and four O $1s^2$), which transform as $5a_1 \oplus 4t_2$, and twenty-five valence orbitals. These latter originate from the five orbitals of the filled 3d shell of Ni, which transform as $e \oplus t_2$, and four sets of valence orbitals of the CO molecules (each containing five filled MOs: σ_{CO}, a π_{CO} pair, lone pairs n on C and O). The twenty valence orbitals of the four CO molecules transform as $3a_1 \oplus e \oplus t_1 \oplus 4t_2$. From these, the e and t_1 shells originate exclusively from the four pairs of π_{CO} MOs of the ligands (which contribute a set of t_2 MOs as well). In the Abelian subgroup D_2, the molecular electron configuration of the electronic ground state of tetracarbonylnickel corresponds to the following types and numbers of filled shells: $12a \oplus 10b_1 \oplus 10b_2 \oplus 10b_3$. After repetition of the SCF calculation in point group C_1, subsequent orbital localization was successfully achieved by the iterative algorithm for the Boys localization as implemented in MOLPRO (17 iteration steps), but this required the grouping of the canonical valence MOs into three different disjoint subsets: a set of five MOs of mainly d character at Ni ($e \oplus t_2$ in T_d), a set of eight MOs of mainly π character at the CO ligands ($e \oplus t_1 \oplus t_2$ in T_d), and a set of all remaining valence MOs. A subset of the resulting Boys localized MOs is shown in Fig. 3.24. In addition to the set of five d orbitals on Ni, shown in the lower row, one can readily identify in the upper row a nonbonding orbital on O (at the distant end of the ligand), a set of three bonding orbitals of CO (σ_{CO} and a π_{CO} pair), and an orbital of mainly σ donor character (at the proximal end of the ligand).

The molecules of tris(acetylacetonato)aluminium exist in two enantiomeric forms, both shown in Fig. 3.23. These two forms differ in the way in which the three bidentate anionic acetylacetonate ligands ($C_5H_7O_2^-$) are arranged around the central Al^{3+} ion, resembling either a right-handed or a left-handed screw. The molecular point group is D_3. For both forms, the electron configuration in their electronic ground state can be derived from the superposition (or combination) of an Al^{3+} ion with Ne-like closed-shell electron configuration and of the filled MOs of the three $C_5H_7O_2^-$

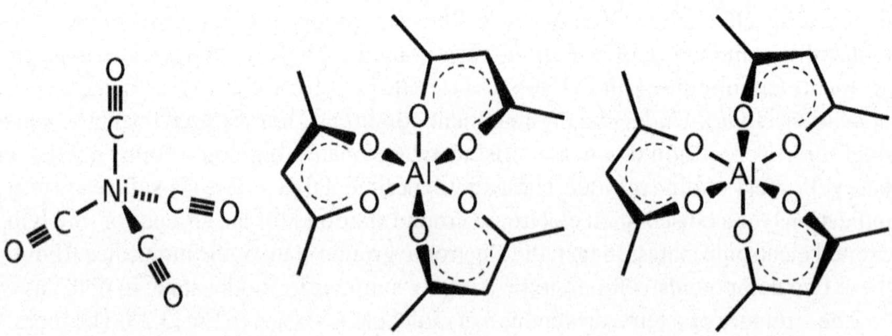

$[Ni(CO)_4]$ (T_d) Λ-$[Al(acac)_3]$ (D_3) Δ-$[Al(acac)_3]$ (D_3)

Fig. 3.23 Structural formulas for the molecules of tetracarbonylnickel, $[Ni(CO)_4]$, and of the two enantiomeric forms of tris(acetylacetonato)aluminium, $[Al(acac)_3]$

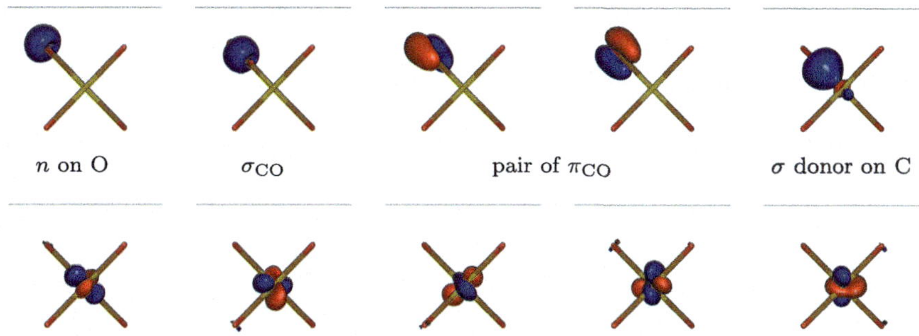

Fig. 3.24 A full set of localized valence orbitals on one of the four CO ligands (upper row) and localized orbitals of mainly d character on the central atom (lower row) in tetracarbonylnickel, $[Ni(CO)_4]$ (T_d; view along the S_4 axis)

ligands being in closed-shell electron configurations as well, see the Appendix 2 for the summarized final result. The electron configuration of $[Al(acac)_3]$ is built up from twenty-six atomic core orbitals (Al $1s^2$ $2s^2$ $2p^6$, fifteen C $1s^2$, and six O $1s^2$), which transform as $6a_1 \oplus 4a_2 \oplus 8e$, and sixty valence orbitals. Each of the three $C_5H_7O_2^-$ ligands contributes twenty valence orbitals, which exist in the space spanned by seven σ_{CH} orbitals, four σ_{CC} orbitals, two σ_{CO} orbitals, four non-bonding orbitals (n) for free electron pairs on the O atoms of the carbonyl groups, and three π orbitals delocalized over the chain of atoms O-C-CH-C-O. These sixty valence orbitals transform as $10a_1 \oplus 10a_2 \oplus 20e$. In the Abelian subgroup C_2, the molecular electron configuration of the electronic ground state of tris(acetylacetonato)aluminium corresponds to the following types and numbers of filled shells: $44a \oplus 42b$. After repetition of the SCF calculation in point group C_1, subsequent orbital localization was successfully achieved by the iterative algorithm for the Boys localization as implemented in MOLPRO (23 iteration steps). No grouping of canonical valence orbitals into different disjoint subsets was necessary. A subset of the resulting Boys localized MOs, the twenty localized valence MOs on one of the three acetylacetonato ligands, is shown in Fig. 3.25. Most interesting here are the following four points: (i) all non-hydrogen atoms in each of the three ligands are found to lie almost in one plane (this is also found in the crystal structures of the different known polymorphs [62]); (ii) the ligand acts as a bidentate chelating ligand with the two O atoms of the carbonyl groups behaving as σ donors; (iii) the bonding orbitals of the carbonyl groups (one σ orbital and one π orbital on each of these groups) are mixed, so that they appear as pairs of banana-shaped orbitals, as one might have expected (since the Boys localization algorithm was used without any further restrictions); (iv) a reasonably well-localized π orbital is found at and in the neighborhood of the central methylidyne group (CH) of each acetylacetonate ligand.

We have shown here, with two coordination compounds as examples, that common orbital localization techniques can be successfully applied also to cases outside the

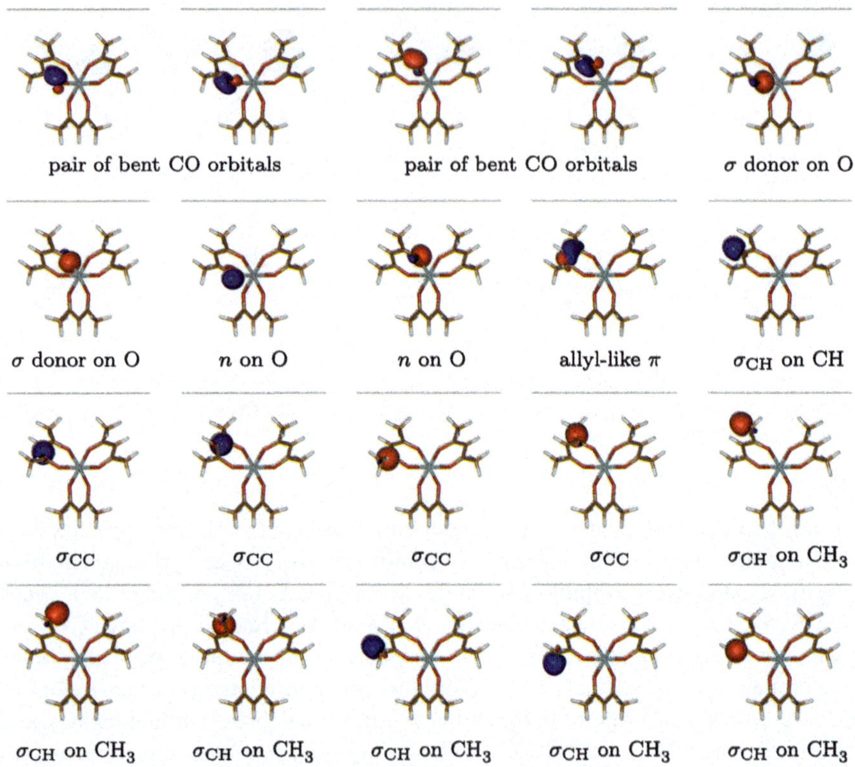

Fig. 3.25 A full set of localized valence orbitals on one of the three ligands in tris(acetylacetonato)-aluminium, [Al(acac)$_3$] (D_3; view along the C_3 axis)

class of compounds that are built exclusively from covalently linked atoms of light main-group elements.

3.4.5 Large Systems: Polycyclic Aromatic Hydrocarbons

As seen in a previous subsection, it is well possible to localize not only the occupied π orbitals, but also the virtual π orbitals in small aromatic hydrocarbon molecules, e.g., in the case of benzene (see Fig. 3.17). Each of the localized π orbitals thus obtained can be easily assigned to a particular pair of neighboring carbon nuclei. In benzene, the resulting complete set of six localized π orbitals may be used even to study π–π^* excitations, at least in first approximation, since it spans the same part of function space as does the complete set of the six canonical π orbitals.

We turn now to Polycyclic Aromatic Hydrocarbons (PAHs), a class of compounds that includes large molecular systems. Well-known families of PAHs are the homologous sequences of the [n]acenes and the [n]phenacenes (with a common chemical formula $C_{4n+2}H_{2n+4}$, see Figs 3.26 and 3.27) as well as the [n]coronenes ($C_{6n^2}H_{6n}$, see Fig. 3.28). In the limit $n \to \infty$, the former sequences turn into carbon nanoribbons of smallest possible width (with limit formula $^{1}_{\infty}(C_4H_2)$), whereas the coronenes approach graphene (two-dimensional infinite hexagonal lattice of carbon atoms, with two carbon atoms per unit cell, $^{2}_{\infty}(C_2)$). A common feature of the electronic structure of all PAHs is the presence of carbon p orbitals oriented normal (or perpendicular) to the surface containing the atomic nuclei, which form the so-called π orbitals. The surface mentioned, a two-dimensional manifold embedded in three-dimensional Euclidean space, does not need to be a plane, see, e.g., the [n]cyclacenes, *cyclo*-$(C_4H_2)_n$, where this surface is a part of a right circular cylinder with carbon p orbitals normal to the surface pointing radially out- and inwards.

It has been properly stated above that orbital localization methods can destroy orbital symmetry, because these methods form new localized functions by linearly combining canonical orbitals of different symmetry behavior (too) freely. Mixing of σ and π orbitals would be a problem in PAHs too, but this problem can be resolved, as shown below.

However, before we attempt to localize the canonical orbitals in PAHs in such a way that their symmetry with respect to the surface containing the atomic nuclei is conserved, we need to make sure that the correct (or desired) electronic state has been selected (irrespective of whether the point group of the PAH has been, or is going to be, fully exploited or not—we remind the reader that the localization process forces us to use point group C_1). As shown in Table 3.1 with three examples taken from the [n]acene sequence ($n = 10$ or 20 or 30; point group D_{2h}), an erroneous choice or selection of the electron configuration for a state as simple as a totally symmetric closed-shell singlet state (^1A_g) easily leads to failure in finding the solution with the lowest possible total energy. In comparison with the electron configuration selected

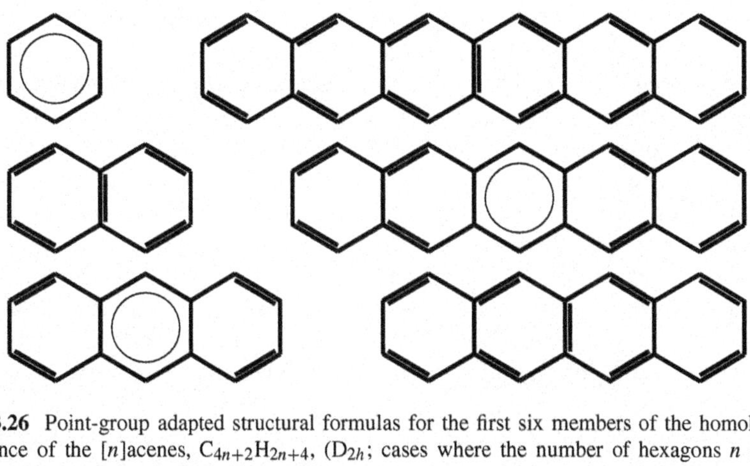

Fig. 3.26 Point-group adapted structural formulas for the first six members of the homologous sequence of the [n]acenes, $C_{4n+2}H_{2n+4}$, (D_{2h}; cases where the number of hexagons n is odd require the use of the single Robinson–Clar π sextet that can be drawn here, being placed in the central hexagon)

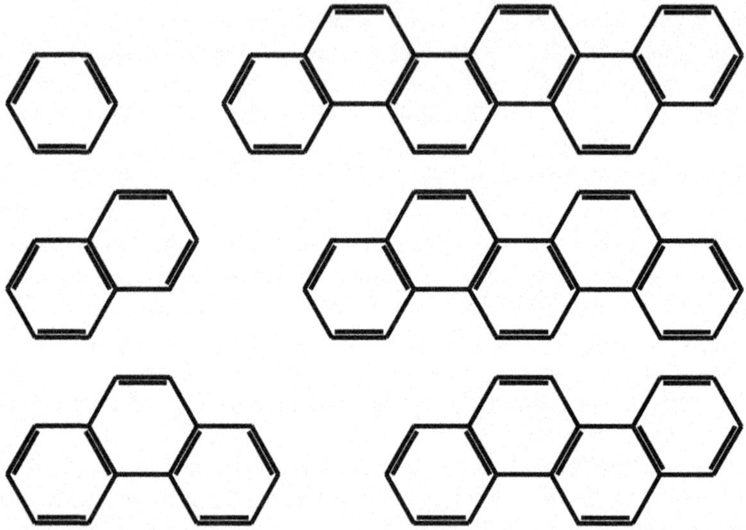

Fig. 3.27 Point-group adapted structural formulas for the first six members of the homologous sequence of the [n]phenacenes, $C_{4n+2}H_{2n+4}$ (n odd: C_{2v}, n even: C_{2h}; the use of Robinson–Clar π sextets for the purpose of symmetry adaption is not necessary here)

in some automatized way by the software applied, the correct choice leads to a lower total energy and to a lower (higher) orbital energy for the HOMO (LUMO), thus ensuring a larger HOMO-LUMO gap. The two totally symmetric electron configurations involved in each example simply differ in the selection of type and number of occupied π orbitals. In the cases of [10]- and [20]acene, a simple orbital rotation, exchanging the role of HOMO and LUMO, corrects the fault. For [30]acene, in contrast, the initial guess for the number of fully occupied π orbitals is wrong for all four irreducible representations (irreps) available for π orbitals.

As it appears now, it may seem to be difficult to select the π orbital part of the electron configuration in such a way that it gives the totally symmetric closed-shell singlet state with the lowest total energy for the larger members of the homologous sequences of PAHs considered here ([n]acenes, [n]phenacenes, and [n]coronenes), but this is wrong. The sought-after electron configuration can be straightforwardly calculated from the reducible representation carried by the occupied localized orbitals (both σ and π) seen in or represented by the point-group adapted structural formulas already shown in Figs. 3.26, 3.27, and 3.28. As shown, these structural formulas sometimes require the inclusion of Robinson–Clar π sextets. Each such π sextet is to be understood then as representing three occupied π orbitals (which are well known from benzene, in point group D_{6h}, as the $1a_{2u}$ orbital and the $1e_{1g}$ orbital pair). In each six-membered ring of the PAH decorated with a Robinson–Clar π sextet, these three π orbitals then simply contribute to the character of the reducible representation in the usual way. After reduction of the reducible representation, the electron configuration for the PAHs considered here in their lowest-energy totally symmet-

Table 3.1 Electron configuration (number of filled MOs per irrep), total SCF energy (single point[a], RHF/cc-pVDZ, in Hartree), and frontier orbital energies (in Hartree) for low-lying totally symmetric closed-shell singlet states ^1A_g (D_{2h}) of three larger [n]acenes ($n = 10, 20, 30$)

— irreps for σ MOs —				— irreps for π MOs —				remark	E_{SCF} or ΔE_{SCF}	ε_{HOMO}[c]	ε_{LUMO}[c,d]
a_g	b_{3u}	b_{2u}	b_{1g}	b_{1u}	b_{2g}	b_{3g}	a_u				
s	x	y	xy	z	xz	yz	xyz				
(a) [10]acene, $C_{42}H_{24}$ (276 electrons):											
33	27	31	26	6	5	6	4	automatic choice	−1604.4514	−0.1525 ($6b_{3g}$)	−0.0563 ($5a_u$)
33	27	31	26	6	5	5	5	corrected[b]	−1604.5031	−0.1730 ($5a_u$)	−0.0324 ($6b_{3g}$)
				energy gain due to change of electron configuration:					−0.0517		
(b) [20]acene, $C_{82}H_{44}$ (536 electrons):											
63	52	61	51	11	10	11	9	automatic choice	−3130.8708	−0.1539 ($11b_{3g}$)	−0.0571 ($10a_u$)
63	52	61	51	11	10	10	10	corrected[b]	−3130.9003	−0.1578 ($10a_u$)	−0.0496 ($11b_{3g}$)
				energy gain due to change of electron configuration:					−0.0295		
(c) [30]acene, $C_{122}H_{64}$ (796 electrons):											
93	77	91	76	17	14	16	14	automatic choice	−4657.2375	−0.1525 ($17b_{1u}$)	−0.0584 ($15b_{2g}$)
93	77	91	76	16	15	15	15	corrected[b]	−4657.2974	−0.1544 ($15a_u$)	−0.0536 ($16b_{3g}$)
				energy gain due to change of electron configuration:					−0.0599		

[a] all nuclei in the xy plane, all CC bond lengths equal to 142.1 pm, all CH bond lengths equal to 108.4 pm, all bond angles equal to 120°.
[b] confirmation of a corresponding local minimum on the potential energy hypersurface was subsequently achieved by optimization of molecular structure followed by normal mode analysis, see Fig. 3.29 and text for more details.
[c] orbital irreps given in parentheses.
[d] negative value of ε_{LUMO} at this level (RHF-SCF) hints at possible existence of either stable mono-anion or low-lying open-shell state(s)

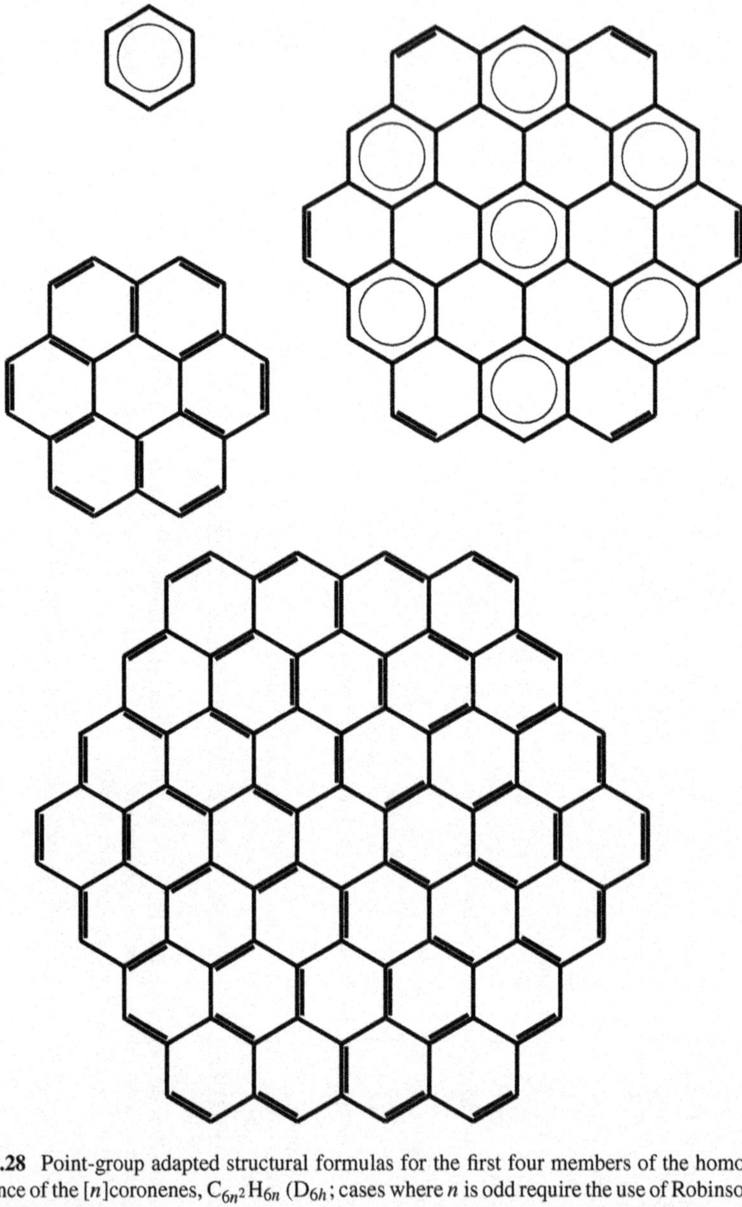

Fig. 3.28 Point-group adapted structural formulas for the first four members of the homologous sequence of the [*n*]coronenes, $C_{6n^2}H_{6n}$ (D_{6h}; cases where *n* is odd require the use of Robinson–Clar π sextets)

ric closed-shell singlet states is determined (details are given in the Appendix 2). The group theoretical approach works because it extracts information that is invariant under orthogonal transformation from one set of orbitals (like filled canonical orbitals) to another, equivalent set (like filled localized orbitals).

The problem discussed here is the problem of choice of an initial guess for the electron configuration of a totally symmetric closed-shell singlet state for molecules from homologous sequences of PAHs. When the HOMO-LUMO gap is small, which happens quickly upon an increase in PAH size, the typical algorithms and methods used to accomplish this task (based, e.g., on application of extended Hückel theory or on superposition of atomic densities) very likely fail sooner or later. Of course, it is not the software that is to be blamed for such failures.

After having solved the problem of electron configuration and electronic state, we show in Fig. 3.29 that consistent and converging results are obtained for several quantities of interest for the series of members of the homologous sequences of [n]acenes and [n]phenacenes up to $n = 30$. Figure 3.29a shows the convergence of the energy per monomer unit, $\Delta E(C_4H_2)$, as derived from data obtained both in single-point calculations at the initially assumed molecular structure (labeled ASP) and after full optimization of molecular structure (labeled OPT). Convergence of this quantity to better than 10^{-6} Hartree (much smaller than the unit shown in the figure) for the fully optimized molecular structures required to go up to [30]acene in the acene sequence, while it was achieved in the phenacene sequence already at [16]phenacene. Figure 3.29b shows corresponding data for the HOMO-LUMO gap. Upon optimization of molecular structure, the HOMO-LUMO gap opens up significantly, as expected, and it does so much more strongly for the [n]acenes than for the [n]phenacenes. Figure 3.29c shows the lowest normal mode frequency found among the rod-bending out-of-plane modes for the fully optimized molecular structures of the [n]acenes and [n]phenacenes. As $n \to \infty$, i.e., as the length of these PAH molecules tends to infinity, one expects that this frequency approaches zero. To be sure, normal mode frequencies for a finite-sized system in a local minimum of its potential energy hypersurface have to be positive. The figure clearly shows that this criterion is fulfilled. As a final statement on Fig. 3.29, we remark that any incorrect choice of electronic state made for members in these homologous sequences of PAHs would have lead to severe "kinks" or "steps" in the curves shown here.

Now that the question of the correct choice of the electronic state of the PAH molecules has been settled, we can eventually turn to the task of determination of localized orbitals for them. This orbital localization usually takes place within point-group symmetry C_1, i.e., the symmetry recognition of the software tools is or has to be switched off. However, one must not forget that the set of nuclear Cartesian coordinates (frequently dubbed "molecular geometry") still carries the full point-group symmetry into the electronic structure calculation even when its use has been switched off by intention. The check for the correctness of results obtained from a calculation done without the use of symmetry for a system that does have symmetry requires the availability of results obtained from a previous calculation where that symmetry has been fully exploited.

As for all other systems, orbital localization in PAHs can be achieved either by a posteriori methods or by a priori methods. The usually unwanted mixing of σ and π orbitals that may occur when a posteriori methods of orbital localization are applied can be safely avoided by preparation of two disjoint subsets of canonical orbitals, a subset of σ orbitals and a subset of π orbitals, and subsequent application of orbital

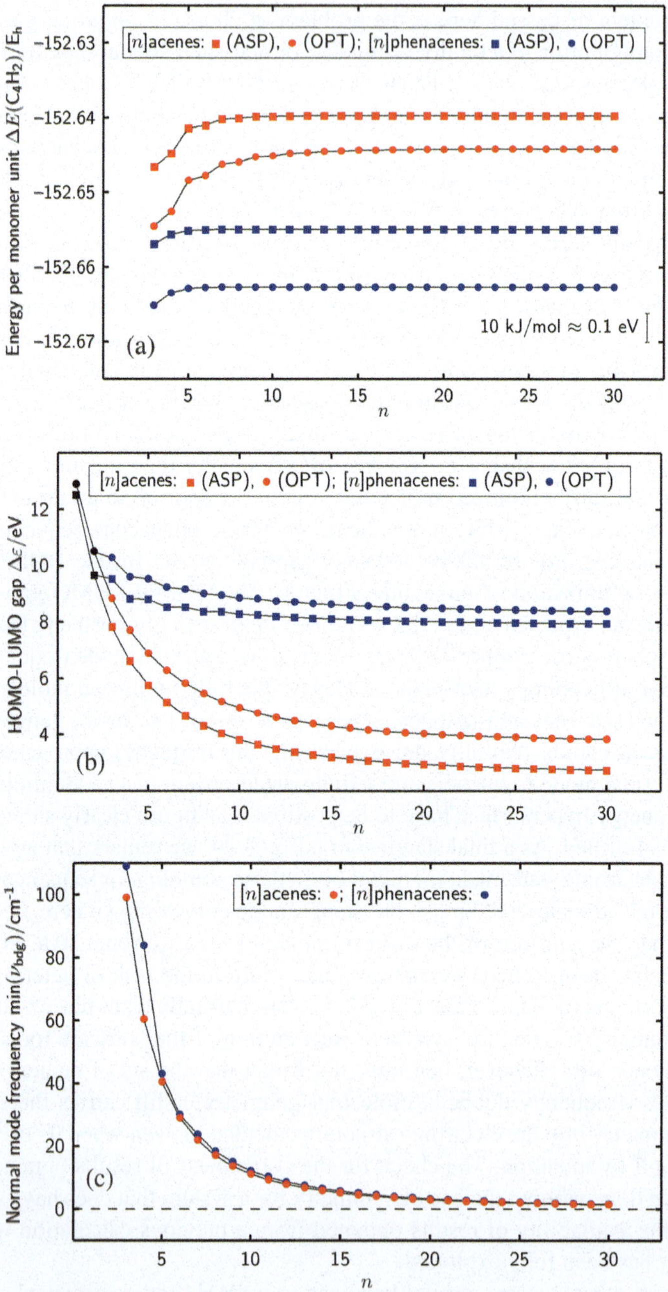

Fig. 3.29 Results from RHF-SCF calculations (basis set cc-pVDZ) for [*n*]acenes (in red) and [*n*]phenacenes (in blue) at assumed molecular structure (ASP: squares) and after full optimization (OPT: circles): **a** energy per monomer unit C_4H_2; **b** HOMO-LUMO gap; **c** lowest frequency of rod-bending normal modes

localization separately within each of these two subsets. It has to be emphasized again that both the SCF calculation, which precedes the orbital localization step and generates the canonical orbitals, as well as the orbital localization step itself are done completely without exploitation of point-group symmetry, i.e., they use point group C_1 only.

We first show in Fig. 3.30 two sets of localized π-MOs for the six molecules representing the first members from the homologous sequences of the [n]acenes and the [n]phenacenes ($n = 1$ to 4). Benzene and naphthalene serve as possibly somewhat untypical first members of both sequences. The group of panels associated with each molecule shows in its upper half a set of localized π-MOs obtained from the Boys localization, whereas the lower half presents a set of localized π-MOs obtained by the DoLo method. In all cases shown, for both the [n]acenes and the [n]phenacenes, the Boys localized π-MOs are found at those positions where localized π-MOs were drawn in the point-group adapted structural formulas presented in Figs. 3.26 and 3.27. In the cases of benzene and anthracene, a set of three Boys localized π-MOs is found to be associated with the (central) six-membered ring, thus representing the Robinson–Clar π sextet, as expected. It is worth to mention here in passing that in the case of benzene, C_6H_6 (point group D_{6h}), the three localized π-MOs can be rotated continuously around the hexagon [63, 64].

We turn now to the localized π-MOs obtained by the DoLo method. Since this localization method is an a priori method, the freedom of choice of unitary (orthogonal) transformation, here within the set of occupied π-orbitals, can be exploited by the user in order to generate a new set of occupied localized π-orbitals with user-specified properties. For phenanthrene and chrysene, the localized π-MOs obtained by the DoLo method are identical to the ones obtained from the Boys localization. However, in the case of the first members of the [n]acene sequence, the DoLo method offers the possibility to choose (or to enforce) localization of a subset of $n + 1$ π-orbitals (out of a total of $2n + 1$) at the $n + 1$ CC σ bonds that connect the two (upper and lower) zigzag chains of carbon atoms seen in these molecules. This attempt is successful indeed, and leads (i) to these $n + 1$ localized π-MOs, as intended, and (ii) to n π-MOs localized across each of the n six-member rings and having an additional nodal surface perpendicular to the molecular mirror plane.

The Figs. 3.31 and 3.32 show complete sets of localized π-MOs for coronene and for circumcoronene, two members of the [n]coronene sequence with $n = 2$ and $n = 3$. Comparison with the point-group adapted structural formulas seen in Fig. 3.28 shows for both molecules a close similarity between the localized π-MOs and the symbols (lines and circles) used to represent the π-system in these structural formulas. For all twelve localized π-MOs of coronene, one finds a one-to-one correspondence between a localized π-MO in Fig. 3.31 and a line formally indicating a double bond in Fig. 3.28. The twenty-seven localized π-MOs of circumcoronene can be grouped into seven sets, each containing three MOs associated with one six-membered ring, and six other π-MOs located at the periphery of this disk-shaped molecule. Figure 3.32 shows these seven three-membered sets of localized π-MOs in its upper part and the remaining six π-MOs in its lower part. The three-membered sets of π-MOs correspond in a one-to-one fashion to the circles drawn in the for-

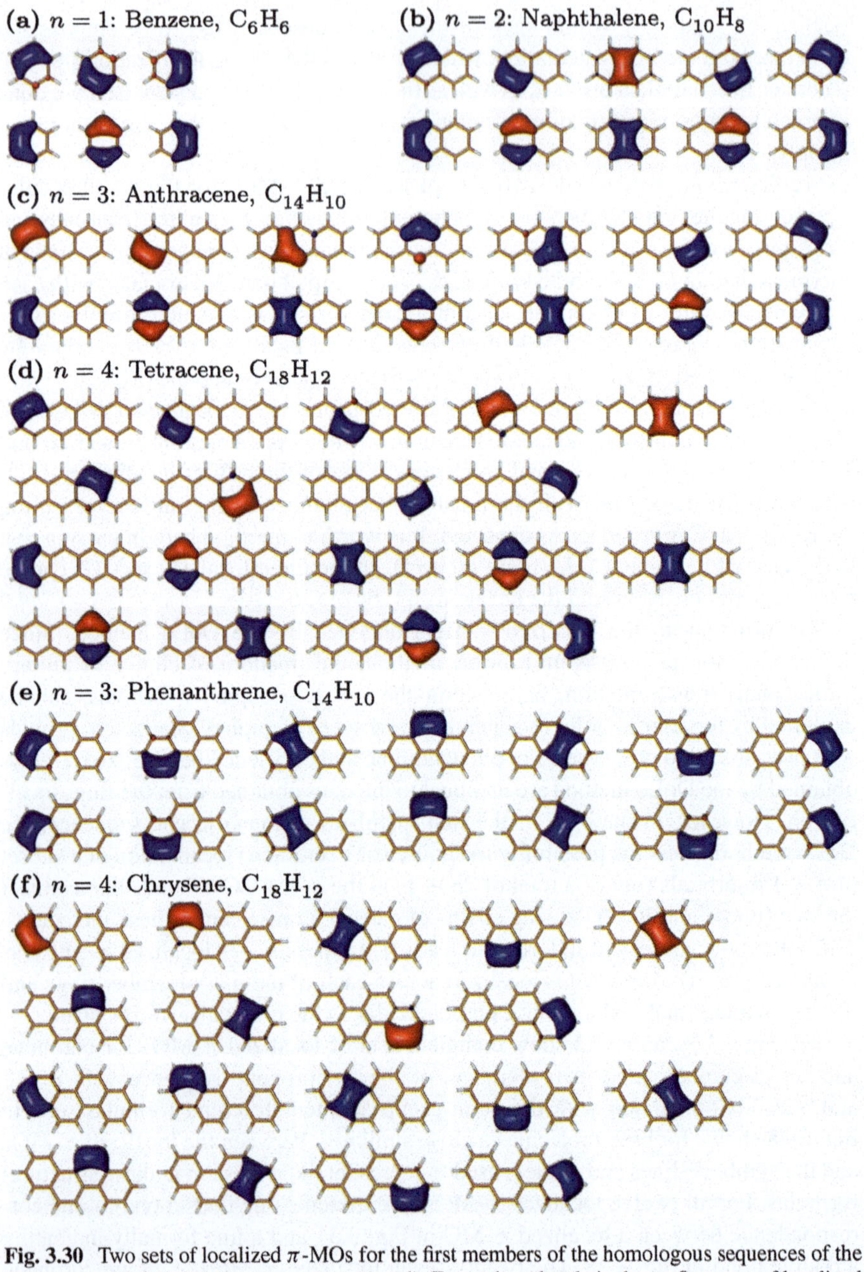

Fig. 3.30 Two sets of localized π-MOs for the first members of the homologous sequences of the [n]acenes and of the [n]phenacenes ($n = 1$ to 4). For each molecule is shown first a set of localized π-MOs obtained from the Boys localization (upper half of panels) and, second, a set of localized π-MOs obtained from the DoLo method (lower half of panels)

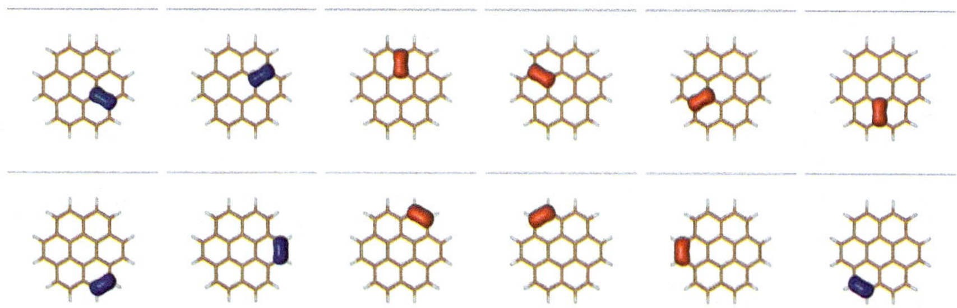

Fig. 3.31 Localized π-MOs for coronene ([2]coronene), $C_{24}H_{12}$, as obtained from the Boys localization applied to the set of occupied canonical π-MOs

mula for circumcoronene in Fig. 3.28. Each of the localized π-MOs at the periphery corresponds to a line formally indicating a double bond, exactly in the same way as seen before for coronene.

It has been demonstrated now with the examples considered so far in this subsection that the generation of localized occupied σ-MOs (not shown) and π-MOs in PAHs can be achieved, in more than one way, without problems. In order to be successful, the Boys localization method requires the identification and specification of the canonical σ- and π-MOs, in terms of lists to be processed, prior to its application. The DoLo method, on the other hand, gives much freedom to the user and can generate also quite unconventional, but equally valid sets of localized π-MOs, as shown above for the early members of the acene sequence.

All PAH molecules discussed up to now belong to the class of alternating aromatic hydrocarbons. Our last example, the comparative study of π-MO localization in naphthalene and azulene, includes a non-alternating aromatic hydrocarbon. Structural formulas for these two isomers of $C_{10}H_8$ are shown in Fig. 3.33. The next figure, Fig. 3.34, shows the five canonical occupied π-MOs ψ_i and the five localized occupied π-MOs ϕ_k for both molecules. The panels are arranged in a way that facilitates comparison. In the middle, one sees the canonical π-MOs, those for naphthalene in Fig. 3.34b and those for azulene in Fig. 3.34c, arranged from left to right according to the increase of orbital energy. The localized π-MOs, as obtained from the Boys localization, are shown in Fig. 3.34a for naphthalene and in Fig. 3.34d for azulene. As for the canonical π-MOs, there is again a close similarity and correspondence between localized π-MOs for naphthalene and localized π-MOs for azulene. However, despite the fact that the iterative Boys localization for azulene eventually converged (to very good quality, but only after 489 iterations), the Boys localized π-MOs for azulene are less well localized than those for naphthalene: they extend over three (ϕ_{31} to ϕ_{34}) to four (ϕ_{30}) carbon nuclei and sometimes exhibit comparatively large orthogonalization tails (see ϕ_{30}). In Figs. 3.35 and 3.36, we show two different sets of localized π-MOs for azulene obtained with the DoLo method. In addition, both figures include the unoccupied antibonding localized π-MOs in the upper row. The localized π-MOs in Fig. 3.35 have been generated by imposing that a localized occupied π-MO is found at the bond common to both rings (this is the fourth π-MO

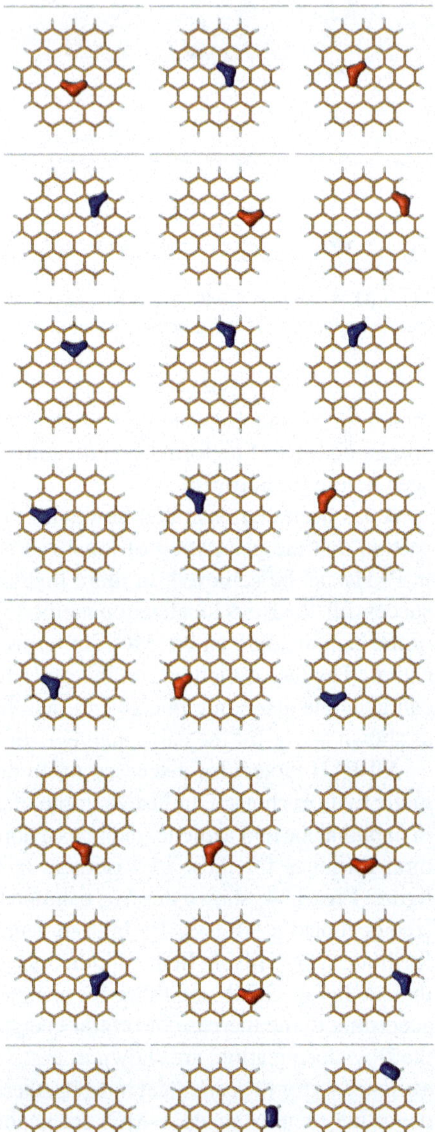

Fig. 3.32 Localized π-MOs for circumcoronene ([3]coronene), $C_{54}H_{18}$, as obtained from the Boys localization applied to the set of occupied canonical π-MOs

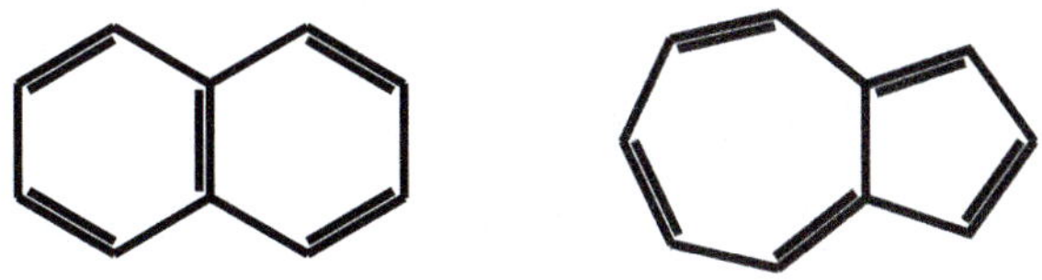

Fig. 3.33 Structural formulas for the molecules of naphthalene ($C_{10}H_8$, D_{2h}) and azulene ($C_{10}H_8$, C_{2v})

from the left in the lower row). As seen before for the Boys localization, some of the π-MOs in this set extend over the three carbon nuclei (the first and the last in the lower row). It is worth to point out that the localized π-MOs seen in the last row of Fig. 3.36 perfectly correspond to the lines drawn in the structural formula for azulene in Fig. 3.33 as formal indicators of a double bond.

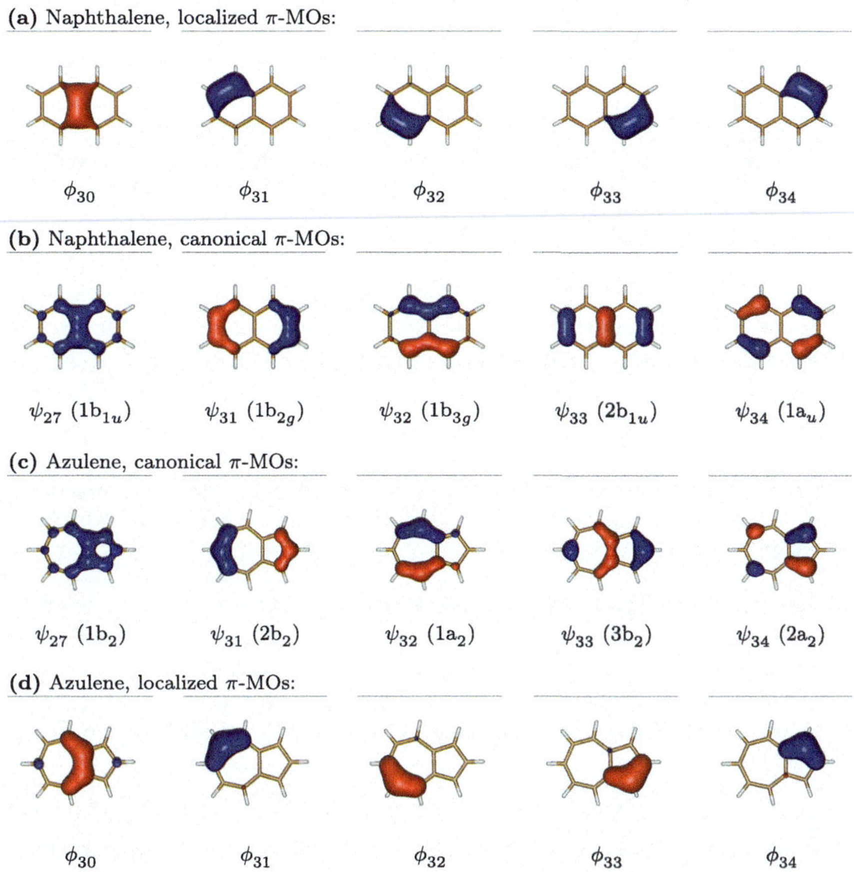

Fig. 3.34 Canonical π-MOs (second and third row) and the Boys localized π-MOs (first and last row) for naphthalene and azulene

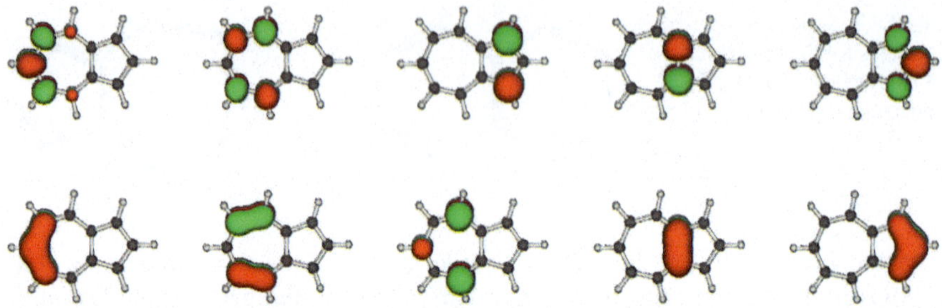

Fig. 3.35 Localized π orbitals on each cycle of the azulene molecule

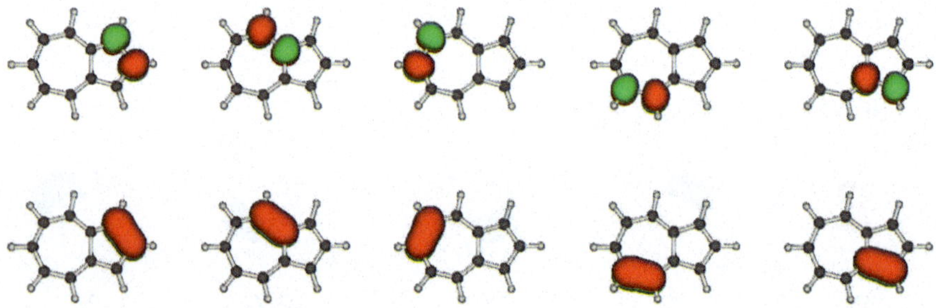

Fig. 3.36 Localized π orbitals on bonds of the azulene molecule

As an example demonstrating the usefulness of localized π-MOs in PAH molecules, we consider the calculation of CCSD correlation energies within the π-system using either canonical π-MOs or localized π-MOs. The calculation of correlation energies with CMOs is done as usual, whereas LMOs require the application of the method of increments [65, 66]. The relevant data for seven small aromatic hydrocarbon molecules are shown in Table 3.2. We can draw the following conclusions with respect to the usefulness of localized π-MOs for calculating correlation energies:

(1) the sum of single electron pair or one-body contributions e_1^π gives at least 60 % of the correlation energy of the π-system in alternating PAHs, but only about 56 % in the case of the non-alternating hydrocarbon azulene;
(2) inclusion of the two-body terms ($e_1^\pi + e_2^\pi$) increases this fraction to at least 99 % in alternating PAHs, and to the slightly smaller value of 98 % in the case of azulene.

Future studies will show whether these findings can be generalized to larger PAH molecules or not. Inclusion of localized σ-MOs and study of the relative contributions to correlation energy from π-only, σ-only, and mixed π/σ parts also require further study.

The data in Table 3.2 also allow us to compare total energies between isomers. For naphthalene and azulene, the two isomers of $C_{10}H_8$, the former is more stable

Table 3.2 Total energy at RHF level (basis set: cc-pVDZ) and various correlation energies at CCSD level (in Hartree) for some polycyclic aromatic hydrocarbons

Compound[a]	Benzene	Naphthalene	Anthracene	Tetracene	Phenanthrene	Chrysene	Azulene
Molecular formula	C_6H_6	$C_{10}H_8$	$C_{14}H_{10}$	$C_{18}H_{12}$	$C_{14}H_{10}$	$C_{18}H_{12}$	$C_{10}H_8$
$N_{core}/N_\sigma/N_\pi$ [b]	6/12/3	10/19/5	14/26/7	18/33/9	14/26/7	18/33/9	10/19/5
E_{RHF} [c]	−230.716167	−383.371777	−536.018362	−688.661379	−536.028684	−688.683101	−383.303824
E^π_{corr} [d]	−0.108098	−0.184584	−0.263737	−0.344314	−0.259434	−0.335108	−0.187560
e^π_1 [e]	−0.065836	−0.117347	−0.162564	−0.209505	−0.159906	−0.206156	−0.104903
e^π_1/E^π_{corr}	60.9%	63.6%	61.6%	60.8%	61.6%	61.5%	55.9%
e^π_2 [f]	−0.041238	−0.066721	−0.099657		−0.099540		−0.079191
$(e^\pi_1 + e^\pi_2)/E^\pi_{corr}$	99.1%	99.7%	99.4%		100.0%		98.2%
$E^{(\pi)}_{tot} = E_{RHF} + E^\pi_{corr}$	−230.824264	−383.556361	−536.282099	−689.005693	−536.288118	−689.018209	−383.491384
$E^{\pi+\sigma}_{corr}$ [g]	−0.828104	−1.353339	−1.880633	−2.408871	−1.879221	−2.405850	−1.358196
$E^{(\pi+\sigma)}_{tot} = E_{RHF} + E^{\pi+\sigma}_{corr}$	−231.544271	−384.725116	−537.898996	−691.070251	−537.907905	−691.088951	−384.662019

[a] all nuclei in the xy plane, all CC bond lengths equal to 142.1 pm, all CH bond lengths equal to 108.4 pm,n-membered rings as regular n-gons with CH bonds pointing radially outwards.
[b] number of electron pairs in the core, in the σ-system (excl. core), and in the π-system ($N_\pi = N_{core}/2$).
[c] total energy at Hartree–Fock level.
[d] canonical correlation energy of the π-system at CCSD level.
[e] sum of single electron pair (one-body) contributions to the correlation energy of the π-system.
[f] sum of non-additive increments of pairs of electron pairs (two-body) contributions to the correlation energy of the π-system.
[g] canonical correlation energy of π- and σ-system (excl. core) at CCSD level

than the latter at all levels of calculation, as expected. In the cases of anthracene / phenanthrene (isomers of $C_{14}H_{10}$) and tetracene / chrysene (isomers of $C_{18}H_{12}$), one finds that the acene compound is less stable than the phenacene compound. This is well known and had already been stated above with respect to the RHF energies. We find for all three pairs of isomers that the less stable molecule always exhibits the larger correlation energy (in all situations studied, i.e., in both π-only correlation and π-and-σ correlation).

3.5 Conclusion

In the vast majority of quantum chemistry methods, the system is described through mathematical objects (wave functions, densities, etc.) that are defined with respect to a set of Molecular Orbitals. At the end of some (usually iterative) procedure, one ends up with a set of Molecular Orbitals that are usually strongly delocalized on several atoms, in some cases even over the entire system. This is true, in particular, every time the orbitals are obtained via a diagonalization procedure that acts on a Hamiltonian matrix defined on a global set of Atomic Orbitals. The orbitals obtained in such a way are often named as "Canonical" for a given method, since in general every method produces a single set of Canonical Orbitals (usually, they will depend on the method). The orbitals that are obtained in such a way, however, have little in common with the orbitals that a chemist would expect to see, since the latter are associated, with a few exceptions, with the notion of bonds between specific pairs of atoms in the molecule. This fact leads to some difficulties in the interpretation of the obtained results, and a difficulty in reading the physical nature of a molecular structure, whose essence is dispersed among an overwhelming number of contributions coming from the different orbitals.

For this reason, starting from the very beginning of Quantum Chemistry, a large number of numerical procedures have been proposed to transform the Canonical Molecular Orbitals into a set of *equivalent* Local Molecular Orbitals. We presented and discussed here some of the different techniques that can be used to obtain these Local Orbitals. Generally speaking, all these methods present advantages and drawbacks, and the choice of the method may sometimes crucially depend on the type of treatment one intends to perform for a given system.

In the more recent years, an additional reason for the use of Localized Orbitals has emerged, related to computational efficiency. In fact, through the use of Localized Orbitals, it is possible to take advantage of the locality of the interaction, which is something almost universal in the physical world. In fact, with the partial exception of gravity, the interactions between the constituents of matter have a rather short-range nature. Even the Coulomb force between charged particles at distance R that decays, in principle, as R^{-2} like gravitational force, gives rise to effective interactions having a much weaker and shorter-ranging interaction because of the presence of an equal number of charges of opposite sign. In order to exploit from a computational point of view, this locality of the interaction, and to obtain the methods that belong to the family of Linear-Scaling methods, it is crucial to approach the study of the system by using local orbitals.

It should be noticed, however, that the Localization methods that are needed to implement the Linear-Scaling approaches should essentially concentrate the orbitals in restricted regions of the space, without an absolute need of obtaining, for instance, orbitals having a bond nature. In other words, they are somehow less demanding than the methods that localize the orbitals in order to give a better chemical interpretation. From this point of view, methods that at first sight could seem formally less satisfactory can be used with good results.

Even though the "perfect localization method" has never been proposed, and probably it simply does not exist, very efficient and versatile procedures are nowadays available. In such a way, a larger and larger number of systems have become not only computationally treatable, but it is also possible to understand the precise nature of their structure in terms of local entities.

Acknowledgements The Toulouse group would like to thank for financial support the French CNRS and the "Programme Investissements d'Avenir" under the program ANR-11-IDEX-0002-02, reference ANR-10-LABX-0037-NEXT. The calculations of this work have been partly performed by using the resources of the HPC center CALMIP under the grant 2016-p1048. D. A. most thankfully acknowledges the generous support from the "Programme Investissements d'Avenir" under the program ANR-11-IDEX-0002-02, reference ANR-10-LABX-0037-NEXT, which made possible a one-month visit to Toulouse. The support received from the high-performance computing facilities at Zentraleinrichtung für Datenverarbeitung (ZEDAT), Freie Universität Berlin, is gratefully acknowledged as well.

Appendix 1: A Priori Orbitals: Input Data (DoLo)

The a priori local orbitals are obtained thanks to the DoLo program. DoLo gives a guess of non-orthogonal local orbitals. Usually, two steps follow: (i) the projection of this guess of local orbitals onto the SCF or CASSCF orbitals; the quality of the guess of local orbitals is then improved (`proj_scf` code) and (ii) the orthogonalization of the obtained orbitals (`schmudort` code). Finally, this multi-step process provides a set of local and orthogonal orbitals, with SCF or CASSCF quality. Input data for the first step (DoLo) are reported here for several examples presented in the chapter. Some explanations are given here and more details can be found in the manual. The source and manual are available on https://github.com/LCPQ/Cost_package (Tables 3.3, 3.4, 3.5 and 3.6).

Table 3.3 C_2H_2: Localization of the SCF Orbitals. The core orbitals are defined using CORE_. In this example, the star character (C*) is used to define the core of all C atoms in the same time. To define the CH bond orbitals with a distance lower than 1.2 Å, SIGMA_ data is used. BOND_ allows to define the σ CC bond and DELOC_ the π orbitals. In all cases, the atomic orbitals used to construct the local orbitals are defined. * at the beginning of the line corresponds to a comment

```
* CORE ORBITALS
CORE_ C* 1S(1)
* σ CH BONDS
SIGMA_ C* 1S(2) 2P(1) : H* 1S(1) DMAX = 1.2 ANGSTROM
* σ CC BONDS
BOND_ chain = 'C1-C2',bas = 2*'1S(2) 2pz(1)',noc = 1,nvirt = 1
* π CC BONDS
DELOC_ chain = 'C1-C2',bas = 2*'2px(1) 2py(1)',noc = 2,nvirt = 2
```

Table 3.4 C_2F_2: Localization of the SCF Orbitals. The core orbitals are defined using CORE_. The σ CC bond is defined with BOND_. In all cases, the atomic orbitals used to construct the local orbitals are defined.* at the beginning of the line corresponds to a comment

```
* CORE ORBITALS
CORE_ C* 1S(1)
CORE_ F* 1S(1)
* σ F-C and CC BONDS and axis Fluorine lone pair (LP1 stands for one lone pair)
BOND_ BASREF = 'C:1S(2) 2PZ(1),F:1S(2) 2PZ(1)'
BOND_ chain = 'F1(LP1)-C1-C2-F2(LP1)
* π CC BONDS
DELOC_ chain = 'C1-C2',bas = 2*'2px(1) 2py(1)',noc = 2,nvirt = 2
* Fluorine lone pairs(LP = 2 stands for two lone pairs)
BOND_ chain = 'F1',BAS = '2PX(1) 2py(1)',LP = 2
BOND_ chain = 'F2',BAS = '2PX(1) 2py(1)',LP = 2
```

Table 3.5 **Benzene:** Localization of the CAS(6,6)SCF Orbitals. The core orbitals are defined using CORE_. To define the CH and CC bond orbitals, SIGMA_ data is used. The π bonds are defined on all the cycles using the DELOC_ data. In all cases, the atomic orbitals used to construct the local orbitals are defined. LABEL='A' means that the local orbitals correspond to active orbitals. * at the beginning of the line corresponds to a comment. Inside the BOND or the SIGMA data, * is used when all the occurrence of an atom are considered. For example, C* means all C atoms

```
* CORE ORBITALS
CORE_ C* 1S(1)
* σ CH BONDS
SIGMA_ C* 1S(2) 2P(1) : H* 1S(1) DMAX = 1.2 ANGSTROM
* σ CC BONDS
SIGMA_ C* 1S(2) 2P(1) : C* 1S(2) 2P(1) dmax = 1.5 ANGSTROM
* π CC BONDS
DELOC_ chain = 'C1-C2-C3-C4-C5-C6',bas = 6*'2pz(1)',noc = 3,nvirt = 3,LABEL = 'A'
```

Table 3.6 *Trans*-**stilbene:** Localization of the π SCF orbitals. The input data of the *trans*-stilbene molecule is very similar to the benzene. The π bonds are defined on each cycle and on the central bond using the DELOC_ data

```
* CORE ORBITALS
CORE_ C* 1S(1)
* σ CH BONDS
SIGMA_ C* 1S(2) 2P(1) : H* 1S(1) DMAX = 1.2 ANGSTROM
* σ CC BONDS
SIGMA_ c* 1S(2) 2P(1) : c* 1S(2) 2P(1) dmax = 1.5 ANGSTROM
* π CC BONDS
DELOC_ chain = 'C1-C8',bas = 2*'2pz(1)',noc = 1,nvirt = 1
DELOC_ chain = 'C2-C3-C4-C5-C6-C7',bas = 6*'2pz(1)',noc = 3,nvirt = 3
DELOC_ chain = 'C9-C10-C11-C12-C13-C14,bas = 6*'2pz(1)',noc = 3,nvirt = 3
```

Appendix 2: Molecular Electronic States and Electron Configurations

By intention, all small molecules selected for discussion in this work on orbital localization have a closed-shell totally symmetric singlet electronic ground state. For larger molecules (for example, acenes and phenacenes), such an electronic state can still always be easily defined and fully characterized, so that it may serve at least as a reference state for other subsequent purposes. As a consequence, the Configuration State Function (CSF) of the electronic state is a single Slater determinant. Both the CSF and the total energy associated with it (that is the expectation value of the Born–Oppenheimer Hamiltonian taken over the CSF) are invariant with respect to unitary transformations among the set of doubly occupied Molecular Orbitals (MOs). Orbital localization then simply takes advantage of the fact that all observables of such molecular electronic states, in particular their total energy, are invariant with respect to a unitary transformation carried out within the space of doubly occupied MOs (in fact, an orthogonal transformation suffices, as only real-valued quantities are involved).

The following list provides for each molecule its point group, the symmetry label of the closed-shell totally symmetric singlet state considered here, and the corresponding electron configuration, i.e., symmetry label and number of all the completely filled shells (an entry like "e 1-3" is to be understood as $1e^4\ 2e^4\ 3e^4$).

1. Ethane, C_2H_6 (18 electrons) $\widetilde{X}\ ^1A_{1g}$ (D_{3d}) / $^1A_1$ (D_3) / $^1A_1'$ (D_{3h}), electron configuration: a_{1g} 1–3, e_g 1, a_{2u} 1–2, e_u 1 (D_{3d}) / a_1 1–3, a_2 1–2, e 1–2 (D_3) / a_1' 1–3, e′ 1, a_2'' 1–2, e″ 1 (D_{3h})
2. Ethene (ethylene), C_2H_4 (16 electrons) $\widetilde{X}\ ^1A_g$ (D_{2h}), electron configuration: a_g 1–3, b_{1g} 1, b_{1u} (π) 1, b_{2u} 1–2, b_{3u} 1
3. Ethyne (acetylene), C_2H_2 (14 electrons) $X\ ^1\Sigma_g^+$ ($D_{\infty h}$), electron configuration: σ_g 1–3, σ_u 1–2, π_u 1

4. Monofluoroethyne, C_2HF (22 electrons) $X\ ^1\Sigma^+$ ($C_{\infty v}$), electron configuration: σ 1–7, π 1–2
5. Difluoroethyne, C_2F_2 (30 electrons) $X\ ^1\Sigma_g^+$ ($D_{\infty h}$), electron configuration: σ_g 1–5, σ_u 1–4, π_g 1, π_u 1–2
6. Benzene, C_6H_6 (42 electrons) $\widetilde{X}\ ^1A_{1g}$ (D_{6h}), electron configuration: a_{1g} 1–3, e_{1g} (π) 1, e_{2g} 1–3, a_{2u} (π) 1, b_{1u} 1–2, b_{2u} 1, e_{1u} 1–3
7. (*E*)-1,2-diphenylethene (*trans*-stilbene), $C_{14}H_{12}$ (96 electrons) $\widetilde{X}\ ^1A_g$ (C_{2h}), electron configuration (planar conformation): a_g 1–21, b_g (π) 1–3, a_u (π) 1–4, b_u 1–20
8. Tetracarbonylnickel, $[Ni(CO)_4]$ (84 electrons) $\widetilde{X}\ ^1A_1$ (T_d), electron configuration: a_1 1–8, e 1–2, t_1 1, t_2 1–9
9. Tris(acetylacetonato)aluminium, $[Al(acac)_3]$ (172 electrons) $\widetilde{X}\ ^1A_1$ (D_3), electron configuration: a_1 1–16, a_2 1–14, e 1–28
10. [*n*]acenes, $C_{4n+2}H_{2n+4}$ ($26n + 16$ electrons) ^1A_g (D_{2h}; all nuclei in the plane $z = 0$, largest molecular extension oriented along the y-axis), electron configuration for $n = 2k + 1$ ($k = 0$: benzene, $k = 1$: anthracene, ...): a_g $6k + 6$, b_{1g} $5k + 3$, b_{2g} (π) $k + 1$, b_{3g} (π) $k + 1$, a_u (π) k, b_{1u} (π) $k + 1$, b_{2u} $6k + 4$, b_{3u} $5k + 5$, electron configuration for $n = 2k$ ($k = 1$: naphthalene, $k = 2$: tetracene, ...): a_g $6k + 3$, b_{1g} $5k + 1$, b_{2g} (π) k, b_{3g} (π) k, a_u (π) k, b_{1u} (π) $k + 1$, b_{2u} $6k + 1$, b_{3u} $5k + 2$
11. [*n*]phenacenes, $C_{4n+2}H_{2n+4}$ ($26n + 16$ electrons, all nuclei in the plane $z = 0$, largest molecular extension oriented along the y-axis), $^1A_1$ (C_{2v}, n odd) or ^1A_g (C_{2h}, n even), electron configuration for $n = 2k + 1$ ($k = 0$: benzene, $k = 1$: phenanthrene, ...): a_1 $11k + 10$, a_2 (π) $2k + 1$, b_1 $11k + 8$, b_2 (π) $2k + 2$, electron configuration for $n = 2k$ ($k = 1$: naphthalene, $k = 2$: chrysene, ...): a_g $11k + 4$, b_g (π) $2k$, a_u (π) $2k + 1$, b_u $11k + 3$
12. [*n*]coronenes, $C_{6n^2}H_{6n}$ ($6n(6n + 1)$ electrons, number of hexagons $N_{\text{hex}} = 3n(n - 1) + 1$, all nuclei in the plane $z = 0$, central hexagon oriented such that two of its carbon nuclei lie on the y-axis), $^1A_{1g}$ (D_{6h}), electron configuration for $n = 2k + 1$ ($k = 0$: benzene, $k = 1$: circumcoronene, ...): a_{1g} $(k + 1)(5k + 3)$, a_{2g} $k(5k + 3)$, b_{1g} (π) $k(k + 1)$, b_{2g} (π) $k(k + 1)$, e_{1g} (π) $2k^2 + 2k + 1$, e_{2g} $(2k + 1)(5k + 3)$, a_{1u} (π) k^2, a_{2u} (π) $k^2 + 2k + 1$, b_{1u} $5k^2 + 5k + 1$, b_{2u} $5k^2 + 6k + 2$, e_{1u} $(2k + 1)(5k + 3)$, e_{2u} (π) $2k(k + 1)$, electron configuration for $n = 2k$ ($k = 1$: coronene, ...): a_{1g} $k(5k + 3)$, a_{2g} $k(5k - 2)$, b_{1g} (π) k^2, b_{2g} (π) k^2, e_{1g} (π) $2k^2$, e_{2g} $k(10k + 1)$, a_{1u} (π) $k(k - 1)$, a_{2u} (π) $k(k + 1)$, b_{1u} $5k^2$, b_{2u} $k(5k + 1)$, e_{1u} $k(10k + 1)$, e_{2u} (π) $2k^2$
13. Azulene, $C_{10}H_8$ (68 electrons) $\widetilde{X}\ ^1A_1$ (C_{2v}; all nuclei in the plane $y = 0$), electron configuration: a_1 1–17, a_2 (π) 1–2, b_1 1–12, b_2 (π) 1–3

All results summarized above have been confirmed in several ways. First, the group theoretical information given above, which is an invariant of the closed-shell singlet electronic state, is in conformity with (or can be derived from) the molecular structural formula, if the latter is adapted to the molecular point group (this may require extension/inclusion of Robinson–Clar π sextets for some polycyclic aromatic hydrocarbons). Examples of such structural formulas are shown in Figs. 3.26, 3.27, and 3.28.

Second, the molecular structures shown have been confirmed as representing local minima of the total energy hypersurface at the SCF level of theory, via full optimization of molecular structure followed by normal mode analysis with full exploitation of molecular point-group symmetry. In the case of the [n]acenes and [n]phenacenes, this has been done for all n up to and including $n = 30$; in the case of the [n]coronenes for all n up to and including $n = 6$.

Third, the electron configurations and the corresponding total energies were vindicated in calculations where the use of molecular symmetry has been completely and intentionally switched off (so that convergence to a state lower in energy may occur, but was never observed). Since the geometric structure of the nuclear framework (the "molecular geometry") still carries the full point-group symmetry information into the calculation running now under point group C_1, all symmetry information is still present (and has to be so).

Both types of calculations (those with as well as those without the exploitation of point-group symmetry) are important, since orbital localization typically and necessarily ignores point-group symmetry. In general, one needs to be able to orient oneself within the set of occupied CMOs, in order, e.g., to select and sort these orbitals into different groups prior to localization. An example is provided by the acenes, where separation of σ CMOs from π CMOs appears to be (an easily achievable) step before any attempt to localize the orbitals.

References

1. Einstein A, Podolsky B, Rosen N (1935) Can quantum-mechanical description of physical reality be considered complete? Phys Rev 47(10):777. https://doi.org/10.1103/PhysRev.47.777
2. Schrödinger E (1926) An undulatory theory of the mechanics of atoms and molecules. Phys Rev 28(6):1049. https://doi.org/10.1103/PhysRev.28.1049
3. Lewis GN (1916) The atom and the molecule. J Am Chem Soc 38(4):762. https://doi.org/10.1021/ja02261a002
4. Heitler W, London F (1927) Wechselwirkung neutraler Atome und homöopolare Bindung nach der Quantenmechanik. Zeitschrift für Physik 44(6–7):455. https://doi.org/10.1007/BF01397394
5. Daudey J (1974) Direct determination of localized SCF orbitals. Chem Phys Lett 24(4):574. https://doi.org/10.1016/0009-2614(74)80185-5
6. Rubio J, Povill A, Malrieu JP, Reinhardt P (1997) Direct determination of localized Hartree Fock orbitals as a step toward N scaling procedures. J Chem Phys 107(23):10044. https://doi.org/10.1063/1.474161
7. Weinstein H, Pauncz R, Cohen M (1971) Localized molecular orbitals. Adv Atomic Molecular Phys 7:97. https://doi.org/10.1016/S0065-2199(08)60359-0
8. Foster JM, Boys SF (1960) Canonical Configurational interaction procedure. Rev Mod Phys 32(2):300. https://doi.org/10.1103/RevModPhys.32.300
9. Edmiston C, Ruedenberg K (1963) Localized atomic and molecular orbitals. Rev Mod Phys 35(3):457. https://doi.org/10.1103/RevModPhys.35.457

10. Pipek J, Mezey PG (1989) A fast intrinsic localization procedure applicable for ab initio and semiempirical linear combination of atomic orbital wave functions. J Chem Phys 90(9):4916. https://doi.org/10.1063/1.456588
11. Aquilante F, Bondo Pedersen T, Sánchez de Merás A, Koch H (2006) Fast noniterative orbital localization for large molecules. J Chem Phys 125(17):174101. https://doi.org/10.1063/1.2360264
12. Leonard JM, Luken WL (1984) Calculation of localized molecular orbitals. Int J Quantum Chem 25(2):355. https://doi.org/10.1002/qua.560250208
13. Rajzmann M, Brenier B, Purcell KF (1987) Localized virtual and occupied molecular orbitals. Theor Chim Acta 72:13. https://doi.org/10.1007/BF00526550
14. Boys SF (1960) Construction of some molecular orbitals to be approximately invariant for changes from one molecule to another. Rev Mod Phys 32:296. https://doi.org/10.1103/RevModPhys.32.296
15. Bartha F, Bogár F, Kapuy E (1990) Localization of virtual orbitals. Int J Quantum Chem 38(2):215. https://doi.org/10.1002/qua.560380216
16. Magnasco V, Perico A (1967) Uniform localization of atomic and molecular orbitals. I. J Chem Phys 47(3):971. https://doi.org/10.1063/1.1712065
17. Magnasco V, Perico A (1968) Uniform localization of atomic and molecular orbitals. II. J Chem Phys 48(2):800. https://doi.org/10.1063/1.1668714
18. Resta R (2006) Kohn's theory of the insulating state: a quantum-chemistry viewpoint. J Chem Phys 124(10):104104. https://doi.org/10.1063/1.2176604
19. Resta R, Sorella S (1999) Electron localization in the insulating state. Phys Rev Lett 82(2):370
20. Resta R (2005) Electron localization in the quantum Hall regime. Phys Rev Lett 95(19). https://doi.org/10.1103/PhysRevLett.95.196805
21. Resta R (2006) Polarization fluctuations in insulators and metals: new and old theories merge. Phys Rev Lett 96(13). https://doi.org/10.1103/PhysRevLett.96.137601
22. El Khatib M, Leininger T, Bendazzoli GL, Evangelisti S (2014) Computing the position-spread tensor in the CAS-SCF formalism. Chem Phys Lett 591:58. https://doi.org/10.1016/j.cplett.2013.10.080
23. Høyvik IM, Jansík B, Jørgensen P (2012) Orbital localization using fourth central moment minimization. J Chem Phys 137(22):224114. https://doi.org/10.1063/1.4769866
24. Jansík B, Høst S, Kristensen K, Jørgensen P (2011) Local orbitals by minimizing powers of the orbital variance. J Chem Phys 134. https://doi.org/10.1063/1.3590361
25. Edmiston C, Ruedenberg K (1965) Localized atomic and molecular orbitals. II. J Chem Phys 43(10):S97. https://doi.org/10.1063/1.1701520
26. von Niessen W (1972) Density localization of atomic and molecular orbitals. I. J Chem Phys 56(9):4290. https://doi.org/10.1063/1.1677859
27. Lehtola S, Jónsson H (2014) Pipek-Mezey orbital localization using various partial charge estimates. J Chem Theory Comput 10(2):642. https://doi.org/10.1021/ct401016x
28. Cioslowski J (1991) Partitioning of the orbital overlap matrix and the localization criteria. J Math Chem 8(1):169. https://doi.org/10.1007/BF01166933
29. Alcoba DR, Lain L, Torre A, Bochicchio RC (2006) An orbital localization criterion based on the theory of "fuzzy" atoms. J Comput Chem 27(5):596. https://doi.org/10.1002/jcc.20373
30. Knizia G (2013) Intrinsic atomic orbitals: an unbiased bridge between quantum theory and chemical concepts. J Chem Theory Comput 9(11):4834. https://doi.org/10.1021/ct400687b
31. Pipek J (1989) Localization measure and maximum delocalization in molecular systems. Int J Quantum Chem 36(4):487. https://doi.org/10.1002/qua.560360405
32. Reed AE, Weinhold F (1983) Natural bond orbital analysis of near-Hartree-Fock water dimer. J Chem Phys 78(6):4066. https://doi.org/10.1063/1.445134
33. Reed AE, Weinhold F (1985) Natural localized molecular orbitals. J Chem Phys 83(4):1736. https://doi.org/10.1063/1.449360
34. Reed AE, Weinstock RB, Weinhold F (1985) Natural population analysis. J Chem Phys 83(2):735. https://doi.org/10.1063/1.449486

35. Glendening ED, Badenhoop JK, Reed AE, Carpenter JE, Bohmann JA, Morales CM, Karafiloglou P, Landis CR, Weinhold F (2018) NBO 7.0, Theoretical Chemistry Institute, University of Wisconsin, Madison, available from https://nbo7.chem.wisc.edu
36. Bhattacharjee J, Waghmare UV (2006) Localized orbital description of electronic structures of extended periodic metals, insulators, and confined systems: density functional theory calculations. Phys Rev B 73(12). https://doi.org/10.1103/PhysRevB.73.121102
37. Høyvik IM, Olsen J, Jørgensen P (2017) Generalising localisation schemes of orthogonal orbitals to the localisation of non-orthogonal orbitals. Mol Phys 115(1):16. https://doi.org/10.1080/00268976.2016.1173733
38. Maynau D, Evangelisti S, Guihéry N, Calzado CJ, Malrieu JP (2002) Direct generation of local orbitals for multireference treatment and subsequent uses for the calculation of the correlation energy. J Chem Phys 116(23):10060. https://doi.org/10.1063/1.1476312
39. Leininger T, Angeli C, Evangelisti S, Cimiraglia R, Maynau D (2003) Geometry optimization within a localized CAS-SCF approach. Chem Phys Lett 371(1):49. https://doi.org/10.1016/S0009-2614(03)00224-0
40. Maynau D, NOSCF, a development of Laboratoire de Chimie et Physique Quantiques de Toulouse, available on https://github.com/lcpq/cost_package
41. Zulfikri H, Amovilli C, Filippi C (2016) Multiple-resonance local wave functions for accurate excited states in quantum monte carlo. J Chem Theory Comput 12(3):1157. https://doi.org/10.1021/acs.jctc.5b01077
42. Ma D, Li Manni G, Gagliardi L (2011) The generalized active space concept in multiconfigurational self-consistent field methods. J Chem Phys 135(4):044128. https://doi.org/10.1063/1.3611401
43. Vogiatzis KD, Li Manni G, Stoneburner SJ, Ma D, Gagliardi L (2015) Systematic expansion of active spaces beyond the CASSCF limit: a GASSCF/SplitGAS benchmark study. J Chem Theory Comput 11(7):3010. https://doi.org/10.1021/acs.jctc.5b00191
44. Ivanic J (2003) Direct configuration interaction and multiconfigurational self-consistent-field method for multiple active spaces with variable occupations. I. method. J Chem Phys 119(18):9364. https://doi.org/10.1063/1.1615954
45. Krah T, Ben Amor N, Maynau D, Berger J, Robert V (2014) A rational reduction of CI expansions: combining localized molecular orbitals and selected charge excitations. J Mol Model 20(7):1
46. Mezey PG (1987) The shape of molecular charge distributions: Group theory without symmetry. J Comput Chem 8:462. https://doi.org/10.1002/jcc.540080426
47. Bories B, Maynau D, Bonnet ML (2007) Selected excitation for CAS-SDCI calculations. J Comput Chem 28(3):632. https://doi.org/10.1002/jcc.20588
48. Ben Amor N, Bessac F, Hoyau S, Maynau D (2011) Direct selected multireference configuration interaction calculations for large systems using localized orbitals. J Chem Phys 135(1):014101. https://doi.org/10.1063/1.3600351
49. Chang C, Calzado CJ, Ben Amor N, Marin JS, Maynau D (2012) Multi-scale multireference configuration interaction calculations for large systems using localized orbitals: partition in zones. J Chem Phys 137(10):104102
50. Widmark PO, Malmqvist P, Roos BO (1990) Density matrix averaged atomic natural orbital (ANO) basis sets for correlated molecular wave functions: I. First row atoms. Theor Chim Acta 77(5):291. https://doi.org/10.1007/BF01120130
51. Dunning TH (1989) Gaussian basis sets for use in correlated molecular calculations. I. The atoms boron through neon and hydrogen. J Chem Phys **90**(2):1007 (1989). https://doi.org/10.1063/1.456153
52. Aquilante F, Pedersen TB, Veryazov V, Lindh R (2013) MOLCAS-a software for multiconfigurational quantum chemistry calculations: MOLCAS. Wiley Interdiscip Rev Comput Mol Sci 3(2):143. https://doi.org/10.1002/wcms.1117
53. Werner HJ, Knowles PJ, Knizia G, Manby FR, Schütz M, Celani P, Györffy W, Kats D, Korona T, Lindh R, Mitrushenkov A, Rauhut G, Shamasundar KR, Adler TB, Amos RD, Bernhardsson A, Berning A, Cooper DL, Deegan MJO, Dobbyn AJ, Eckert F, Goll E, Hampel

C, Heßelmann A, Hetzer G, Hrenar T, Jansen G, Köppl C, Liu Y, Lloyd AW, Mata RA, May AJ, McNicholas SJ, Meyer W, Mura ME, Nicklass A, O'Neill DP, Palmieri P, Peng D, Pflüger K, Pitzer R, Reiher M, Shiozaki T, Stoll H, Stone AJ, Tarroni R, Thorsteinsson T, Wang M (2015) MOLPRO, version 2015.1, a package of ab initio programs
54. TURBOMOLE V6.2 2010, a development of University of Karlsruhe and Forschungszentrum Karlsruhe GmbH, 1989–2007, TURBOMOLE GmbH, since 2007; available from http://www.turbomole.com
55. Maynau D, DoLo, a development of Laboratoire de Chimie et Physique Quantiques de Toulouse, available on https://github.com/lcpq/cost_package
56. Veryazov V (2014) GV, available on https://www.molcas.org/gv
57. Avogadro: an open-source molecular builder and visualization tool. version 1.xx., available on http://avogadro.cc/
58. Schaftenaar G, Noordik JH (2000) Molden: a pre- and post-processing program for molecular and electronic structures. J Comput Aided Mol Des 14(2):123. https://doi.org/10.1023/A:1008193805436
59. Schaftenaar G, Vlieg E, Vriend G (2017) Molden 2.0: quantum chemistry meets proteins. J Comput Aided Mol Des 31(9):789. https://doi.org/10.1007/s10822-017-0042-5
60. Zimmerman PM, Molina AR, Smereka P (2015) Orbitals with intermediate localization and low coupling: Spanning the gap between canonical and localized orbitals. J Chem Phys 143. https://doi.org/10.1063/1.4923084
61. Ben Amor N, Hoyau S, Maynau D, Brenner V (2018) Low-lying excited states of model proteins: Performances of the CC2 method versus multireference methods. J Chem Phys 148(18):184105. https://doi.org/10.1063/1.5025942
62. von Chrzanowski LS, Lutz M, Spek AL (2007) α-Tris(2,4-pentanedionato-κ^2 O, O$'$)aluminium(III) at 240, 210, 180, 150 and 110K: a new δ phase at 110 K. Acta Crystallogr C 63(3):m129. https://doi.org/10.1107/S0108270107005707
63. Scheurer P, Schwarz WHE (2000) Externally Localized Molecular Orbitals: A Numerical Investigation of Localization Degeneracy. Int J Quantum Chem. 76(3):420. https://doi.org/10.1002/(SICI)1097-461X (2000) 76:3 < 420::AID-QUA10>3.0.CO;2-8
64. Scheurer P, Schwarz WHE (2000) Continuous Degeneracy of Sets of Localized Orbitals. Int J Quantum Chem 76(3):428. https://doi.org/10.1002/(SICI)1097-461X (2000)76:3<428::AID-QUA11>3.0.CO;2-Y
65. Stoll H (1992) The correlation energy of crystalline silicon. Chem Phys Lett 191:548. https://doi.org/10.1016/0009-2614(92)85587-Z
66. Paulus B (2006) The method of increments—a wavefunction-based ab initio correlation method for solids. Phys Rep 428:1. https://doi.org/10.1016/j.physrep.2006.01.003
67. Mulliken RS (1955) Electronic population analysis on LCAO-MO molecular wave functions. II. Overlap populations, bond orders, and covalent bond energies. J Chem Phys 23(10):1841. https://doi.org/10.1063/1.1740589
68. Griffith MG, Goodman L (1967) Sigma and pi electronic reorganization in acetylene. J Chem Phys 47(11):4494. https://doi.org/10.1063/1.1701659
69. Switkes E, Stevens RM, Lipscomb WN, Newton MD (1969) Localized Bonds in SCF Wave-functions for Polyatomic Molecules. I. Diborane. J Chem Phys 51(5):2085. https://doi.org/10.1063/1.1672303
70. Schlosser H (1971) Localized orbitals for polyatomic systems (I). Phys Status Solidi B 44(1):183. https://doi.org/10.1002/pssb.2220440118
71. Daudey JP, Diner S (1972) PCILON. perturbative configuration interaction using localized orbitals and numerical integration. I. Numerical integration techniques for the calculation of Hamiltonian matrix elements between localized orbitals. Int J Quantum Chem 6(3):575. https://doi.org/10.1002/qua.560060318
72. von Niessen W (1974) Localized molecular orbitals for aromatic molecules: Mono- and disubstituted benzenes. Theor Chim Acta 33(3):185. https://doi.org/10.1007/BF00551254
73. Coffey P (1974) Virtual orbital transformation prior to configuration interaction with localized orbitals. Int J Quantum Chem 8(5):777. https://doi.org/10.1002/qua.560080512

74. Kleier DA, Halgren TA, Hall JH, Lipscomb WN (1974) Localized molecular orbitals for polyatomic molecules. I. A comparison of the Edmiston-Ruedenberg and Boys localization methods. J Chem Phys 61(10):3905. https://doi.org/10.1063/1.1681683
75. Barr R, Basch H (1975) Improved convergence in orbital localization methods. Chem Phys Lett 32(3):537. https://doi.org/10.1016/0009-2614(75)85235-3
76. Haddon RC, Williams GR (1976) Calculation of localised molecular orbitals with the Foster-Boys criterion. Chem Phys Lett 42(3):453. https://doi.org/10.1016/0009-2614(76)80651-3
77. Daudel R, Stephens ME, Csizmadia IG, Kozmutza C, Kapuy E, Goddard JD (1977) Theory of lone pairs. II. A moment analysis of localized molecular orbitals in ten-electron hydrides. Int J Quantum Chem 11(4):665. https://doi.org/10.1002/qua.560110410
78. Brown LD, Kleier DA, Lipscomb WN (1978) ChemInform abstract: Localized molecular orbitals for carbon dioxide and carbonate (2-). A comparison of localization types and a comment on isoelectronic structures. Chem Informationsdienst 9(2). https://doi.org/10.1002/chin.197802064
79. Shei TC, Pan KC (1979) Transferability of Electronic Interaction Energies Between Localized Orbitals. J Chin Chem Soc 26(2):43. https://doi.org/10.1002/jccs.197900009
80. Surjan PR, Mayer I (1981) Delocalization corrections to the strictly localized molecular orbitals: A linearized SCF approximation. Theor Chim Acta 59(6):603. https://doi.org/10.1007/BF00552853
81. Böhm MC (1981) A simple extension of the external Magnasco-Perico localization procedure to the virtual MO-space. Theor Chim Acta 59(6):609. https://doi.org/10.1007/BF00552854
82. Leonard JM, Luken WL (1982) Quadratically convergent calculation of localized molecular orbitals. Theor Chim Acta 62(2):107. https://doi.org/10.1007/BF00581477
83. Luken WL, Beratan DN (1982) Localized orbitals and the Fermi hole. Theor Chim Acta 61(3):265. https://doi.org/10.1007/BF00550971
84. Laidig WD, Purvis GD, Bartlett RJ (1982) Localized orbitals in the coupled cluster singles and doubles model. Int J Quantum Chem 22:561. https://doi.org/10.1002/qua.560220848
85. Cizek J, Förner W, Ladik J (1983) Localization of the filled and virtual orbitals in the nucleotide bases. Theor Chim Acta 64:107. https://doi.org/10.1007/BF00550325
86. Mayer I (1983) Charge, bond order and valence in the AB initio SCF theory. Chem Phys Lett 97(3):270. https://doi.org/10.1016/0009-2614(83)80005-0
87. Kapuy E, Csépes Z, Kozmutza C (1983) Application of the many-body perturbation theory by using localized orbitals. Int J Quantum Chem 23(3):981. https://doi.org/10.1002/qua.560230321
88. Pipek J (1984) Approximate upper bound for two-electron integrals of molecular orbitals. Chem Phys Lett 111(4):430. https://doi.org/10.1016/0009-2614(84)85535-9
89. Kapuy E, Csépes Z, Pipek J (1984) Application of the many-body perturbation theory based on localized orbitals to cyclic polyenes. Acta Phys Hung 55:365. https://doi.org/10.1007/BF03155948
90. Pipek J (1985) Controlled orthogonalization of localized orbitals. Int J Quantum Chem. 27(5):527. https://doi.org/10.1002/qua.560270503
91. Förner W, Cizek J, Otto P, Ladik J, Steinborn OE (1985) Coupled-cluster studies. I. Application to small molecules, basis set dependences. Chem Phys 97:235. https://doi.org/10.1016/0301-0104(85)87034-8
92. Förner W, Ladik J, Otto P, Cizek J (1985) Coupled-cluster studies. II. The role of localization in correlation calculations on extended systems. Chem Phys 97:251. https://doi.org/10.1016/0301-0104(85)87035-X
93. Förner W (1987) Coupled cluster studies. IV. Analysis of the correlated wavefunction in canonical and localized orbital basis for ethylene, carbon monoxide, and carbon dioxide. Chem Phys 114:21. https://doi.org/10.1016/0301-0104(87)80016-2
94. Sano T, Narita S, I'Haya YJ (1987) Optimization with a direct search for orbital localization. Chem Phys Lett 138(4):291. https://doi.org/10.1016/0009-2614(87)80386-X
95. Pipek J (1988) Long-range behavior of the off-diagonal Fock matrix elements of localized molecular orbitals. Chem Phys Lett 143(3):293. https://doi.org/10.1016/0009-2614(88)87383-4

96. Pipek J, Mezey PG (1988) Dependence of MO shapes on a continuous measure of delocalization. Int J Quantum Chem 34:1. https://doi.org/10.1002/qua.560340804
97. Kapuy E, Bartha F, Bogár F, Csépes Z, Kozmutza C (1990) Applications of the MBPT in the localized representation. Int J Quantum Chem 38(2):139. https://doi.org/10.1002/qua.560380209
98. Kozmutza C, Kapuy E (1990) Localized orbitals for the description of molecular interaction. Int J Quantum Chem 38(5):665. https://doi.org/10.1002/qua.560380508
99. Kozmutza C, Kapuy E (1991) Counterpoise corrected calculations at the correlated level: A simplified method using LMOs. J Comput Chem 12(8):953. https://doi.org/10.1002/jcc.540120808
100. Förner W (1992) Formulation of the coupled cluster theory with localized orbitals in correlation calculations on polymers. Int J Quantum Chem 43(2):221. https://doi.org/10.1002/qua.560430206
101. Surjan PR, Pipek J, Paizs B (1993) Localization maps by orbital partitioning of the electron density. Theor Chim Acta 86(5):379. https://doi.org/10.1007/BF01122430
102. Kozmutza C, Kapuy E, Evleth E (1993) Calculation of the interaction energy in a localized representation for a trimer (ne3) system. J Comput Chem 14(10):1136. https://doi.org/10.1002/jcc.540141003
103. Kapuy E, Bogár F, Tfirst E (1994) Application of many-body perturbation theory in the localized representation for the all-trans conjugated polyenes. Int J Quantum Chem 52(1):127. https://doi.org/10.1002/qua.560520113
104. Perera SA, Bernholdt DE, Bartlett RJ (1994) Localized Hartree product orbitals in correlated studies of molecules. Int J Quantum Chem 49(5):559. https://doi.org/10.1002/qua.560490502
105. Pipek J, Varga I (1994) Mathematical characterization and shape analysis of localized, fractal, and complex distributions in extended systems. Int J Quantum Chem 51(6):539. https://doi.org/10.1002/qua.560510619
106. Kozmutza C, Kapuy E, Evleth EM, Pipek J, Trézl L (1996) Application of the localized representation for studying interaction energies. Int J Quantum Chem 57(4):775. https://doi.org/10.1002/(SICI)1097-461X(1996)57:4<775::AID-QUA25>3.0.CO;2-Y
107. Moyano A, Paniagua JC (1996) Distortive properties of σ- and σ-electrons and aromaticity: a semiempirical localized molecular orbital approach. J Mol Struct: Theochem 369(1):39. DOI:https://doi.org/10.1016/S0166-1280(96)04599-X
108. Reinhardt P, Malrieu JP, Povill A, Rubio J (1998) Localized orbitals in nonmetallic ring systems. Int J Quantum Chem 70(1):167. https://doi.org/10.1002/(SICI)1097-461X(1998)70:1<167::AID-QUA14>3.0.CO;2-C
109. Pipek J, Varga I (1998) Scaling behavior of energy functionals of highly complex electron distributions. Int J Quantum Chem 70(1):125. https://doi.org/10.1002/(SICI)1097-461X(1998)70:1<125::AID-QUA10>3.0.CO;2-E
110. Jug K, Hiberty PC, Shaik S (2001) σ-π energy separation in modern electronic theory for ground states of conjugated systems. Chem Rev 101(5):1477. https://doi.org/10.1021/cr990328e
111. Whitten JL (2003) Localized orbital interactions: d-electron exchange and correlation. Int J Quantum Chem 95(6):758. https://doi.org/10.1002/qua.10608
112. Evangelisti S, Guihéry N, Leininger T, Malrieu JP, Maynau D, Pitarch Ruiz JV, Suaud N, Angeli C, Cimiraglia R, Calzado CJ (2004) Local orbitals for quasi-degenerate systems. J Mol Struct: Theochem 709(1):1. https://doi.org/10.1016/j.theochem.2003.12.054
113. Pitarch-Ruiz J, Evangelisti S, Maynau D (2004) Localized molecular orbitals for excited states of polyenals, polyendials, and polyenones. Int J Quantum Chem 97(1):688. https://doi.org/10.1002/qua.10766
114. Santos JC, Andres J, Aizman A, Fuentealba P (2005) An Aromaticity Scale Based on the Topological Analysis of the Electron Localization Function Including σ and π Contributions. J Chem Theory Comput 1(1):83. https://doi.org/10.1021/ct0499276
115. Subotnik JE, Dutoi AD, Head-Gordon M (2005) Fast localized orthonormal virtual orbitals which depend smoothly on nuclear coordinates. J Chem Phys 123(11):114108. https://doi.org/10.1063/1.2033687

116. Zoboki T, Mayer I (2011) Extremely localized nonorthogonal orbitals by the pairing theorem. J Comput Chem 32(4):689. https://doi.org/10.1002/jcc.21654
117. Guo Y, Li W, Li S (2011) An efficient linear scaling procedure for constructing localized orbitals of large molecules based on the one-particle density matrix. J Chem Phys 135(13):134107. https://doi.org/10.1063/1.3644893
118. Silvi B, Reinhardt P (2012) ChemInform abstract: localization and localizability in quantum organic chemistry: localized orbitals and localization functions. ChemInform 43(13). https://doi.org/10.1002/chin.201213275
119. Beran GJO, Hirata S (2012) Fragment and localized orbital methods in electronic structure theory. Phys Chem Chem Phys 14(21):7559. https://doi.org/10.1039/C2CP90072F
120. Høyvik IM, Jansík B, Jørgensen P (2012) Trust region minimization of orbital localization functions. J Chem Theory Comput 8(9):3137. https://doi.org/10.1021/ct300473g
121. Høyvik IM, Jansík B, Jørgensen P (2013) Pipek-Mezey localization of occupied and virtual orbitals. J Comput Chem 34(17):1456. https://doi.org/10.1002/jcc.23281
122. Høyvik IM, Jansík B, Kristensen K, Jørgensen P (2013) Local Hartree-Fock orbitals using a three-level optimization strategy for the energy. J Comput Chem 34(15):1311. https://doi.org/10.1002/jcc.23256
123. Abarenkov IV, Boyko MA, Sushko PV (2013) Localized directed orbitals representing chemical bonds in ion-covalent crystals. Int J Quantum Chem. 113(14):1868. https://doi.org/10.1002/qua.24367
124. Lehtola S, Jónsson H (2013) Unitary Optimization of Localized Molecular Orbitals. J Chem Theory Comput 9(12):5365. https://doi.org/10.1021/ct400793q
125. Suaud N, Ruamps R, Malrieu JP, Guihéry N (2014) Singly Occupied MOs in Mono- and Diradical Conjugated Hydrocarbons: Comparison between Variational Single-Reference, π-Fully Correlated and Hückel Descriptions. J Phys Chem A 118(31):5876. https://doi.org/10.1021/jp4120892
126. Hohenstein EG, Luehr N, Ufimtsev IS, Martínez TJ (2015) An atomic orbital-based formulation of the complete active space self-consistent field method on graphical processing units. J Chem Phys 142(22):224103. https://doi.org/10.1063/1.4921956
127. Meyer B, Guillot B, Ruiz-Lopez MF, Genoni A (20163) Libraries of extremely localized molecular orbitals. 1. Model molecules approximation and molecular orbitals transferability. J Chem Theory Comput 12(3):1052. DOI:https://doi.org/10.1021/acs.jctc.5b01007
128. Høyvik IM, Jørgensen P (2016) Characterization and generation of local occupied and virtual Hartree-Fock orbitals. Chem Rev 116(5):3306. https://doi.org/10.1021/acs.chemrev.5b00492
129. Heßelmann A (2016) Local Molecular Orbitals from a Projection onto Localized Centers. J Chem Theory Comput 12(6):2720. https://doi.org/10.1021/acs.jctc.6b00321
130. Tenti L, Giner E, Malrieu JP, Angeli C (2017) Strongly localized approaches for delocalized systems. I. Ground state of linear polyenes. Comput Theor Chem 1116:102. https://doi.org/10.1016/j.comptc.2017.01.021

Chapter 4
An Introduction to Discretization Error Analysis for Computational Chemists

Eric Cancès

4.1 Introduction

Validation and certification of numerical results is a key issue in all fields involving computer simulation. The error between the exact and computed values of a given physical quantity of interest (QOI), e.g. the dissociation energy of a molecule, has several origins [1]: a model error (resulting from the choice of a computationally tractable, but not extremely accurate, model, e.g. Kohn-Sham with PBE functional), a discretization error (resulting from the choice of a finite basis set or a grid), an algorithmic error (due to the choice of stopping criteria in self-consistent field and other iterative algorithms), an implementation error (due to possible bugs or uncontrolled round-off errors), a computing error (due to random hardware failures). Quantifying these different sources of errors is of major importance for two reasons. First, guaranteed estimates on these five components of the error would allow one to supplement the computed value of the QOI returned by the numerical simulation with guaranteed error bars (certification of the result). Second, this would allow one to choose the parameters of the simulation (approximate model, discretization parameters, algorithm and stopping criteria, data structures) in an optimal way in order to minimize the computational effort required to reach the target accuracy (error balancing).

In contrast with the current situation in other fields, such as computational mechanics and engineering sciences [2], neither fully guaranteed error bounds nor black-box error balancing schemes are available yet for molecular simulation. However, recent progress has been made on the analysis of the different sources of errors for various electronic structure models, see e.g. [1, 3–16] and references therein, and in particular on discretization error, which is the matter of the present chapter.

For the sake of clarity and brevity, we will restrict ourselves to the analysis of the plane-wave approximation of the Gross-Pitaevskii model. This model was introduced

E. Cancès (✉)
CERMICS, Ecole des Ponts and Inria Paris, 6 and 8 avenue Blaise Pascal, 77455 Marne-la-Vallée, France
e-mail: cances@cermics.enpc.fr

E. Perlt (ed.), *Basis Sets in Computational Chemistry*, Lecture Notes in Chemistry 107,
https://doi.org/10.1007/978-3-030-67262-1_4

in the early 60s to describe the ground state of Bose-Einstein condensates [17]. From a mathematical point of view, it can be seen as a simplified version of the Kohn-Sham model, involving a single orbital and a very simple mean-field potential. The discretization error cancellation phenomenon, which plays a crucial role in electronic structure calculation, will be analyzed in Sect. 4.4. Beforehand, we will introduce in Sect. 4.2 the key concepts of QOI-related a priori and a posteriori error estimators leading to post-processing methods, and asymptotic expansions leading to extrapolation methods.

We will omit the proofs of the rigorous mathematical results mentioned in this contribution, but we will comment on these results in detail.

4.2 Basic Concepts in Discretization Error Analysis

To clarify what error analysis is about, consider a reference model for which the ground-state energy is obtained by solving a minimization problem of the form

$$E_0 = \inf\left\{\mathscr{E}(v),\ v \in \mathscr{X},\ c(v) = 0\right\}, \tag{4.1}$$

where $\mathscr{E} : \mathscr{X} \to \mathbb{R}$ is an energy functional defined on some infinite-dimensional function space $\mathscr{X}$, and $c : \mathscr{X} \to \mathscr{Y}$ represents the constraints on the admissible states ($\mathscr{Y}$ is a finite- or infinite-dimensional vector space). Hartree-Fock, Kohn-Sham, multi-configuration self-consistent field (MCSCF), and many other models are of the generic form (4.1). For instance, the restricted Hartree-Fock problem for the helium atom can be written, in atomic units, as (4.1) with

$$\mathscr{E}(v) = \int_{\mathbb{R}^3} |\nabla v|^2 - 4\int_{\mathbb{R}^3} \frac{v(\mathbf{r})^2}{|\mathbf{r}|}\,d\mathbf{r} + 2\int_{\mathbb{R}^3}\int_{\mathbb{R}^3} \frac{v(\mathbf{r})^2\, v(\mathbf{r}')^2}{|\mathbf{r}-\mathbf{r}'|}\,d\mathbf{r}\,d\mathbf{r}',$$

$\mathscr{X} = H^1(\mathbb{R}^3)$, $\mathscr{Y} = \mathbb{R}$ and $c(v) = \int_{\mathbb{R}^3} v(\mathbf{r})^2\,d\mathbf{r} - 1$, where $H^1(\mathbb{R}^3)$ is the Sobolev space of real-valued functions of $\mathbb{R}^3$ which are square integrable and whose gradient is square integrable as well:

$$H^1(\mathbb{R}^3) := \left\{ v : \mathbb{R}^3 \to \mathbb{R} \mid \|v\|_{H^1}^2 := \int_{\mathbb{R}^3} v(\mathbf{r})^2\,d\mathbf{r} + \int_{\mathbb{R}^3} |\nabla v(\mathbf{r})|^2\,d\mathbf{r} < \infty \right\}. \tag{4.2}$$

Likewise, the restricted Kohn-Sham LDA model for a non-magnetic molecular system with N_p electron pairs can be written as (4.1) with $\mathscr{X} = (H^1(\mathbb{R}^3))^{N_p}$, $\mathscr{Y}$ the space of real, symmetric, $N_p \times N_p$ matrices, and for all $v = (\phi_1, \cdots, \phi_{N_p}) \in (H^1(\mathbb{R}^3))^{N_p}$,

$$\mathscr{E}(v) = \sum_{i=1}^{N_p} \int_{\mathbb{R}^3} |\nabla\phi_i|^2 + \int_{\mathbb{R}^3} \rho_v V_{\rm nuc} + \frac{1}{2}\int_{\mathbb{R}^3}\int_{\mathbb{R}^3} \frac{\rho_v(\mathbf{r})^2\,\rho_v(\mathbf{r}')^2}{|\mathbf{r}-\mathbf{r}'|}\,\mathbf{dr}\,\mathbf{dr}' + E_{\rm xc}^{\rm LDA}(\rho_v),$$

$$\text{with}\quad \rho_v(\mathbf{r}) = 2\sum_{i=1}^{N_p} |\phi_i(\mathbf{r})|^2, \quad\text{and}\quad [c(v)]_{ij} = \int_{\mathbb{R}^3} \phi_i\phi_j - 1.$$

Here $V_{\rm nuc}$ is the electrostatic potential generated with the nuclei, and $E_{\rm xc}^{\rm LDA}$ the local density approximation of the exchange-correlation functional [18].

4.2.1 Variational Approximations

A variational approximation of (4.1) is obtained by choosing a finite-dimensional subspace $\mathscr{X}_{\mathscr{N}}$ of $\mathscr{X}$ and in considering

$$E_{0,\mathscr{N}} = \inf\left\{\mathscr{E}(v_{\mathscr{N}}),\ v_{\mathscr{N}} \in \mathscr{X}_{\mathscr{N}},\ c(v_{\mathscr{N}}) = 0\right\}. \tag{4.3}$$

Obviously, since $\mathscr{X}_{\mathscr{N}} \subset \mathscr{X}$, we have $E_{0,\mathscr{N}} \geq E_0$: the approximate ground-state energy $E_{0,\mathscr{N}}$ is an upper bound of the exact ground-state energy E_0.

A particularly important QOI is the ground-state energy E_0. It is, therefore, natural to try and estimate the error $E_{0,\mathscr{N}} - E_0$ and compare it to other characteristic energies of the problem (e.g. to $k_B T$) to determine whether the discretization error is sufficiently small or not. In other cases, the QOI is a function of the minimizer u of (4.1) (e.g. the dipolar momentum of a neutral molecule is obtained from the ground-state electronic density, which is itself computed from the Kohn-Sham orbitals). In such cases, the exact value of the QOI is $q(u)$ while the computed value is $q(u_{\mathscr{N}})$, where $q : \mathscr{X} \to \mathbb{R}$ is a given function, u a minimizer of (4.1), and $u_{\mathscr{N}}$ a minimizer of (4.3). The error on the QOI to be estimated then is $q(u_{\mathscr{N}}) - q(u)$.

4.2.2 A Priori Error Analysis

For systematically improvable discretization methods, such as plane-waves (PW) [19–21], finite-elements [22, 23], or wavelets [24], we can construct a sequence of approximation spaces $(\mathscr{X}_{\mathscr{N}})_{\mathscr{N}>0}$ such that

1. for $\mathscr{N} < \mathscr{N}'$, $\mathscr{X}_{\mathscr{N}} \subsetneq \mathscr{X}_{\mathscr{N}'}$, that is $\mathscr{X}_{\mathscr{N}}$ gets larger and larger when $\mathscr{N}$ grows;
2. any function of $\mathscr{X}$ can be approximated arbitrarily well by some function of $\mathscr{X}_{\mathscr{N}}$ provided $\mathscr{N}$ is large enough:

$$\forall v \in \mathscr{X}, \qquad \min_{v_{\mathscr{N}} \in \mathscr{X}_{\mathscr{N}}} \|v - v_{\mathscr{N}}\|_{\mathscr{X}} \underset{\mathscr{N}\to\infty}{\to} 0,$$

where $\|\cdot\|_{\mathscr{X}}$ is the norm of the function space $\mathscr{X}$.

This can be done by refining the mesh in finite-element methods, and by increasing the resolution in wavelet methods, or the energy cut-off in PW methods. In the latter case, $\mathcal{N}$ is usually the wave-vector cut-off, which is related to the energy cut-off $E_{\rm co,\mathcal{N}}$ by the relation $E_{\rm co,\mathcal{N}} = \frac{\mathcal{N}^2}{2}$ (in atomic units).

A priori error estimators are results assessing that the computed value of the QOI converges to the exact value of the QOI when $\mathcal{N}$ goes to infinity, and providing in addition convergence rates. A typical such result (which holds true [3] for the PW discretization of the periodic Kohn-Sham LDA model with pseudopotentials [25–27], for well-chosen minimizers $u_{\mathcal{N}}$ of (4.3)) is the existence of positive constants s, c_-, c_+ and c_s such that for all $\mathcal{N}$,

$$c_-\|u_{\mathcal{N}} - u\|_{\mathcal{X}}^2 \leq E_{0,\mathcal{N}} - E_0 \leq c_+\|u_{\mathcal{N}} - u\|_{\mathcal{X}}^2 \tag{4.4}$$

and

$$\|u_{\mathcal{N}} - u\|_{\mathcal{X}} \leq \frac{c_s}{\mathcal{N}^s}. \tag{4.5}$$

This result implies that, on the one hand, the error on the energy goes to zero at the same speed as the square of the error on the orbitals (measured in $\mathcal{X}$-norm), and that, on the other hand, the $\mathcal{X}$-norm error on the orbitals goes to zero as $\mathcal{N}^{-s}$. Gathering (4.4) and (4.5), we obtain

$$0 \leq E_{0,\mathcal{N}} - E_0 \leq \frac{c_+c_s^2}{\mathcal{N}^{2s}} = \frac{c_+c_s^2}{2^s E_{\rm co,\mathcal{N}}^s}. \tag{4.6}$$

The admissible values of s in (4.5)–(4.6) can usually be obtained explicitly. Typically, estimate (4.5) will hold true for any $s < s_{\max}$, but not for $s > s_{\max}$, where the value of $s_{\max}$ is an explicit outcome of the mathematical analysis. As a matter of example [3], $s_{\max} = \frac{7}{2}$ for PW discretizations of periodic Kohn-Sham LDA models with Troullier-Martins pseudopotentials [26]. Note that $-s_{\max}$ is basically the slope of the convex hull of the log-log plot of the discretization error $E_{0,\mathcal{N}} - E_0$ as the function of the energy cut-off $E_{\rm co,\mathcal{N}}$. The higher $s_{\max}$, the faster the asymptotic convergence of the computed ground-state energy towards the exact value for the considered model.

The main interest of a priori error estimators is that they allow to get quantitative insight on the difficulty of getting an accurate approximation of a given quantity of interest with a given numerical method. Indeed, the value of $s_{\max}$ for which

$$|q(u_{\mathcal{N}}) - q(u)| \leq \frac{C_s}{E_{\rm co,\mathcal{N}}^s} \tag{4.7}$$

for any $s < s_{\max}$, but not for $s > s_{\max}$ heavily depends on the QOI q. If $s_{\max}$ is “large” (say $s_{\max} = 3$) doubling the energy cut-off will typically increase the accuracy by a factor 8, while if $s_{\max}$ is “small” (say $s_{\max} = 1$) doubling the energy cut-off will only double the accuracy. Again for PW discretizations of periodic Kohn-Sham models with Troullier-Martins pseudopotentials, we have seen that $s_{\max} = \frac{7}{2}$ if the QOI is the energy; on the other hand, $s_{\max} = \frac{3}{2}$ is the QOI is the value of the ground-state density at a particular point of the simulation cell, which makes the latter QOI much more difficult to converge than the former one.

This argument also allows one to clearly understand one of the main roles of pseudopotentials, namely smoothing out the Coulomb singularities generated by point nuclei. Indeed, as we have seen before, if the QOI is the ground-state energy, $s_{\max} = \frac{7}{2}$ for Troullier-Martins pseudopotentials, while we only have $s_{\max} = \frac{3}{2}$ for point-like nuclei.

It is also interesting to reformulate the above results in terms of the computational time $\mathrm{CPU}_\varepsilon^q$ necessary to reach a given accuracy ε for the QOI q. For PW Kohn-Sham LDA calculations, the computational time typically scales as $E_{\mathrm{co},\mathcal{N}}^{3/2} \log E_{\mathrm{co},\mathcal{N}}$ (using preconditioned gradient methods and Fast Fourier Transforms, see e.g. [28] and references therein). A simple calculation shows that if (4.7) is satisfied for any $s < s_{\max}$, but not for $s > s_{\max}$, then

$$\log \mathrm{CPU}_\varepsilon^q \sim -\frac{3}{2s_{\max}} \log \varepsilon. \tag{4.8}$$

We will see later that a priori error estimates can also be useful to design new, efficient, numerical schemes.

A priori error estimators, however, suffer from two severe limitations. First, the optimal value of the constant C_s in (4.7) is usually unknown. The constant C_s derived from the mathematical analysis is most often dramatically overestimated, sometimes by several orders of magnitude. In addition, it usually depends on the *exact* solution u to the problem, which is unknown. The constant C_s does not appear in (4.8) because this relation is in log-log scales, but an estimation of the optimal value of C_s would of course be of major interest for practical purposes. The second limitation is that an inequality such as (4.7) is only useful when the right-hand side is small enough, that is in the asymptotic regime when the cut-off energy $E_{\mathrm{co},\mathcal{N}}$ is large enough.

4.2.3 A Posteriori Error Estimators and Post-Processing

A posteriori estimates are very different in nature from a priori error estimates. An a posteriori discretization error estimator for the QOI q is a pair of inequalities of the form

$$\eta_{\mathrm{l.b.}}^q(u_{\mathcal{N}}) \le q(u_{\mathcal{N}}) - q(u) \le \eta_{\mathrm{l.b.}}^q(u_{\mathcal{N}}) \tag{4.9}$$

(where we recall that u is a minimizer of (4.1) and $u_{\mathcal{N}}$ is a minimizer of (4.3), and where l.b. and u.b. stand for lower bound and upper bound respectively), which, *ideally*, satisfy the following properties:

1. the estimator is *guaranteed*, in the sense that inequalities (4.9) can be established with full mathematical rigour;

2. the lower and upper bounds $\eta^q_{\rm l.b.}(u_{\mathcal N})$ and $\eta^q_{\rm l.b.}(u_{\mathcal N})$ are *fully computable* from the approximate solution $u_{\mathcal N}$ and the data of the problem; in particular, they do not involve the exact solution u, in contrast with the bounds resulting from a priori error estimates;
3. $\eta^q_{\rm l.b.}(u_{\mathcal N})$ and $\eta^q_{\rm l.b.}(u_{\mathcal N})$ are cheap to compute: their numerical values can be obtained with a negligible, or small enough, computational cost;
4. the estimates are accurate, in the sense that $\eta^q_{\rm l.b.}(u_{\mathcal N})$, $q(u_{\mathcal N}) - q(u)$, and $\eta^q_{\rm l.b.}(u_{\mathcal N})$ are of the same order of magnitude for generic values of $\mathcal N$ (note however that $|q(u_{\mathcal N}) - q(u)|$ can be, by chance, much smaller than $|\eta^q_{\rm l.b.}(u_{\mathcal N})|$ and $|\eta^q_{\rm l.b.}(u_{\mathcal N})|$ for some specific values of $\mathcal N$);
5. the estimates give insights on what to do to improve the quality of the approximation.

Let us clarify the last point. Finite-element methods, as well as wavelet or some hierarchical tensor methods, have more flexibility than PW discretization methods. While in PW method, the user only controls a single discretization parameter, namely the wave-vector cut-off $\mathcal N$, or equivalently, the energy cut-off $E_{{\rm co},\mathcal N} = \frac{\mathcal N^2}{2}$, the quality of a finite-element approximation space can be improved by *locally* refining the mesh in the regions of the simulation cell where the field u strongly varies. In many cases, it is possible to construct lower and upper bounds $\eta^q_{\rm l.b.}(u_{\mathcal N})$ and $\eta^q_{\rm l.b.}(u_{\mathcal N})$ as a sum of localized contributions to the error [29], each of them being obtained by solving a small-size local problem. The advantage of such a decomposition is twofold: first the computation of these local contributions can be easily parallelized on a large number of processors; second, it paves the way to adaptive finite-element methods, where the mesh is refined only in the regions of the simulation cell where the local error is significant. This can be done with a black-box algorithm and can dramatically reduce the overall computational effort necessary to reach a given accuracy (compared to brute force, uniform, mesh-refinement methods).

Let us emphasize that the above five properties of ideal a posteriori error estimators are usually not completely fulfilled by most of the a posteriori error estimators available in practice. Indeed,

1. Inequalities (4.9) are sometimes only satisfied for large enough values of $\mathcal N$. In this case, it is interesting to have at our disposal checkable conditions allowing one to know whether the bounds are reliable or not. Such conditions can take the form
$$\text{if } c_q(u_{\mathcal N}) > 0, \text{ then (9) hold true,}$$
where $c_q(u_{\mathcal N})$ is a real number computable from the approximate solution $u_{\mathcal N}$ at low cost;
2. The lower and upper bounds may not be fully computable in the sense that they are in fact a function of the (known) approximate solution $u_{\mathcal N}$ *and* of the exact (unknown) solution u, but nevertheless decomposable as
$$\eta^q_{\star{\rm .b.}}(u_{\mathcal N}, u) = \eta^q_{\star{\rm .b.},1}(u_{\mathcal N}) + \eta^q_{\star{\rm .b.},2}(u_{\mathcal N}, u),$$

where $\eta^q_{\star.\text{b.},1}(u_{\mathcal{N}})$ is fully computable and $\eta^q_{\star.\text{b.},2}(u_{\mathcal{N}}, u)$ small compared to $\eta^q_{\star.\text{b.},1}(u_{\mathcal{N}})$, at least when $\mathcal{N}$ is large enough. A priori error estimates can be called out to justify the smallness of $\eta^q_{\star.\text{b.},2}(u_{\mathcal{N}}, u)$;

3. Computing $\eta^q_{\text{l.b.}}(u_{\mathcal{N}})$ and $\eta^q_{\text{l.b.}}(u_{\mathcal{N}})$ may require solving an auxiliary problem of the same complexity as the original problem (4.1), which may double or triple the cost of the calculation. In engineering sciences, simulations are more and more substitutes to experiments and prototypes in the design process; it is then worth paying a significant extra-cost to guarantee the quality of the simulation results;
4. Quite often, the relative quality of the lower and upper bounds increases with $\mathcal{N}$. In the case when the QOI is the ground-state energy, a posteriori error estimates are of the form

$$0 < \eta^E_{\text{l.b.}}(u_{\mathcal{N}}) \leq E(u_{\mathcal{N}}) - E(u) \leq \eta^E_{\text{l.b.}}(u_{\mathcal{N}}), \tag{4.10}$$

and we can define the efficiency factors of the lower and upper bounds as

$$1 \leq I^{\text{l.b.}}_{\mathcal{N}} = \frac{E(u_{\mathcal{N}}) - E(u)}{\eta^E_{\text{l.b.}}(u_{\mathcal{N}})} \quad \text{and} \quad 1 \leq I^{\text{u.b.}}_{\mathcal{N}} = \frac{\eta^E_{\text{u.b.}}(u_{\mathcal{N}})}{E(u_{\mathcal{N}}) - E(u)}.$$

The closer $I^{\text{l.b.}}_{\mathcal{N}}$ and $I^{\text{u.b.}}_{\mathcal{N}}$ to 1, the better. The a posteriori estimate (4.10) is called asymptotically exact if both $I^{\text{l.b.}}_{\mathcal{N}}$ and $I^{\text{u.b.}}_{\mathcal{N}}$ converge to 1 when $\mathcal{N}$ goes to infinity. Note that if, for instance, $I^{\text{l.b.}}_{\mathcal{N}}$ goes to 1 when $\mathcal{N}$ goes to infinity, then for $\mathcal{N}$ large enough, the post-processed approximation of the ground-state energy

$$\widetilde{E}(u_{\mathcal{N}}) = E(u_{\mathcal{N}}) - \eta^E_{\text{l.b.}}(u_{\mathcal{N}})$$

is more accurate than the original one $E(u_{\mathcal{N}})$.

4.2.4 Asymptotic Expansions and Extrapolation

In some specific cases, it is possible to expand the error $q(u_{\mathcal{N}}) - q(u)$ in terms of simple functions of $\mathcal{N}$ in the limit when $\mathcal{N}$ goes to infinity, and obtain, as a matter of illustration—this is just an example—, asymptotic expansions of the form

$$q(u_{\mathcal{N}}) - q(u) = \frac{a_1}{\mathcal{N}^{2/3}} + \frac{a_2}{\mathcal{N}} + o\left(\frac{1}{\mathcal{N}^{4/3}}\right). \tag{4.11}$$

The main interest of asymptotic expansions is that they allow extrapolations. Indeed, assuming a result such as (4.11), one can combine the values of $q(u_{\mathcal{N}})$ for several correlated values of $\mathcal{N}$, and obtain, for instance,

$$(\alpha q(u_{\mathcal{N}}) + \beta q(u_{2\mathcal{N}}) + \gamma q(u_{3\mathcal{N}})) - q(u) = O\left(\frac{1}{\mathcal{N}^{4/3}}\right),$$

where the weights α, β, γ are obtained by solving the linear system

$$\begin{pmatrix} 1 & 1 & 1 \\ 1 & \frac{1}{2^{2/3}} & \frac{1}{3^{2/3}} \\ 1 & \frac{1}{2^{2/3}} & \frac{1}{3^{2/3}} \end{pmatrix} \begin{pmatrix} \alpha \\ \beta \\ \gamma \end{pmatrix} = \begin{pmatrix} 1 \\ 1 \\ 1 \end{pmatrix}.$$

In other words, one can obtain a much better convergence rate by linear combinations of a few calculations performed with different values of $\mathcal{N}$.

Extrapolation methods are very appealing. Unfortunately, the situations where the error on the QOI of interest is known to admit an asymptotic expansion are not so common in the field of electronic structure calculation. An interesting example is a Makov-Payne correction for computing the energy of charge defects in insulators and semiconductors [30]. It has indeed been proved in [31] that the Makov-Payne correction corresponds to the leading term of the asymptotic expansion of the error on the ground-state energy when the discretization parameter is the size L of the supercell.

The second limitation of extrapolation methods based on asymptotic expansions of the error is that they are only efficient for $\mathcal{N}$ "large enough". It is usually not clear how to check whether the asymptotic regime has been reached without running a number of calculations with different values of $\mathcal{N}$ covering a large range and check whether the results match the prediction of the asymptotic expansion.

4.3 Periodic Gross-Pitaevskii and Kohn-Sham Models

We now turn to the analysis of discretization errors for self-consistent quantum problems. For pedagogical reasons, we will mainly deal with the (relatively simple) Gross-Pitaevskii model, and the existing results on the Kohn-Sham model will only be mentioned. Still, for pedagogical results, we will focus on the periodic versions of these models, and on plane-wave discretization methods.

For simplicity, we assume that the periodic simulation cell is $\Omega = (0, 2\pi)^d$ ($d \leq 3$), but all the results below can easily be extended to the generic case of a d-dimensional periodic cell of any shape. The fundamental Hilbert space for periodic Gross-Pitaevskii and Kohn-Sham models is

$$L^2_\#(\Omega) := \left\{u \in L^2_{\rm loc}(\mathbb{R}^d, \mathbb{R}) \mid u \ 2\pi\mathbb{Z}^d\text{-periodic}\right\}, \quad \langle u|v\rangle_{L^2_\#} = \int_\Omega u\, v,$$

where $L^2_{\rm loc}(\mathbb{R}^d, \mathbb{R})$ is the space of locally square-integrable real-valued functions on $\mathbb{R}^d$. We will make extensive use of the periodic Sobolev spaces (see e.g. [32])

$$H^s_\#(\Omega) := \left\{ v = \sum_{\mathbf{k}\in\mathbb{Z}^d} \widehat{v}_{\mathbf{k}} e_{\mathbf{k}},\ v \text{ real valued},\ \|v\|^2_{H^s_\#} := \sum_{\mathbf{k}\in\mathbb{Z}^d} \left(1+|\mathbf{k}|^2\right)^s |\widehat{v}_{\mathbf{k}}|^2 < \infty \right\},$$

$s \in \mathbb{R}$, where $e_{\mathbf{k}} = |\Omega|^{-1/2} e^{i\mathbf{k}\cdot\mathbf{r}}$ is the Fourier mode with wave-vector $\mathbf{k} \in \mathbb{Z}^d$, which, endowed with the inner product

$$\langle u|v\rangle_{L^2_\#} = \sum_{\mathbf{k}\in\mathbb{Z}^d} \left(1+|\mathbf{k}|^2\right)^s \overline{\widehat{u}_{\mathbf{k}}}\, \widehat{v}_{\mathbf{k}},$$

are also Hilbert spaces. Note in particular that $H^0_\#(\Omega) = L^2_\#(\Omega)$ and that

$$H^1_\#(\Omega) = \left\{ v \in L^2_{\rm loc}(\mathbb{R}^d, \mathbb{R}) \mid \nabla v \in (L^2_{\rm loc}(\mathbb{R}^d, \mathbb{R}))^d,\ v\ 2\pi\mathbb{Z}^d\text{-periodic} \right\}.$$

4.3.1 Plane-Wave Discretization of the Gross-Pitaevskii Model

The d-dimensional periodic Gross-Pitaevskii model is defined as

$$I = \inf \left\{ E(v),\ v \in H^1_\#(\Omega),\ \int_\Omega v^2 = 1 \right\}, \tag{4.12}$$

where the Gross-Pitaevskii energy functional is given by

$$E(v) = \int_\Omega |\nabla v|^2 + \int_\Omega V v^2 + \frac{\mu}{2} \int_\Omega v^4.$$

Here, the trapping potential V is a $2\pi\mathbb{Z}^d$-periodic real-valued continuous function, and the mean-field interaction parameter μ is chosen positive (repulsive interaction). The mathematical properties of the minimization problem (4.12) are well-understood:

- Equation (4.12) has exactly two minimizers u (with $u > 0$ in Ω) and $-u$;
- There exists a unique real number $\lambda \in \mathbb{R}$ such that (λ, u) satisfies the nonlinear Schrödinger equation

$$-\Delta u + V u + \mu u^3 = \lambda u, \qquad \|u\|_{L^2_\#} = 1. \tag{4.13}$$

  Physically, λ is the chemical potential of the condensate. Mathematically, it is the Lagrange multiplier of the equality constraint $\int_\Omega v^2 = 1$ in (4.12);
- λ is the lowest eigenvalue of the self-consistent Hamiltonian

$$H_u = -\Delta + V + \mu u^2.$$

We refer to the appendix of [4] for detailed proofs of these standard results.

In plane-wave discretization methods, the approximation spaces are defined as

$$X_{\mathcal{N}} = \left\{ v_{\mathcal{N}} = \sum_{|\mathbf{k}| \le \mathcal{N}} \widehat{v}_{\mathbf{k}} e_{\mathbf{k}}, \; v_{\mathcal{N}} \text{ real valued} \right\},$$

where $\mathcal{N}$ is the cut-off parameter. The Galerkin approximation of (4.12) in $X_{\mathcal{N}}$ consists in searching $u_{\mathcal{N}} \in X_{\mathcal{N}}$ satisfying the constraint $\int_\Omega |u_{\mathcal{N}}|^2 = 1$, and such that

$$I_{\mathcal{N}} = E(u_{\mathcal{N}}) = \inf \left\{ E(v_{\mathcal{N}}), \; v_{\mathcal{N}} \in X_{\mathcal{N}}, \; \int_\Omega |v_{\mathcal{N}}|^2 = 1 \right\}, \quad (u_{\mathcal{N}}, 1)_{L^2_\#} \ge 0. \tag{4.14}$$

The additional requirement $(u_{\mathcal{N}}, 1)_{L^2_\#} \ge 0$ ensures that $u_{\mathcal{N}}$ approximates the positive solution u to (4.12) (and not the other solution, $-u$).

Relying on the fact that the operator $-\Delta$ commutes with the projection operator $\Pi_{\mathcal{N}}$, we obtain that the function $u_{\mathcal{N}}$ satisfies the Euler-Lagrange equation

$$-\Delta u_{\mathcal{N}} + \Pi_{\mathcal{N}}(V + \mu u_{\mathcal{N}}^2)\Pi_{\mathcal{N}} u_{\mathcal{N}} = \lambda_{\mathcal{N}} u_{\mathcal{N}}, \tag{4.15}$$

where $\lambda_{\mathcal{N}}$ is the Lagrange multiplier of the $L^2_\#$-normalization constraint. It can be shown that, except perhaps for very small values of $\mathcal{N}$, $\lambda_{\mathcal{N}}$ is the lowest eigenvalue of the operator $-\Delta + \Pi_{\mathcal{N}}(V + \mu u_{\mathcal{N}}^2)\Pi_{\mathcal{N}}$ on $L^2_\#(\Omega)$.

From a geometrical point of view, the situation is as depicted in Fig. 4.1. The positive solution u to (4.12) is not in general in the approximation space $X_{\mathcal{N}}$. The best approximation of u in $X_{\mathcal{N}}$ for a given norm $\|\cdot\|_{H^s_\#}$ is the orthogonal projection of u on $X_{\mathcal{N}}$ for the inner product of $H^s_\#$. An interesting property is that this orthogonal projector is independent of s: it is simply the Fourier truncation operator $\Pi_{\mathcal{N}}$ defined by

$$\Pi_{\mathcal{N}} \left(\sum_{\mathbf{k} \in \mathbb{Z}^d} \widehat{v}_{\mathbf{k}} e_{\mathbf{k}} \right) = \sum_{|\mathbf{k}| \le \mathcal{N}} \widehat{v}_{\mathbf{k}} e_{\mathbf{k}}.$$

Indeed, for all $s \in \mathbb{R}$ and all $\mathcal{N} \in \mathbb{N}$, $X_{\mathcal{N}} \subset H^s_\#(\Omega)$, and for all $v \in H^s_\#(\Omega)$,

$$\Pi_{\mathcal{N}} v \in X_{\mathcal{N}},$$

$$\|v - \Pi_{\mathcal{N}} v\|_{H^s_\#} = \min_{w_{\mathcal{N}} \in X_{\mathcal{N}}} \|v - w_{\mathcal{N}}\|_{H^s_\#} = \left(\sum_{|\mathbf{k}| > \mathcal{N}} \left(1 + |\mathbf{k}|^2\right)^s |\widehat{v}_{\mathbf{k}}|^2 \right)^{1/2}.$$

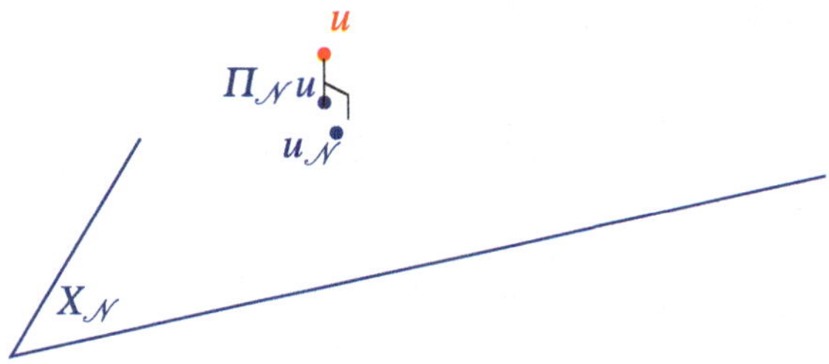

Fig. 4.1 Graphical representation of the best approximation $\Pi_{\mathcal{N}} u$ in the discretization space $X_{\mathcal{N}}$ of the exact solution u to (4.12), and of the approximation $u_{\mathcal{N}}$ obtained by the variational method (4.14)

Note that the Galerkin approximation $u_{\mathcal{N}}$ of u obtained by solving (4.14) is not the best approximation $\Pi_{\mathcal{N}} u$ of u in $X_{\mathcal{N}}$ (see Fig. 4.1). The best we can hope for is that $u_{\mathcal{N}}$ will be close to $\Pi_{\mathcal{N}} u$ for the various norms of interest.

4.3.2 A Priori Error Analysis

The following result has been proved in [4]. It is an extension of classical results for linear eigenvalue problems (see [33] and references therein) to the nonlinear setting of the Gross-Pitaevskii model. The case of Kohn-Sham LDA models is dealt with in [3] for PW discretizations and in [7] for other systematically improvable discretization methods.

Theorem 4.1 *Let u be the unique positive minimizer of* (4.12) *and $u_{\mathcal{N}}$ a minimizer of* (4.14), *which is unique for $\mathcal{N}$ large enough . Then, there exists $0 < c \leq C < \infty$ such that for all $\mathcal{N} \in \mathbb{N}$,*

$$\|u - \Pi_{\mathcal{N}} u\|_{H^1_\#} \leq \|u - u_{\mathcal{N}}\|_{H^1_\#} \leq C\|u - \Pi_{\mathcal{N}} u\|_{H^1_\#} \underset{\mathcal{N}\to 0}{\longrightarrow} 0, \tag{4.16}$$

$$c\|u - u_{\mathcal{N}}\|^2_{H^1_\#} \leq I_N - I = E(u_{\mathcal{N}}) - E(u) \leq C\|u - u_{\mathcal{N}}\|^2_{H^1_\#}. \tag{4.17}$$

Assume that $V \in H^\sigma_\#(\Omega)$ for some $\sigma > d/2$. Then,

- *$(u_{\mathcal{N}})_{\mathcal{N}\in\mathbb{N}}$ converges to u in $H^{\sigma+2}_\#(\Omega)$;*
- *there exists positive constants C and C_s such that*

$$\forall -\sigma \leq s < \sigma + 2, \quad \|u - u_{\mathcal{N}}\|_{H^s_\#} \leq \frac{C_s}{\mathcal{N}^{\sigma+2-s}}, \quad |\lambda - \lambda_{\mathcal{N}}| \leq \frac{C}{\mathcal{N}^{2(\sigma+1)}}. \tag{4.18}$$

According to estimate (4.17), the error on the ground-state energy behaves as the square of the H^1-norm of the error on the eigenfunction, and according to estimate (4.16), the latter goes to zero when $\mathcal{N}$ goes to infinity.

If, in addition, the external periodic potential V is regular enough, more precisely if V belongs to the Sobolev space $H^\sigma_\#(\Omega)$ for some $\sigma > d/2$, then (4.18) provides

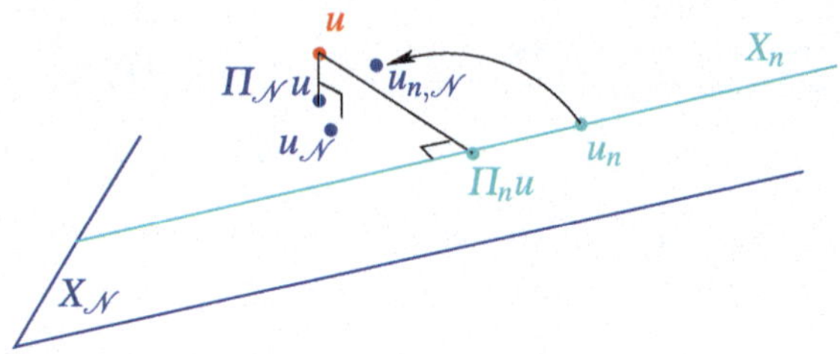

Fig. 4.2 Graphical representation of the two-grid method

optimal a priori convergence rates for both the Lagrange multiplier λ and the ground-state eigenfunction (the optimality has been checked numerically [4]). The estimates $\|u - u_{\mathcal{N}}\|_{H^s_\#} \leq \frac{C_s}{\mathcal{N}^{\sigma+2-s}}$ are valid for the whole hierarchy of Sobolev spaces $H^s_\#(\Omega)$, $-\sigma \leq s < \sigma + 2$, and therefore allow one to derive optimal convergence rates for any differentiable observable $q : H^s(\Omega) \to \mathbb{R}$ with $-\sigma \leq s < \sigma + 2$. For instance, the value of the ground-state density at some point $\mathbf{r}_0 \in \mathbb{R}^d$ is defined as $q_{\mathbf{r}_0}(u) = u(\mathbf{r}_0)^2$. Using Sobolev embedding theorems (see e.g. [32]), we obtain that $q_{\mathbf{r}_0}$ is a differentiable functional on the Sobolev space $H^s_\#(\Omega)$ for all $s > d/2$. It follows that for all $s < \sigma + 2 - d/2$, there exists $C_s \in \mathbb{R}_+$ such that for all $\mathcal{N}$,

$$|u_{\mathcal{N}}(\mathbf{r}_0)^2 - u(\mathbf{r}_0)^2| = |q_{\mathbf{r}_0}(u_{\mathcal{N}}) - q_{\mathbf{r}_0}(u)| \leq \frac{C_s}{\mathcal{N}^s}.$$

In addition to providing optimal convergence rates for various QOI, a priori error estimates can also be used to design computational cost reduction methods based on neglecting terms with higher convergence rates. For instance, two-grid methods consist in finding in a first stage a solution u_n to the full problem in a coarse variational space X_n, and in a second stage a solution $u_{n,\mathcal{N}}$ to a simpler problem parameterized by u_n in a finer approximation space $X_{\mathcal{N}}$ (see Fig. 4.2). For a well-chosen value of n, it is possible to obtain in this way, at a much lower cost, the same accuracy as if the full problem had been solved in $X_{\mathcal{N}}$. These methods were introduced by Xu and Zhou to solve nonlinear elliptic problems [34], then adapted to linear eigenvalue problems in [35, 36], and to nonlinear eigenvalue problems in [37].

Indeed, solving (4.14) in a fine approximation space $X_{\mathcal{N}}$ is costly since it requires about $\sim \mathcal{K} \mathcal{N}^d \ln \mathcal{N}$ elementary operations, where $\mathcal{K}$ is a constant related to the structure of problem (4.14). In the two-grid method,

1. u_n is computed by solving the full problem on the coarse approximation space X_n, $n \ll \mathcal{N}$, which requires $\sim \mathcal{K} n^d \ln n$ elementary operations,
2. $u_{n,\mathcal{N}}$ can be computed in $\sim \kappa \mathcal{N}^d \ln \mathcal{N}$ elementary operations with $\kappa \ll \mathcal{K}$ since the problem to be solved is much simpler.

Typically, in the present case, the simpler problem can be (i) a linear eigenvalue problem obtained by freezing the mean-field potential to $V + \mu u_n^2$, or (ii) the linear system

$$-\Delta v + (V + \mu u_n^2) v = \lambda_n u_n.$$

Both strategies have been tested numerically and are quite efficient [37]. Using the a priori error estimators (4.18), the following theoretical justification of the efficiency of the first strategy can be given.

Theorem 4.2 *Assume that $V \in H^\sigma_\#(\Omega)$ for some $\sigma > d/2$. Let u_n be a solution to* (4.14) *in a coarse approximation space X_n and $u_{n,\mathcal{N}}$ the variational approximation in $X_\mathcal{N}$ to the ground state of the linear eigenvalue problem*

$$-\Delta v + (V + \mu u_n^2)v = \mu v, \quad \|v\|_{L^2_\#} = 1.$$

Then, there exists $C \in \mathbb{R}_+$ such that for all n and $\mathcal{N}$ with $n \leq \mathcal{N}$,

$$\|u_{n,\mathcal{N}} - u\|_{H^1_\#} \leq C\left(n^{-\sigma-3} + \mathcal{N}^{-\sigma-1}\right),$$
$$0 \leq E(u_{n,\mathcal{N}}) - E(u) \leq C\left(n^{-\sigma-3} + \mathcal{N}^{-\sigma-1}\right)^2.$$

Choosing $n \sim \mathcal{N}^{\frac{\sigma+1}{\sigma+3}}$ in order to balance the error contributions in the right-hand sides of the above inequalities, we obtain same convergence rates as in Theorem 4.1:

$$\|u_{n,\mathcal{N}} - u\|_{H^1_\#} \leq C\mathcal{N}^{-(\sigma+1)}, \qquad 0 \leq E(u_{n,\mathcal{N}}) - E(u) \leq C\mathcal{N}^{-2(\sigma+1)},$$
$$\|u_\mathcal{N} - u\|_{H^1_\#} \leq C\mathcal{N}^{-(\sigma+1)}, \qquad 0 \leq E(u_\mathcal{N}) - E(u) \leq C\mathcal{N}^{-2(\sigma+1)},$$

with a significant gain in CPU time (see the numerical results in [37]).

4.3.3 A Posteriori Error Analysis and Post-Processing

A posteriori error analysis for linear elliptic eigenvalue problems has been the matter of a large number of numerical analysis papers (see [38, 39] and references therein). It turns out that even the simple case of the Laplace operator on a bounded polyhedral domain with Dirichlet boundary conditions is quite challenging (see [38, 40] and references therein). The case of linear and nonlinear Schrödinger operator has been considered in [10, 41–45] (see also the references therein and the appendix in [46]), leading to adaptive discretization procedures with optimal complexity [47–49]. Some results regarding Hartree-Fock and Kohn-Sham models have also been established [12, 15, 22].

As far as PW discretizations of Gross-Pitaevskii models are concerned, post-processing methods can be obtained by a non-standard application of Rayleigh-Schrödinger perturbation theory (RSPT).

Recall that if we have at hand a simple eigenmode (E_0, ψ_0) of a reference Hamiltonian H_0 on $L^2_\#(\Omega)$:

$$H_0\psi_0 = E_0, \quad \|\psi_0\|_{L^2_\#} = 1, \tag{4.19}$$

and if W is a small perturbation of H_0 (small in a sense made precise by Kato [50]), then the perturbed Hamiltonian $H = H_0 + W$ has a unique eigenvalue E in the vicinity of E_0, which is simple. Using first-order perturbation for the eigenvector and second-order perturbation for the eigenvalue, we obtain

$$\begin{aligned} H\psi = E\psi, \quad \text{with } \psi \simeq \psi_0 - \Pi_{\psi_0^\perp}(H_0 - E_0)|_{\psi_0^\perp}^{-1}\Pi_{\psi_0^\perp}(W\psi_0), \\ E \simeq E_0 + \langle \psi_0|W|\psi_0\rangle - \langle \Pi_{\psi_0^\perp}(W\psi_0)|(H_0 - E_0)|_{\psi_0^\perp}^{-1}|\Pi_{\psi_0^\perp}(W\psi_0)\rangle, \end{aligned}$$

where $\Pi_{\psi_0^\perp}$ is the orthogonal projector on the space

$$\psi_0^\perp = \left\{ \phi \in L^2_\#(\Omega) \mid \langle \psi_0|\phi\rangle = \int_\Omega \psi_0\phi = 0 \right\},$$

for the $L^2_\#(\Omega)$ inner product, and where $(H_0 - E_0)|_{\psi_0^\perp}^{-1}$ is the inverse of the restriction of the operator $H_0 - E_0$ to the invariant space $\psi_0^\perp$ (this operator is invertible since E_0 is simple).

As shown in [51], RSPT can be used to derive a posteriori error estimators. The idea is to consider the Euler-Lagrange equation of the variational approximation of (4.12) in X_n, i.e.

$$-\Delta u_n + \Pi_n(V + \mu u_n^2)\Pi_n u_n = \lambda_n u_n, \tag{4.20}$$

as the unperturbed eigenvalue problem, and the Euler-Lagrange equation of (4.12), i.e.

$$-\Delta u + (V + \mu u^2)u = \lambda u \tag{4.21}$$

as the perturbed eigenvalue problem. In other words, we take

$$\begin{aligned} H_0 &= -\Delta + \Pi_n(V + \mu u_n^2)\Pi_n, \quad \psi_0 = u_n, \quad E_0 = \lambda_n, \\ W &= (V + \mu u^2) - \Pi_n(V + \mu u_n^2)\Pi_n. \end{aligned}$$

Note that $\langle \psi_0|W|\psi_0\rangle = 0$: the first-order correction to the eigenvalue vanishes; this is the reason why we need to consider the second-order correction of the eigenvalue. We then notice that since both u_n and Δu_n belong to X_n, (4.20) also reads

$$\Pi_n\left(-\Delta u_n + (V + \mu u_n^2)u_n - \lambda_n u_n\right) = 0,$$

which means that the residual $r_n := -\Delta u_n + (V + \mu u_n^2)u_n - \lambda_n u_n$ is in $X_n^\perp$. Since $u_n \in X_n$ and $r_n \in X_n^\perp$, this implies that

$$\Pi_{u_n^\perp} r_n = r_n,$$

It follows that

$$
\begin{aligned}
\Pi_{\psi_0^\perp}(W\psi_0) &= \Pi_{u_n^\perp}\left(((V+\mu u_n)^2) - \Pi_n(V+\mu u_n^2)\Pi_n)\, u_n\right) \\
&= \Pi_{u_n^\perp}\left(r_n + \mu(u^2 - u_n^2)u_n\right) \\
&= r_n + \mu \Pi_{u_n^\perp}\left((u^2 - u_n^2)u_n\right).
\end{aligned}
$$

Next, we observe that the block representation of H_0 associated with the decomposition $L^2_\#(\Omega) = X_n \oplus X_n^\perp$ reads

$$
H_0 = \left(\begin{array}{c|c} -\Delta|_{X_n} + \Pi_n(V+\mu u_n^2)\Pi_n & 0 \\ \hline 0 & -\Delta_{X_n^\perp} \end{array}\right).
$$

As a consequence,

$$
(H_0 - E_0)|_{\psi_0^\perp}^{-1} \Pi_{\psi_0^\perp}(W\psi_0) = u_n^{(1)} + u_n^{(2)},
$$

with

$$
\begin{aligned}
u_n^{(1)} &= (-\Delta - \lambda_n)|_{X_n^\perp}^{-1} r_n, \\
u_n^{(2)} &= \mu\left(-\Delta + V + \mu u_n^2 - \lambda_n\right)|_{u_n^\perp}^{-1} \Pi_{u_n^\perp}\left((u^2 - u_n^2)u_n\right).
\end{aligned}
$$

Since in PW calculations, functions are stored as vectors of Fourier coefficients, computing a very accurate approximation $u_{n,\mathcal{N}}^{(1)}$ of $u_n^{(1)}$ in a very fine discretization space $X_\mathcal{N}$ with $\mathcal{N} \gg n$ is easy. On the other hand, it can be shown using the a priori error estimates in Theorem 4.1 that $\|u_n^{(2)}\|_{H^1_\#}$ is much smaller than $\|u_n^{(1)}\|_{H^1_\#}$. Introducing

$$
\widetilde{u}_n = u_n + u_n^1 \quad \text{and} \quad \widetilde{\lambda}_n = \lambda_n + (u_n^1, W u_n)_{L^2_\#},
$$

we have

$$
\|u - \widetilde{u}_n\|_{H^1_\#} \le C n^{-2} \|u - u_n\|_{H^1_\#} \quad \text{and} \quad |\lambda - \widetilde{\lambda}_n| \le C n^{-2} |\lambda - \lambda_n|,
$$

for a constant $C \in \mathbb{R}_+$ independent of n. For large enough values of n and for $\mathcal{N} \gg n$, $\widetilde{u}_{n,\mathcal{N}} = u_n + \widetilde{u}_{n,\mathcal{N}}^{(1)}$ therefore represent a much better approximation of u than u_n.

We refer to [4] for an application of this technique to Kohn-Sham LDA models.

4.3.4 Error Balancing

As mentioned in the introduction, discretization error is only one of the various components of the overall error. In this section, we give an example of a numerical scheme automatically balancing discretization and algorithmic error for the Gross-Pitaevskii model. Still for pedagogical reasons, we consider the simplest possible self-consistent algorithm for solving the Gross-Pitaevskii equation, defined at the continuous level by

$$\begin{cases} -\Delta v_k + V v_k + \mu v_{k-1}^2 v_k = \lambda_k v_k, \quad v_k \in H^1_\#(\Omega), \quad \|v_k\|_{L^2_\#} = 1, \quad (v_k, 1)_{L^2_\#} \geq 0, \\ \lambda_k \text{ lowest eigenvalue of } \; -\Delta + V + \mu v_{k-1}^2. \end{cases}$$

The initial guess v_0 can be chosen, for example, as a normalized ground state of the operator $-\Delta + V$ for small values of μ, and as the Thomas-Fermi approximation of the ground state for large values of μ, but many other choices are possible. In this algorithm, the iterate v_k is the $L^2_\#$-normalized positive ground-state (in the weak sense $(v_k, 1)_{L^2_\#} \geq 0$) of the mean-field operator $-\Delta + V + \mu v_{k-1}^2$ constructed from the previous iterate v_{k-1}. In the Hartree-Fock and Kohn-Sham frameworks, this algorithm is referred to as the Roothaan algorithm, and has been analyzed from a mathematical point of view in [52, 53]. It is known in particular that the sequence $(v_k)_{k\geq 0}$

- either converges to the unique positive solution u to the Gross-Pitaevskii equation (4.13);
- or oscillates between two states in the sense that there exist two functions v_e and v_o in $H^1_\#(\Omega)$, with $v_e \neq v_o$ such that

$$\begin{aligned} -\Delta v_e + V v_e + \mu v_o^2 v_e = \lambda_e v_e, \quad \|v_e\|_{L^2_\#} = 1, \quad (v_e, 1)_{L^2_\#} \geq 0, \\ -\Delta v_o + V v_o + \mu v_e^2 v_o = \lambda_o v_o, \quad \|v_o\|_{L^2_\#} = 1, \quad (v_o, 1)_{L^2_\#} \geq 0, \end{aligned}$$

  and

$$v_{2k} \underset{k\to\infty}{\longrightarrow} v_e, \qquad v_{2k+1} \underset{k\to\infty}{\longrightarrow} v_o \quad \text{in } H^1_\#(\Omega).$$

Typically, $(v_k)_{k\geq 0}$ converges if μ is small and oscillates if μ is large. Clearly, this is not an efficient way to solve the Gross-Pitaevskii equation. We focus on this algorithm for pedagogical reasons only, because it is easier to analyse. Note that the oscillatory behaviour can be suppressed by using an optimal damping algorithm [54]. At a discrete level, it is recommended to solve (4.14) using a preconditioned nonlinear conjugate gradient algorithm [55].

The following scheme is a discretized version of the basic self-consistent field algorithm, in which the discretization space depends on k (compare with (4.15)):

$$\begin{cases} -\Delta v_k + \Pi_{n_k}\left(V + \mu v_{k-1}^2\right)\Pi_{n_k} v_k = \lambda_k\, v_k, \quad v_k \in X_{n_k}, \quad \|v_k\|_{L^2_\#} = 1, \quad (v_k,1)_{L^2_\#} \geq 0, \\ \lambda_k = \lambda_{v_{k-1},n_k}, \quad \text{where } \lambda_{v,n} \text{ is the lowest eigenvalue of } \; -\Delta + \Pi_n\left(V + \mu v^2\right)\Pi_n. \end{cases}$$

Intuitively, it is indeed inefficient to compute the first iterates in a very fine discretization space since we are far from convergence. It, therefore, makes sense to increase the size of the discretization space along the iterations when getting closer to the exact solution u. To automatize this process, we need to define a criterion allowing the algorithm to decide when to refine to discretization space. For this purpose, we use the following result [51]:

Proposition 4.1 *Let u be the unique positive minimizer to* (4.12). *Let J be the error criterion defined by*

$$\forall v \in H^1_\#(\Omega) \text{ such that } \|v\|_{L^2_\#} = 1, \quad J(v) = E(v) - E(u) + \frac{1}{2}\|v-u\|^2_{L^2}.$$

Let $v_k \in X_{n_k}$ be the kth iterate of the above algorithm. Then, we have

$$0 \le J(v_k) \le \eta_{\mathrm{d},k} + \eta_{\mathrm{a},k},$$

where the discretization and algorithmic error estimators $\eta_{\mathrm{d},k}$ and $\eta_{\mathrm{a},k}$ are, respectively, defined by

$$\eta_{\mathrm{d},k}=\frac{1}{2}\left(\lambda_{v_k,n_k} - \lambda_{v_k,\infty}\right) \ge 0, \quad \eta_{\mathrm{a},k} = \frac{1}{2}\left(\mu \int_\Omega (v_k^2 - v_{k-1}^2)v_k^2 + \lambda_k - \lambda_{v_k,n_k}\right) \ge 0.$$

We see that $\eta_{\mathrm{d},k} = 0$ if $n_k = \infty$, that is, if the problem at iteration k has been solved in the whole space $H^1_\#(\Omega)$ (no discretization error), and that $\eta_{\mathrm{a},k} = 0$ if $v_{k-1} = v_k$, that is if the SCF iteration has converged in the discretization space X_{n_k} (no algorithmic error). The numerical experiments reported in [51] show that, in practice, the inequalities

$$E(v_k) - E(u) \le J(v_k) \le \eta_{\mathrm{d},k} + \eta_{\mathrm{a},k}$$

are almost equalities; this observation can be theoretically justified in the asymptotic regime using a priori error analysis results. As a consequence, $\eta_{\mathrm{d},k} + \eta_{\mathrm{a},k}$ gives an accurate estimate of the energy error $E(v_k) - E(u)$, which is split into a discretization error and an algorithmic error. A natural strategy to reach a desired accuracy ε in an optimal way from a computational point of view then consists in refining the discretization if $\eta_{\mathrm{d},k} \gg \eta_{\mathrm{a},k}$, and in iterating otherwise in the same discretization space X_{n_k}, until $\eta_{\mathrm{d},k} + \eta_{\mathrm{a},k} \le \varepsilon$.

Note that at iteration k, v_{k-1}, v_k and λ_k are known, but not λ_{v_k,n_k}, whose computation would require solving another eigenvalue problem in X_{n_k}, nor a fortiori $\lambda_{v_k,\infty}$, which is out of reach of numerical methods. It is, therefore, not possible to compute exactly $\eta_{\mathrm{d},k}$. On the other hand, it is possible to obtain very accurate approximations of all these numbers by adapting the approach based on Rayleigh-Schrödinger perturbation theory detailed in the previous section.

In conclusion, $\eta_{\mathrm{d},k}$ and $\eta_{\mathrm{a},k}$ therefore provide relatively cheap and sharp estimators of the discretization and algorithmic errors at iteration k if the quantity of interest is the energy, allowing adaptive error balancing.

We refer to [4] for an extension of this approach to the periodic Kohn-Sham LDA setting.

4.4 Error Cancellation Phenomenon

In many applications in physics, chemistry, materials science, and biology, energy differences are far more important than absolute energies. Consider for instance a simple chemical reaction that can be modelled as a transition from a local minimum of the ground-state Born-Oppenheimer potential energy surface (GS-BO-PES)—the reactants—to another local minimum of the GS-BO-PES—the products—through a well-defined saddle point—the transition state (see Fig. 4.3). According to Arrhenius law, the reaction rate is given by the relation

$$k = \nu_0 \, \exp\left(-E_{\mathrm{a}}/k_{\mathrm{B}}T\right),$$

where ν_0 is a prefactor, E_{a} the activation energy, that is, the difference between the energy E_{ts} of the transition state and the energy E_{re} of the reactants (see Fig. 4.3), k_{B} the Boltzmann constant, and T the temperature. The relevant QOI, therefore, is the *energy difference* $E_{\mathrm{ts}} - E_{\mathrm{re}}$, and not each of the energies E_{ts} and E_{re}. The same is true for the reaction energy, defined as the energy difference $E_{\mathrm{pr}} - E_{\mathrm{re}}$, where E_{pr} is the energy of the products.

Considering two configurations R_1 and R_2 of the system, our goal is to estimate the error

$$\underbrace{(E_{R_1,\mathcal{N}} - E_{R_2,\mathcal{N}})}_{\text{computable quantity}} - \underbrace{(E_{R_1} - E_{R_2})}_{\text{quantity of interest}}$$

where E_{R_j} is the exact ground-state energy for the configuration R_j and $E_{R_j,\mathcal{N}}$ its variational approximation in the discretization space $\mathscr{X}_{\mathcal{N}}$.

It has been observed both in quantum chemistry and computational materials science, that in general,

$$|(E_{R_1,\mathcal{N}} - E_{R_2,\mathcal{N}}) - (E_{R_1} - E_{R_2})| \ll |E_{R_1,\mathcal{N}} - E_{R_1}| + |E_{R_2,\mathcal{N}} - E_{R_2}|.$$

In other words, the error on the energy difference between two configurations is usually much lower than the error on the energy of each configuration, typically by

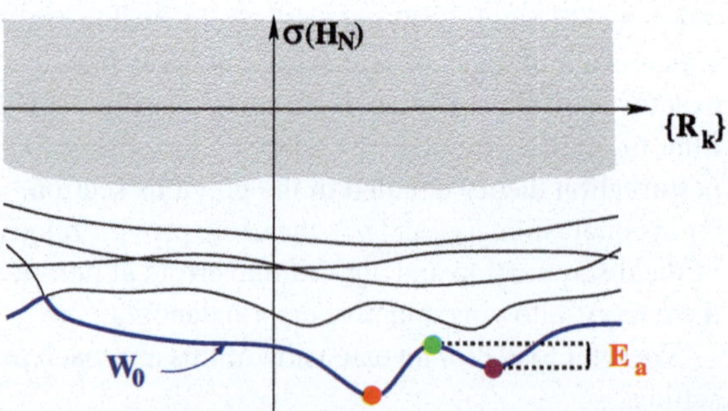

Fig. 4.3 Sketch of a chemical reaction taking place on the ground-state potential energy surface. The activation energy E_{a} of the reaction is the difference between the energy of the reactants and that of the transition state

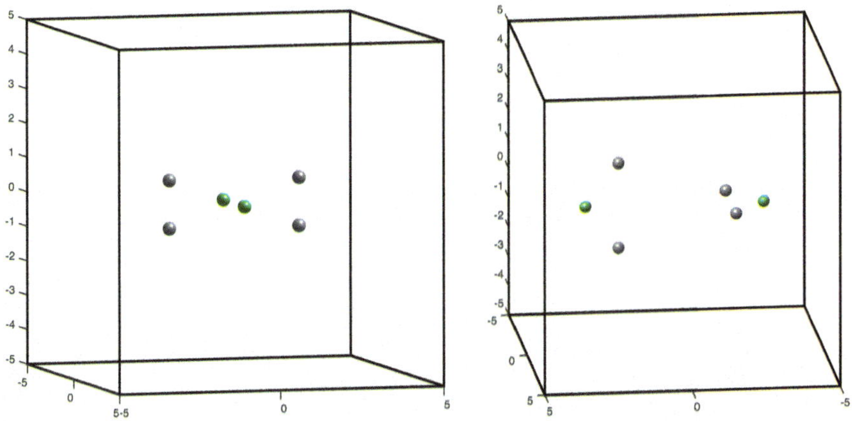

Fig. 4.4 Two different configurations of a system composed of 2 oxygen atoms and 4 hydrogen atoms corresponding to the reactants (left) and the products (right) of the chemical reaction (4.22)

one, sometimes two orders of magnitude. This is the so-called *error cancellation phenomenon*. Chemists and physicists heavily rely on this phenomenon: obtaining an accuracy of 1 kcal/mol (or 1 meV) on an energy difference turns out to be much cheaper in terms of computational effort than obtaining a similar accuracy on a single point energy.

As a matter of example, consider the two different configurations of a system composed of 2 oxygen atoms and 4 hydrogen atoms corresponding, respectively, to the reactants and the products of the chemical reaction (Fig. 4.4).

$$2H_2 + O_2 \longrightarrow 2H_2O. \tag{4.22}$$

The sum and difference of the energy errors

$$S_{\mathcal{N}} := (E_{\text{reactants},\mathcal{N}} - E_{\text{reactants}}) + (E_{\text{products},\mathcal{N}} - E_{\text{products}}), \tag{4.23}$$

$$\begin{aligned} D_{\mathcal{N}} &:= |(E_{\text{reactants},\mathcal{N}} - E_{\text{reactants}}) - (E_{\text{products},\mathcal{N}} - E_{\text{products}})| \qquad (4.24) \\ &= |\underbrace{(E_{\text{reactants},\mathcal{N}} - E_{\text{product},\mathcal{N}})}_{\text{computed value of the QOI}} - \underbrace{(E_{\text{reactants}} - E_{\text{products}})}_{\text{exact value of the QOI}}|, \end{aligned}$$

for PW Kohn-Sham LDA calculations with Troullier-Martins pseudopotentials as a function of the energy cut-off $E_{\mathcal{N}} = \frac{1}{2}\mathcal{N}^2$ are plotted on Fig. 4.5 (top). It can be observed that $D_{\mathcal{N}}$ is indeed smaller than $S_{\mathcal{N}}$ by about two orders of magnitude. In addition, the non-dimensional error cancellation factor

$$0 \leq Q_{\mathcal{N}} := \frac{D_{\mathcal{N}}}{S_{\mathcal{N}}} \leq 1 \tag{4.25}$$

fluctuates about a value close to $Q_\infty \simeq 5 \times 10^{-3}$.

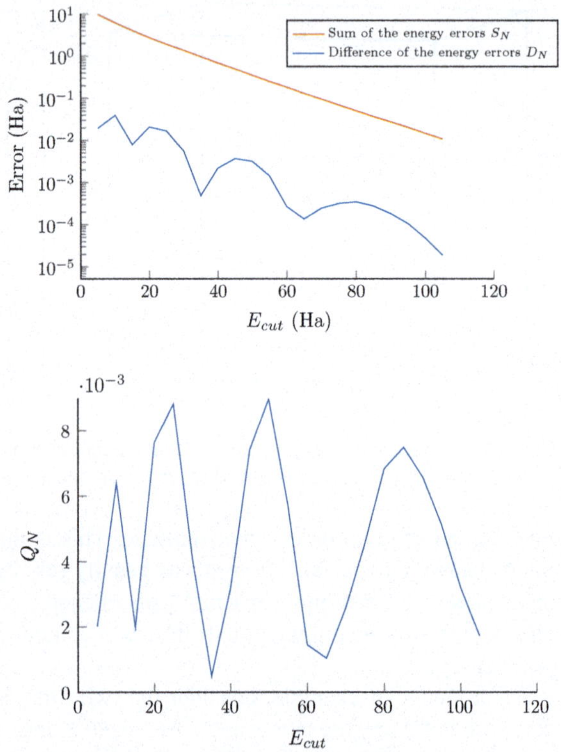

Fig. 4.5 Observed convergence of the quantities $S_{\mathcal{N}}$ and $D_{\mathcal{N}}$ defined by (4.23)–(4.24) as a function of $\mathcal{N}$, and behaviour of the ratio $Q_{\mathcal{N}}$ for the two configurations represented on Fig. 4.4

In order to unravel the origin of the discretization error cancellation phenomenon, let us consider a simple linear 1D model for which explicit calculations can be carried out [1]. In this model, the external potential is periodic (with period $a = 1$) and is a sum of Dirac masses:

$$V_{\mathrm{ext},R} = -\sum_{m\in\mathbb{Z}} z_1\delta_m - \sum_{m\in\mathbb{Z}} z_2\delta_{m+R}.$$

For given values of the charges z_1 and z_2, the configurations are labelled by $R \in (0, 1)$. The ground-state energy and wave-function are obtained by computing the lowest eigenvalue and an associated normalized eigenfunction of the 1D periodic Schrödinger equation

$$\left(-\frac{\mathrm{d}^2}{\mathrm{d}x^2} - \sum_{m\in\mathbb{Z}} z_1\delta_m - \sum_{m\in\mathbb{Z}} z_2\delta_{m+R}\right)\psi_R = E_R\psi_R \quad \text{in } L^2_{\mathrm{per}}(0,1), \tag{4.26}$$

$$\int_0^1 \psi_R^2(x)\mathrm{d}x = 1.$$

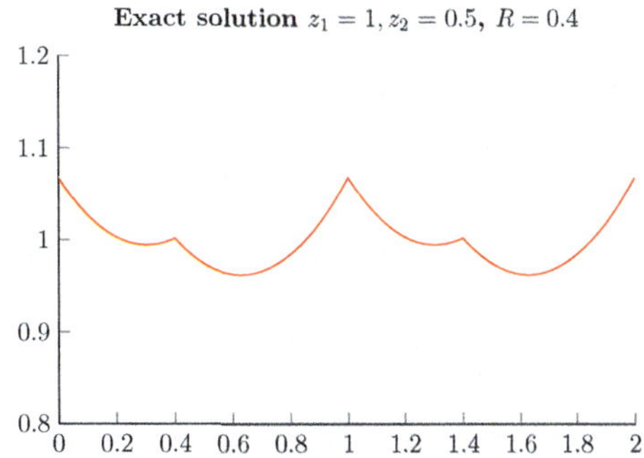

Fig. 4.6 Exact ground-state wave-function of (4.26) for $z_1 = 1$, $z_2 = 0.5$ and $R = 0.4$

As a matter of illustration, the ground-state wave-function for $z_1 = 1$, $z_2 = 0.5$ and $R = 0.4$ is plotted on Fig. 4.6.

The following result is proved in [1].

Theorem 4.3 *Let $z_1, z_2 > 0$ and $R \in (0, 1)$, let E_R be the ground-state energy of (4.26), and $E_{R,\mathcal{N}}$ the variational approximation of E_R in the Fourier approximation space*

$$\mathit{Span}\left\{e^{2i\pi kx},\ k \in \mathbb{Z},\ |k| \leq \mathcal{N}\right\}.$$

Then, we have the asymptotic expansion

$$E_{R,\mathcal{N}} - E_R \doteq \frac{\alpha_R}{\mathcal{N}} - \frac{\alpha_R}{2\mathcal{N}^2} + \frac{\beta^{(1)}_{R,\mathcal{N}}}{\mathcal{N}} + \frac{\gamma_R}{\mathcal{N}}\eta_{R,\mathcal{N}} + o\left(\frac{1}{\mathcal{N}^{3-\varepsilon}}\right), \tag{4.27}$$

where

$$\alpha_R := \frac{z_1^2 u_R(0)^2 + z_2^2 u_R(R)^2}{2\pi^2}, \tag{4.28}$$

$$\gamma_R := \frac{z_1 z_2 u_R(0) u_R(R)}{\pi^2}, \qquad \eta_{R,\mathcal{N}} := \mathcal{N} \sum_{k=\mathcal{N}+1}^{+\infty} \frac{\cos(2\pi k R)}{k^2},$$

$$\beta^{(1)}_{R,\mathcal{N}} := \frac{z_1^2 u_R(0)(u_{R,\mathcal{N}}(0) - u_R(0)) + z_2^2 u_R(R)(u_{R,\mathcal{N}}(R) - u_R(R))}{2\pi^2}.$$

In addition

$$|\eta_{R,\mathcal{N}}| \leq \min\left(1, \frac{2 + \frac{\pi^3}{8}}{|\sin(\pi R)|\mathcal{N}}\right) \quad \textit{and} \quad \forall \varepsilon > 0,\ \exists C_\varepsilon \in \mathbb{C}_+ \ \textit{s.t.}\ |\beta^{(1)}_{R,\mathcal{N}}| \leq \frac{C_\varepsilon}{\mathcal{N}^{1-\varepsilon}}.$$

This result sheds light on the mechanism of discretization error cancellation for the PW discretization of (4.26). First, it implies that errors on energies and errors

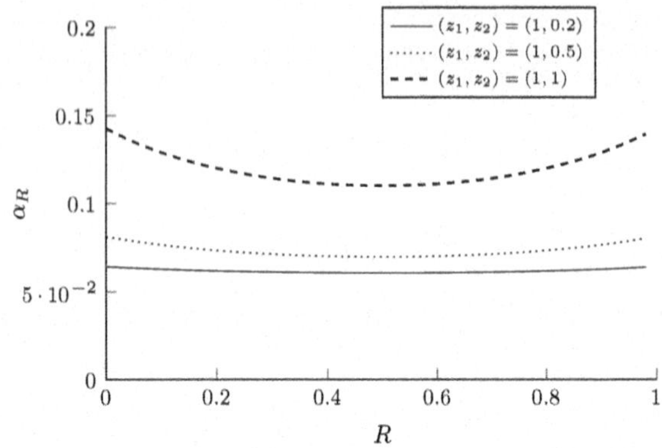

Fig. 4.7 Plot of the function $R \mapsto \alpha_R$ defined by (4.28) for different sets of parameters (z_1, z_2)

on energy differences all scale in N^{-1}, and that discretization error cancellation is a matter of prefactors:

$$E_{R,\mathcal{N}} - E_R \underset{\mathcal{N}\to\infty}{\sim} \frac{\alpha_R}{\mathcal{N}} \quad \text{while} \quad (E_{R_2,\mathcal{N}} - E_{R_1,\mathcal{N}}) - (E_{R_2} - E_{R_1}) \underset{\mathcal{N}\to\infty}{\sim} \frac{\alpha_{R_2} - \alpha_{R_1}}{\mathcal{N}}.$$

Note that α_R only depends on the charges z_1 and z_2 of the Dirac potentials and on the values of the ground-state densities at the positions of the Dirac potentials. Error cancellation is due to the fact that, for $0.1 \leq R_1, R_2 \leq 0.9$, we have

$$|\alpha_{R_2} - \alpha_{R_1}| \ll \max(\alpha_{R_1}, \alpha_{R_2}) \qquad \text{(see Fig. 4.7).} \tag{4.29}$$

We also obtain that the error cancellation factor converges when $\mathcal{N}$ goes to infinity:

$$\lim_{\mathcal{N}\to+\infty} Q_{\mathcal{N}} = \frac{|\alpha_{R_1} - \alpha_{R_2}|}{\alpha_{R_1} + \alpha_{R_2}}.$$

Numerical simulations show that the convergence is monotonous for R_1, R_2 away from the singularities $R = 0$ and $R = 1$ (where the two Dirac combs overlap), and oscillating for R_1 or R_2 close to the singularities (see Fig. 4.8).

Inequality (4.29), which is at the root of error cancellation, can be rewritten as

$$\begin{aligned} &\left|z_1^2\left(\rho_{R_1}(0) - \rho_{R_2}(0)\right) + z_2^2\left(\rho_{R_1}(R_1) - \rho_{R_2}(R_2)\right)\right| \\ &\quad \ll \max(z_1^2\rho_{R_1}(0) + z_2^2\rho_{R_2}(0), z_1^2\rho_{R_1}(R_1) + z_2^2\rho_{R_2}(R_2)), \end{aligned}$$

where $\rho_R(x) = u_R(x)^2$ is the ground-state density at point x in configuration R. For R away from the singularities 0 and 1, the ground-state density at the positions 0 and R of the Dirac potentials does not change much with R.

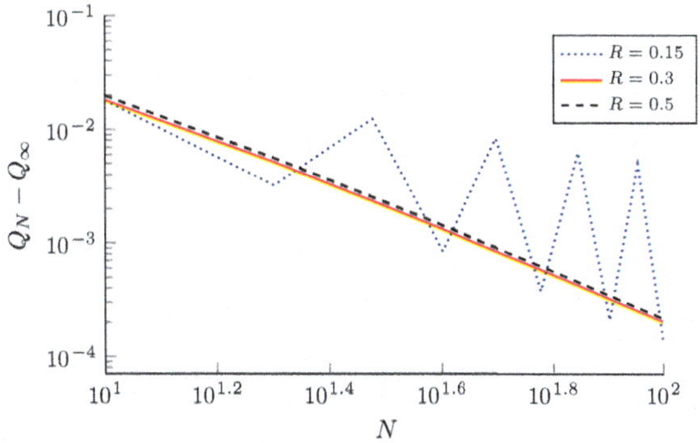

Fig. 4.8 Convergence of $Q_\mathcal{N}$ to Q_∞ for $R_1 = \frac{1}{2}$ and three different values of R_2

Another interesting remark reported in [1] is that

$$\left| \frac{\mathrm{d}E_{R,\mathcal{N}}}{\mathrm{d}R} - \frac{\mathrm{d}E_R}{\mathrm{d}R} \right| \gg \left| \frac{\frac{\mathrm{d}\alpha_R}{\mathrm{d}R}}{\mathcal{N}} \right|.$$

In other words, it is not possible to infer from $E_{R,\mathcal{N}} - E_R \sim_{\mathcal{N}\to\infty} \frac{\alpha_R}{\mathcal{N}}$ a result on the convergence of the forces. In fact, it is observed that

$$\frac{\mathrm{d}E_{R,\mathcal{N}}}{\mathrm{d}R} - \frac{\mathrm{d}E_R}{\mathrm{d}R} \underset{\mathcal{N}\to\infty}{\sim} \frac{\mathrm{d}}{\mathrm{d}R}\left(\frac{\gamma_R}{\mathcal{N}}\eta_{R,\mathcal{N}}\right)$$

and that the function $R \mapsto \frac{\mathrm{d}E_{R,\mathcal{N}}}{\mathrm{d}R} - \frac{\mathrm{d}E_R}{\mathrm{d}R}$ is oscillating more and more when $\mathcal{N}$ becomes large. As a consequence,

1. It is not a good idea to try and compute the energy difference between two configuration R_1 and R_2 by integrating the forces along a path of the configuration space linking R_1 and R_2;
2. Extrapolation methods based on the asymptotic expansion (4.27) can be used to improve the accuracy of the energy, but they will not improve the accuracy of the forces.

4.5 Conclusion

In this chapter, we have introduced basic concepts of mathematically based discretization error analysis: a priori error estimators, a posteriori error estimators and post-processing methods, asymptotic expansions and extrapolation methods, and error cancellation phenomenon. These concepts have been illustrated on the simple examples of plane-wave discretizations of the Gross-Pitaevskii model, and of a 1D periodic Schrödinger equation with Dirac potentials.

Significant progress on discretization error analysis for Kohn-Sham and other electronic structure models has been made in the past few years, and many ongoing works in these directions are in progress in several groups around the world. As witnessed in other fields of science and engineering, rigorously founded error analysis should play a major role in the design of a new generation of electronic structure calculation software, generating numerical results supplemented with error bars, optimizing the available computational resources, and adapted to massively parallel and heterogeneous architectures.

References

1. Cancès E, Dusson G (2017) Discretization error cancellation in electronic structure calculation: a quantitative study. M2AN 51:1617–1636
2. Akin JE (2005) Finite element analysis with error estimators: An introduction to the FEM and adaptive error analysis for engineering students. Elsevier
3. Cancès E, Chakir R, Maday Y (2012) Numerical analysis of the planewave discretization of some orbital-free and Kohn-Sham models. ESAIM: Math Mod Numer Anal 46(2):341–388
4. Cancès E, Dusson G, Maday Y, Stamm B, Vohralík M (2016) A perturbation-method-based post-processing for the planewave discretization of Kohn-Sham models. J Comput Phys 307:446–459
5. Cancès E, Ehrlacher V, Maday Y (2014) Non-consistent approximations of self-adjoint eigenproblems: application to the supercell method. Numer Math 128(4):663–706
6. Chen H, Dai X, Gong X, He L, Zhou A (2014) Adaptive finite element approximations for Kohn-Sham models. SIAM MMS 12(4):1828–1869
7. Chen H, Gong X, He L, Yang Z, Zhou A (2013) Numerical analysis of finite dimensional approximations of Kohn-Sham models. Adv Comput Math 38(2):225–256
8. Chen H, Schneider R (2015) Error estimates of some numerical atomic orbitals in molecular simulations. Commun Comput Phys 18(01):125–146
9. Chen H, Schneider R (2015) Numerical analysis of augmented plane wave methods for full-potential electronic structure calculations. ESAIM: Math Mod Numer Anal 49(3):755–785
10. Dusson G, Maday Y (2016) A posteriori analysis of a nonlinear Gross–Pitaevskii-type eigenvalue problem. IMA J Numer Anal drw001
11. Hanrath M (2008) Wavefunction quality and error estimation of single- and multi-reference coupled-cluster and CI methods: the H4 model system. Chem Phys Lett 466(4–6):240–246
12. Kaye J, Lin L, Yang C (2015) A posteriori error estimator for adaptive local basis functions to solve Kohn-Sham density functional theory. Commun Math Sci 13(7):1741–1773
13. Kutzelnigg W (1991) Error analysis and improvements of coupled-cluster theory. Theoretica Chimica Acta 80(4–5):349–386
14. Kutzelnigg W (2012) Rate of convergence of basis expansions in quantum chemistry. AIP Conf Proc 1504(1):15–30
15. Maday Y, Turinici G (2003) Error bars and quadratically convergent methods for the numerical simulation of the Hartree-Fock equations. Numer Math 94(4):739–770
16. Rohwedder T, Schneider R (2013) Error estimates for the coupled cluster method. ESAIM: Math Mod Numer Anal 47(6):1553–1582
17. Pitaevskii LP, Stringari S (2003) Bose-Einstein condensation. Clarendon Press
18. Kohn W, Sham LJ (1965) Self-consistent equations including exchange and correlation effects. Phys Rev 140(4A):A1133–A1138
19. Giannozzi P, Baroni S, Bonini N, Calandra M, Car R, Cavazzoni C, Ceresoli D, Chiarotti GL, Cococcioni M, Dabo I, Dal Corso A, de Gironcoli S, Fabris S, Fratesi G, Gebauer R,

Gerstmann U, Gougoussis C, Kokalj A, Lazzeri M, Martin-Samos L, Marzari N, Mauri F, Mazzarello R, Paolini S, Pasquarello A, Paulatto L, Sbraccia C, Scandolo S, Sclauzero G, Seitsonen AP, Smogunov A, Umari P, Wentzcovitch RM, QUANTUM ESPRESSO: a modular and open-source software project for quantum simulations of materials. J Phys: Cond Matter 21(39):395502, 2009
20. Gonze X, Amadon B, Anglade P-M, Beuken J-M, Bottin F, Boulanger P, Bruneval F, Caliste D, Caracas R, Côté M, Deutsch T, Genovese L, Ghosez Ph, Giantomassi M, Goedecker S, Hamann DR, Hermet P, Jollet F, Jomard G, Leroux S, Mancini M, Mazevet S, Oliveira MJT, Onida G, Pouillon Y, Rangel T, Rignanese G-M, Sangalli D, Shaltaf R, Torrent M, Verstraete MJ, Zerah G, Zwanziger JW (2009) ABINIT: First-principles approach to material and nanosystem properties. Comput Phys Commun 180(12):2582–2615
21. Kresse G, Furthmüller J (1996) Efficient iterative schemes for ab initio total-energy calculations using a plane-wave basis set. Phys Rev B 54(16):11169–11186
22. Motamarri P, Nowak MR, Leiter K, Knap J, Gavini V (2013) Higher-order adaptive finite-element methods for Kohn-Sham density functional theory. J Comput Phys 253:308–343
23. Pask JE, Sterne PA (2005) Finite element methods in ab initio electronic structure calculations. Model Simul Mater Sci Eng 13(3):R71–R96
24. Mohr S, Ratcliff LE, Boulanger P, Genovese L, Caliste D, Deutsch T, Goedecker S (2014) Daubechies wavelets for linear scaling density functional theory. J Chem Phys 140(20):204110
25. Goedecker S, Teter M, Hutter J (1996) Separable dual-space Gaussian pseudopotentials. Phys Rev B 54(3):1703–1710
26. Troullier N, Martins JL (1990) A straightforward method for generating soft transferable pseudopotentials. Solid State Commun 74:613–616
27. Vanderbilt D (1990) Soft self-consistent pseudopotentials in a generalized eigenvalue formalism. Phys Rev B 41:7892–7895
28. Levitt A, Torrent M (2015) Parallel eigensolvers in plane-wave Density Functional Theory. Comput Phys Commun 187:98–105
29. Ainsworth M, Oden JT (2011) A posteriori error estimation in finite element analysis. Wiley, New York
30. Makov G, Payne MC (1995) Periodic boundary conditions in ab initio calculations. Phys Rev B 51:4014–4022
31. Gontier D, Lahbabi S (2017) Supercell calculations in the reduced Hartree-Fock model for crystals with local defects. Appl Math Res Express (1):1–64
32. Reed M, Simon B (1972) Methods of modern mathematical physics. I. Functional analysis, vol 53. Academic Press Inc., New York
33. Babuška I, Osborn J (1991) Eigenvalue problems. Handbook of numerical analysis, vol. II. North-Holland, Amsterdam, pp 641–787
34. Xu J, Zhou A (1999) Local and parallel finite element algorithms based on two-grid discretizations. Math Comput 69(231):881–909
35. Guo H, Zhang Z, Zhao R (2017) Superconvergent two-grid methods for elliptic eigenvalue problems. J Sci Comput 70:125–148
36. Xu J, Zhou A (1999) A two-grid discretization scheme for eigenvalue problems. Math Comput 70(233):17–25
37. Cancès E, Chakir R, He L, Maday Y, Two–grid methods for a class of nonlinear elliptic eigenvalue problems. IMA J Numer Anal. in press
38. Cancès E, Dusson G, Maday Y, Stamm B, Vohralík M (2017) Guaranteed and robust a posteriori bounds for Laplace eigenvalues and eigenvectors: conforming approximations. SIAM J Numer Anal 55:2228–2254
39. Rannacher R, Westenberger A, Wollner W (2010) Adaptive finite element solution of eigenvalue problems: balancing of discretization and iteration error. J Numer Math 18:303–327
40. Liu X, Oishi S (2013) Verified eigenvalue evaluation for the Laplacian over polygonal domains of arbitrary shape. SIAM J Numer Anal 51:1634–1654
41. Heuveline V, Rannacher R (2001) A posteriori error control for finite approximations of elliptic eigenvalue problems. Adv Comput Math 15(1–4):107–138

42. Larson MG (2000) A posteriori and a priori error analysis for finite element approximations of self-adjoint elliptic eigenvalue problems. SIAM J Numer Anal 38(2):608–625
43. Li H, Ovall JS (2015) A posteriori eigenvalue error estimation for a Schrödinger operator with inverse square potential. Disc Cont Dyn Sys B 20(5):1377–1391
44. Lin L, Stamm B (2017) A posteriori error estimates for discontinuous Galerkin methods using non-polynomial basis functions. Part II: Eigenvalue problems. ESAIM: Math Mod Numer Anal 51(5):1733–1753
45. Mao D, Shen L, Zhou A (2006) Adaptive nite element algorithms for eigenvalue problems based on local averaging type a posteriori error estimates. Adv Comput Math 25(1–3):135–160
46. Cancès E, Dusson G, Maday Y, Stamm B, Vohralík M (2017) Guaranteed and robust a posteriori bounds for Laplace eigenvalues and eigenvectors: a unified framework. ArXiv:hal-01483461
47. Dahmen W, Rohwedder T, Schneider R, Zeiser A (2008) Adaptive eigenvalue computation: complexity estimates. Numer Math 110(3):277–312
48. Dai X, Xu J, Zhou A (2008) Convergence and optimal complexity of adaptive finite element eigenvalue computations. Numer Math 110(3):313–355
49. Mehrmann V, Miedlar A (2011) Adaptive computation of smallest eigenvalues of self-adjoint elliptic partial differential equations. Numer Linear Algebra Appl 18(3):387–409
50. Kato T (1980) Perturbation theory for linear operators, 2nd edn. Springer, Berlin
51. Cancès E, Dusson G, Maday Y, Stamm B, Vohralík M (2014) A perturbation-method-based a posteriori estimator for the planewave discretization of nonlinear Schrödinger equations. C R Math 352(11):941–946
52. Cancès E, Le Bris C (2000) On the convergence of SCF algorithms for the Hartree-Fock equations. M2AN 34:749–774
53. Levitt A (2012) Convergence of gradient-based algorithms for the Hartree-Fock equations. M2AN, 46:1321–1336
54. Dion C, Cancès E (2003) Spectral method for the time?dependent Gross?Pitaevskii equation with harmonic traps. Phys Rev E 67:046706
55. Antoine X, Levitt A, Tang Q (2017) Efficient spectral computation of the stationary states of rotating Bose-Einstein condensates by the preconditioned nonlinear conjugate gradient method. J Comput Phys 343:92–109

Chapter 5
Basis Sets for Correlated Methods

Daniel Claudino and Rodney J. Bartlett

5.1 Introduction

It is an unfortunate fact that virtually all problems of chemical or physical interest do not have a closed-form solution and, therefore, need to be dealt with by a range of different approximations. One of the several approximations underlying most electronic structure efforts is the use of what is commonly known as atomic basis sets.

Simply put, an atomic basis set, or simply basis set or basis, in the context of Quantum Chemistry is a group of known one-particle functions combined in a certain fashion to be further employed in approximating the exact N-particle electronic wave function (or density, in the case of density functional theory) with N being the number of electrons in the system. Two important considerations arise from such a definition: the kind of functions to be used and how many of them. Even though many different classes of functions have been tried to constitute a basis set, none has been more prevalent than Gaussian functions. Unlike the Slater-type orbitals (STOs), which are exponential functions inspired by the solution of hydrogen-like atoms that display the correct cusp and long-range behavior, [1] a Gaussian-type orbital (GTO) suffers from much too rapid decay, meaning that the region spanned by functions' tails is not adequately covered. An even more serious drawback is that they do not comply with Kato's cusp condition, i.e., the electron density does not form a cusp as it approaches the nuclei [2]. Nonetheless, the computational efficiency in turning to GTOs over STOs for polyatomic molecules is so remarkable that it allows for multiple Gaussian functions to extend the reach of the wave function to farther distances from the

D. Claudino (✉) · R. J. Bartlett
University of Florida, Gainesville, FL 32611, USA
e-mail: dclaudino@ufl.edu

R. J. Bartlett
e-mail: bartlett@qtp.ufl.edu

E. Perlt (ed.), *Basis Sets in Computational Chemistry*, Lecture Notes in Chemistry 107,
https://doi.org/10.1007/978-3-030-67262-1_5

nuclei as well as ironing out the cusp region, with computational speedups of orders of magnitude [3].

For the reasons discussed above, basis sets to be used in quantum chemical applications eventually became a synonym for Gaussian-type orbitals, and these terms will be used interchangeably throughout this document. The other question brought up, that is, the number of Gaussians to be employed is rather involved and is closely related to the reasoning behind the development of each basis set.

Before we delve into the specifics of the most popular basis sets, it seems suitable to lay down some important theoretical and computational considerations that are imperative in examining the reasoning adopted as a guide in the construction of such bases.

5.2 The Structure of Gaussian Basis Sets

Even though there are many different families of GTO basis sets, in all instances their radial part can be written in the following general form:

$$\phi(\alpha, \ell, \mathbf{r} - \mathbf{R_A}) = N\mathbf{r}^{\ell} e^{-\alpha|\mathbf{r}-\mathbf{R_A}|^2} \tag{5.1}$$

where N is a normalization constant, ℓ is the angular momentum quantum number associated with the atomic orbital ϕ, $|\mathbf{r} - \mathbf{R_A}|$ represents the distance of the electron with coordinates $\mathbf{r}$ with respect to the Gaussian function centered at the nucleus with coordinates $\mathbf{R_A}$, and α is the Gaussian exponent.

The general GTO shown in Eq. 5.1 is given in the spherical harmonics space, meaning it has a $Y_\ell^m(\theta, \varphi)$ attached to it, where θ and φ are the polar and azimuthal angles. For reasons of visualization, it is more convenient to consider the so-called Cartesian Gaussians. They can be easily transformed to and from spherical harmonics in most quantum chemical codes. Cartesian Gaussians fit the following general formula:

$$\phi(\alpha, a, b, c, x, y, z, \mathbf{r} - \mathbf{R_A}) = N x^a y^b z^c e^{-\alpha|\mathbf{r}-\mathbf{R_A}|^2} \tag{5.2}$$

with x, y, and z being the Cartesian coordinates, $\ell = a + b + c$, and $r^2 = x^2 + y^2 + z^2$, thereby accounting for the spherical harmonic $Y_\ell^m(\theta, \varphi)$ above.

The computational efficiency of resorting to GTOs can be further enhanced by an approach known as contraction. This idea is so widespread in basis set construction that, unless stated otherwise, the term basis sets refers to contracted GTOs, or CTGO for short. In the basis set jargon, a contraction is a basis function formed through a linear combination of GTOs, and the GTOs used in the contraction expansion are named primitives. A contraction can be expressed mathematically by

$$\psi = \sum_n C_n \phi_n \tag{5.3}$$

where the atomic orbital ψ, now a CGTO, is expanded in a linear combination of the ϕ primitive Gaussians, weighed by coefficients C_n.

Simply stated, the objective of contracting a set of primitives is to manipulate fewer functions that incorporate most of the features of the full primitive GTO set.

For a given angular momentum, there are two degrees of freedom in the CGTOs: the Gaussian exponent, α (Eqs. 5.1 and 5.2), and the contraction coefficients, C_n (Eq. 5.3). Each of these two components are discussed in more detail below.

5.2.1 Gaussian Exponents

The exponent α shown in Eqs. 5.1 and 5.2 can be seen as a scaling factor of the distance from the atomic center A. In other words, it controls how the Gaussian functions span the radial space (spherical harmonics are the basis that spans the angular domain). In the early days of basis set quantum chemistry, Gaussians were simply fit to best describe a given STO [4]. But it was known to be a very crude approximation, much constrained by the computational technology of the time and did not embody any correlation effect. The seminal work of Ruedenberg and co-workers showed that Gaussian exponents individually optimized with respect to the minimization of the Hartree-Fock (HF) energy closely followed a series, which was named an even-tempered expansion. Instead of the optimization of a number of parameters equal in size to the primitive set for a certain angular momentum, in turning to an even-tempered expansion all it takes is the determination of two parameters [5]:

$$\alpha_k = \alpha\beta^k \quad , \quad k = 0, \ldots, N \tag{5.4}$$

where $\alpha = \alpha_0$ and β are the parameters to be optimized, and N determines the size of the expansion, with a different expansion for each ℓ-shell to be spanned by the basis set.

This expansion can successfully address sets of primitives of moderate size. As the basis becomes larger, the even-tempered expansion eventually shows significant deviations from the individually optimized exponents. Additionally, the large primitive set devised from the even-tempered expansion usually suffers from linear dependencies, meaning that two (or more) functions are too similar. For these reasons, a number of alternative expansions have been proposed [6–8].

In possession of a primitive set generated from an even-tempered (or other) expansion, it was found that extra flexibility could be attained if the most diffuse functions, i.e., those with the smallest exponents, were included as primitives without being contracted, [9, 10] which became a strategy followed in the development of some of the basis sets discussed below.

With the advance of computational technology and more efficient algorithms that could perform more extensive calculations in a timely manner, it eventually became possible to obtain primitives by optimization of the underlying expansions (or individual exponents) with respect to the energy of a correlated method (or, less frequently, some other property), whose exponents are varied until the lowest energy is achieved.

5.2.2 Contraction Schemes

Given a set of primitive functions, the question is whether there is a way whereby these functions can be combined so that a set of reduced size can effectively and efficiently approximate the larger, more computationally costly set. Such a (linear) combination is commonly referred to as a "contraction".

There are two main contraction types: segmented, [11] in which each contraction has its own set of primitives, and general, [12] with the parent Gaussians being shared by each of the contracted functions. The distinction between the two contraction schemes is illustrated in Fig. 5.1.

In the early days of Quantum Chemistry, segmented contractions were routine due to computational efficiency reasons, but nowadays this is hardly an impediment and many current correlated basis sets are of the general type.

Even though virtually all atomic basis sets fall into one of the two different contraction approaches, the contraction coefficients C_n can be obtained in a variety of ways, which will be discussed in the context of each particular basis set.

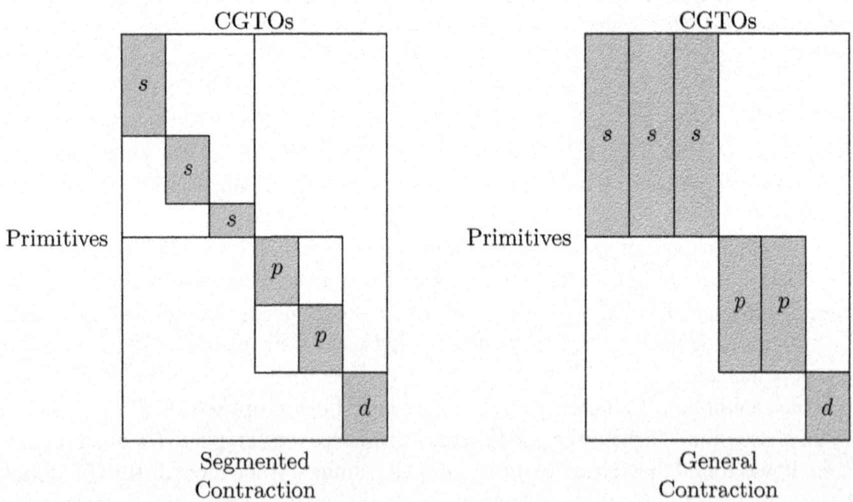

Fig. 5.1 Pictorial representation of the segmented and general contraction schemes of a hypothetical set of primitives contracted to $[3s2p1d]$

5.3 The Correlation Problem

Even though electrons are in constant motion in matter, the early and simplest attempts to address electronic structure relied upon independent-particle models, also known as mean-field theories, where each electron is treated individually as though it is moving in an average field generated by the remaining electrons. This idea has been proposed in several different flavors and yields what is commonly referred to as a reference function. The most famous example is Hartree-Fock, a term often used interchangeably with Self-Consistent Field (SCF) theory. The determination of the HF reference function is a routine first step in the vast majority of quantum chemical calculations, with the notable exception of Density Functional Theory (DFT). In HF, the instantaneous interaction of electrons is not taken into account, and even though it constitutes only a small fraction of the energy, the proper treatment of the so-called "correlation energy" is very often the decisive factor in achieving high-accuracy results in a wide range of problems of chemical and physical relevance.

As defined by Löwdin, the correlation energy is simply the difference between the exact energy and the "restricted" Hartree-Fock energy for a given basis set, [13] but now one usually relaxes the latter to the energy of any mean-field, single determinant like "unrestricted" HF, or other theory based on a single determinant. In that case, the energy correction from the mean-field approximation accounts for the other instantaneous situations absent in the mean field, as can be simply defined as

$$E_{\text{correlation}} = E_{\text{exact}} - E_{\text{HF}} \tag{5.5}$$

That being said, the goal in the development of correlated basis sets is to provide basis functions that favorably bridge this gap between the reference and the exact picture, that is, that describe the many-electron interactions in a satisfactory manner.

5.3.1 *Choice of Correlated Method*

There are quite a few strategies in the treatment of electron correlation, but some of them became more popular than others over the years, which did not necessarily translate in being more associated with basis set development. Within the wave function framework, the correlated wave function, from which the correlation energy (or any other correlated property, in principle) can be evaluated, is obtained via excitations from the occupied orbitals to the virtual orbitals as determined by the reference function, subject to the chosen correlated method. This can be represented mathematically as

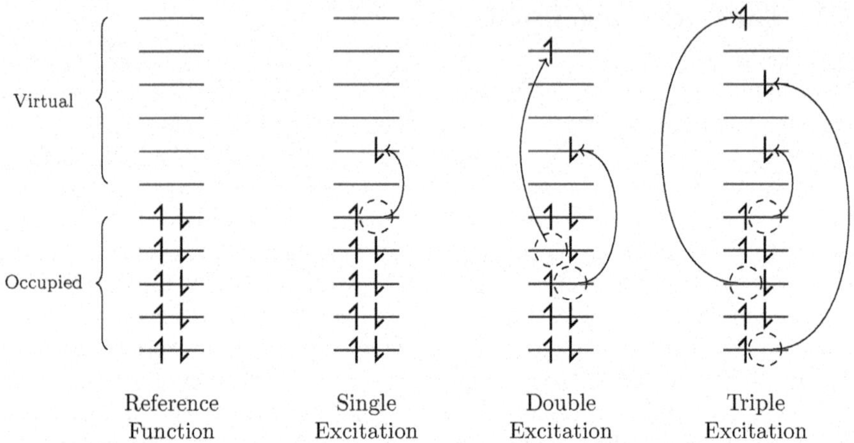

Fig. 5.2 A pictorial way to represent the excitation manifold according to the orbital occupancy

$$\Psi = \hat{C}\Phi_0 = (1 + \hat{\mathbf{C}}_S + \hat{\mathbf{C}}_D + \hat{\mathbf{C}}_T \ldots)\Phi_0 = \underbrace{\Phi_0}_{\text{reference}} + \underbrace{\sum_{ia} C_i^a \Phi_i^a}_{\text{singles}} + \underbrace{\sum_{ijab} C_{ij}^{ab} \Phi_{ij}^{ab}}_{\text{doubles}} + \underbrace{\sum_{ijkabc} C_{ijk}^{abc} \Phi_{ijk}^{abc}}_{\text{triples}} + \ldots \tag{5.6}$$

where $\hat{C}$ is the operator that generates the excitation manifold from the reference function (excitation operator), i, j, and k label occupied orbitals, a, b, and c are associated with virtual orbitals, Φ_S, Φ_D, and Φ_T form the space spanned by determinants obtained from single, double, triple, and so on, excitations from the reference function. The occupation of the orbitals in these determinants can be graphically represented in Fig. 5.2.

The expansion of the wave function presented in Eq. 5.6 is known as Configuration Interaction (CI). This method has the usually desirable property of being variational, which means that the energy of any truncation of the CI expansion is an upper bound to the exact energy. The limiting case in which all possible electronic excitations are included is known as Full CI (FCI), and the exact wave function follows and the total correlation energy is computed. Unfortunately, FCI calculations can seldom be done as the difficulty increases approximately exponentially with the number of basis functions. On the other hand, any truncation of the CI expansion suffers from what is known as a lack of size-extensivity, the best known defect of the CI formalism, due to the presence of terms known as "unlinked diagrams." A method is only size-extensive if its corresponding energy scales correctly with the number of electrons in the system [14].

It is not uncommon that the HF energy is $\approx 99\%$ of the total energy. Given the small magnitude of the correlation energy, it can be instinctively treated by perturbation theory (PT), thus the mean-field picture is seen as the unperturbed, zeroth-order problem, while electron correlation is introduced as a perturbation. Due to Brillouin's theorem, the first energy correction from electron correlation with an HF reference is of second order in the correlation correction, commonly known as second order Møller-Plesset Perturbation Theory (MPPT), or MP2 for short, which is a particular case of Rayleigh-Schrödinger Perturbation Theory (RSPT) with a HF starting point [15]. This result can be viewed as MBPT(2), meaning the first linked diagram approximation given by the Many-Body Perturbation Theory (MBPT) formalism. In the context of MBPT, the energy expressions are size-extensive in all orders, since this method is free of the unlinked diagrams by construction, while this aspect does not arise in MPPT. This approach can be easily applied to non-HF references through generalized MBPT [16].

MBPT taken to infinite-order gives Coupled-Cluster (CC) theory. In CC, infinite-order contributions that are evaluated order by order in PT can be achieved by an exponential excitation operator analogous to the CI operator $\hat{C}$. In the FCI limit, both CC and CI are equivalent, but the major advantage of CC methods (and also MBPT) over CI counterparts is that any truncation short of FCI is inherently size-extensive. As a consequence of the exponential nature of the CC operator, CC methods almost always provide a better approximation to the exact correlation energy and molecular properties because of the presence of the so-called disconnected clusters. For example, CC with single and double excitations, CCSD, scales as $\sim N^6$, where N is the number of basis functions. It has a term $1/2\hat{T}_2^2$, which is an effective quadruple excitation contribution that can only be achieved in CI if one expands the CI wave function up to $\hat{C}_4$, yielding a very computationally expensive approach ($\sim N^{10}$), and would still not be size-extensive for any case with more than four correlated electrons. Schemes combining CC and MBPT are very successful in reaching high accuracy while offering a more affordable alternative. The most notable of such cases is CCSD where the effect of triply excited determinants are incorporated perturbatively, in methods known as CCSD[T], CCSD(T), and ΛCCSD(T) [17–19].

Readers interested in further considerations of the differences between CI and CC and the relations of the latter to perturbation theory are referred to Refs. [14, 20, 21].

While CI, MBPT, and CC are all wave function methodologies, another very common alternative in trying to solve the problem is using Density Functional Theory (DFT) [22]. DFT in its Kohn-Sham form (KS-DFT) is formally an independent-particle theory, but unlike Hartree-Fock, it has a term that attempts to account for correlation effects, known as correlation potential (V_c), or more generally, exchange-correlation potential (V_{xc}) as it is accompanied by the exchange operator (V_x) as well. The functionals that enjoy popularity in Quantum Chemistry, grounded on the Kohn-Sham (KS) formalism, [23] are very often determined in a somewhat *ad hoc* and empirical fashion. Some of interest are the Minnesota functionals, [24, 25], the B97 family of Becke, [26] and the ωB97 series of Head-Gordon and co-workers [27]. Others, like PBE [28, 29] and the QTP family, [30, 31] try to avoid extensive and arbitrary parameterization.

5.3.2 Partitioning of the Orbital Space

When considering the excitations from a single determinant, there are other aspects when it comes to the determinants to be used in correlated calculations. Regardless of the choice of reference function, they all share in common the fact that their final product is a set of orbitals. It is instructive to further partition the orbital space beyond simply occupied and virtual because different orbitals exhibit different requirements with respect to the correlation problem, as depicted in Fig. 5.3.

Core orbitals tend to be virtually unaffected by the correlation phenomenon. It does not mean that they have negligible contribution to the correlation energy. Quite the opposite, the correlation effect in the core is rather large, but such correlation is nearly constant throughout the potential energy surface for most chemical situations, and usually cancels out if the end goal is the evaluation of some kind of energy difference [32]. Since the computational expense of correlated methods scales rather steeply with the number of orbitals to be taken into account, core orbitals are usually left out of the correlation calculation, an approach known as the frozen-core approximation. Basis sets developed under and for such conditions are named valence basis sets, as they target the so-called valence correlation.

However, some cases do require the inclusion of the innermost orbitals, which are often of higher complexity than the valence counterparts, and the frozen-core approximation is no longer valid. Correlation contributions arising from both valence and core orbitals are termed core-valence correlation, while those that are solely from core orbitals comprise the core (or core-core) correlation energy, and are both treated by basis sets named accordingly.

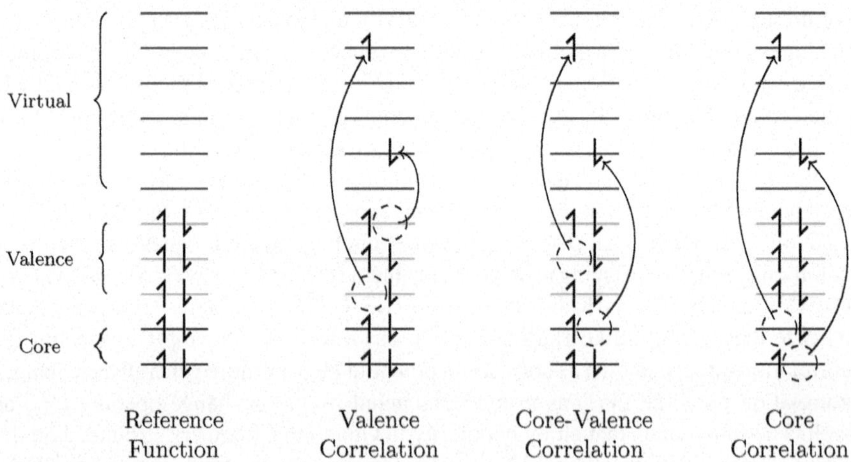

Fig. 5.3 A pictorial way to classify the orbitals and the different types of correlation according to the orbitals that are populated by the correlated electrons

Due to the orthogonality of the spherical harmonics for atoms, basis functions associated with different angular momentum quantum numbers do not mix. For practical purposes, it means that basis functions of higher ℓ-number than those of the occupied orbitals have very little effect on such orbitals. It does not mean, however, that higher angular momentum functions are not important. Since the correlation energy comes from excitations out of the reference function, many such excitations target unoccupied orbitals pertaining to higher ℓ, and an appropriate treatment of these orbitals is crucial to proper recovery of the correlation energy. In the basis set terminology, these functions are termed polarization functions, for they allow the electronic cloud to be distorted and more easily accommodate different chemical environments.

5.3.3 The "Zeta" (ζ) Classification and the Complete Basis Set Limit

A convenient and systematic approach to name basis sets according to their size became known as n-tuple-ζ, or, nZ. In principle, the zeta multiplicity refers to the number of basis functions to be employed in each AO. For instance, in a single zeta (SZ) basis set, each AO is associated with only one basis function; in a double zeta (DZ) basis, there are two basis functions per each AO, and so on. Due to the form that the polarization functions are added in most basis sets, the zeta classification also became a synonym for how the basis set varies with an increment in the highest ℓ number that the basis sets spans. So, the zeta multiplicity is closely related to the function associated with the maximum angular momentum quantum number, ℓ_{max}, in the basis set.

In the most general case, a one-particle basis like those in question here would have to be comprised of an infinite number of functions in order to be an exact representation of an N-particle full CI wave function. Finite basis sets are devised aiming at functions that can be taken as suitable truncations of infinite sets.

Even though the answer from an exact basis set is technically inaccessible, it can still be estimated with excellent accuracy. Fortunately, the increase in the size of a basis set comes with a consistently diminishing improvement. In practical terms, it means that finite basis sets can be extended in a way to asymptotically approach the complete basis set (CBS) limit. Several basis sets discussed below are constructed in a hierarchical fashion providing smooth convergence to the CBS limit. Studies performed in different contexts have shown that the correlation energy varies as some inverse function of the ℓ_{max} [33–39]. This dependency is usually ℓ_{max}^{-3} for typical correlated methods, but is dramatically improved to ℓ_{max}^{-7} for explicitly correlated approaches. Since the energy in SCF and DFT, the most common single determinant theories, displays an exponential dependence on both ℓ_{max} and the number of functions for a given ℓ, it converges much more rapidly to the CBS limit than correlated properties with respect to the basis set [40–43]. That being said, it is advised that

Table 5.1 Some of the most common CBS extrapolation formulas, where $E(X)$ is the correlation energy for a basis set of X-tuple zeta size, E_{CBS} is the extrapolated energy to the CBS limit to be determined along with the parameters A and B, and E^{total} is the total energy

Extrapolation formula	Fitting parameters	References
$E(X) = E_{CBS} + AX^{-3}$	E_{CBS}, A	[44]
$E(X) = E_{CBS} + Ae^{-BX}$	E_{CBS}, A, B	[41]
$E(X) = E_{CBS} + Ae^{-(X-1)} + Be^{-(X-1)^2}$	E_{CBS}, A, B	[46]
$E(X) = E_{CBS} + A\,(X + 1/2)^{-4}$	E_{CBS}, A	[48]
$E(X) = E_{CBS} + AX^{-B}$	E_{CBS}, A, B	[50, 53]
$E(X) = E_{CBS}\left[1 + A(1 + B/X)X^{-3}\right]$	E_{CBS}, A, B	[51]
$E(X) = E_{CBS} + (X + A)^{-3}$	E_{CBS}, A	[52]
$E(X) = E_{CBS} + AE_X^{total}X^{-3}$	A	[54]
$E(X) = E_{CBS} + Ae^{-a\sqrt{X}}$	A	[55]

the SCF energy and its associated correlation correction be treated separately when CBS limit estimates are sought. Several accounts report approaches for extrapolation of energies and other properties to the CBS limit, [44–55] with some of the most commonly used reported in Table 5.1.

This is certainly not an extensive list of all extrapolation approaches reported in the literature. These formulas are usually conceived based on a specific set of parameters, such as method, basis set family, and molecular property, so they can display strikingly different behavior from case to case. The specifics of each formula can be found in the corresponding references in Table 5.1 and more involving discussions regarding performance considerations have been reported by Feller and co-workers [56, 57].

5.3.4 *Diffuse Functions*

The plain valence correlation basis sets have been shown to be suitable for a wide range of applications that can be understood as rising from the valence shell. However, situations where basis functions limited to just the valence region provide an insufficient picture are not unusual. One such case is when the phenomenon under investigation has a long-range nature, which takes place beyond the valence region, and a well-established approach to addressing this deficiency is the inclusion of Gaussian functions whose exponents are rather small, so that their tail is relatively slowly decaying. These are usually referred to as diffuse functions, and the respective basis sets are commonly named augmented, as they are extended sets with added diffuse functions on pre-existent bases.

One common problem where one needs to resort to augmented basis sets would be anions and the corresponding electron affinities. The extra electron in anionic species is usually loosely bound in comparison to the corresponding neutral, and for that reason, tend to dwell in regions rather distant from the nuclei, with diffuse functions being required to describe the spatial domain occupied by these electrons. Another important context where the use of augmented basis sets is crucial is the accurate determination of Rydberg states, whose distinct characteristic is that electrons are excited to orbitals located well beyond the valence shell, usually tens or hundreds of Å away from the nuclei. This is very often spatially inaccessible for Gaussians obtained from calculations on the neutral atoms, as their tails die off much closer to the nuclei. Finally, a myriad of problems of central interest in Chemistry rises from dispersive van der Walls forces. If one is interested in the moments of a charge distribution, one would expect to need more diffuse functions. Both constitute classic examples of the demand for basis sets with additional diffuse functions in order to achieve high accuracy.

5.4 Correlated Basis Sets

With the most important concepts pertinent to basis sets explained in the previous sections, we can now delve into the correlated basis sets themselves and be able to explain their main features and differences.

5.4.1 Pople-Style Basis Sets

The exponents and contraction coefficients in the first incarnation of the basis sets proposed by Pople and co-workers were initially determined at the HF level, meaning they were not designed in order to include electron correlation [58]. In fact, they fall into the category of minimal basis sets (MBS), i.e., the number of basis functions equals the number of occupied orbitals of the corresponding atoms. The HF energy is dependent only on the occupied orbitals, hence the MBS being unable to account for correlation effects. However, bond formation in molecules require a significant charge polarization, an ingredient that is missing in the atomic MBS. This was remedied through the addition of polarization functions. It turns out that these extra functions, while obviously not the best functions for correlation purposes, can capture some electron correlation upon the fact that they extend the determinant space to include virtual orbitals. In general, the Pople-style basis sets are labeled m-ijG, meaning that the core orbitals are represented by a contraction of m Gaussians and that the valence orbitals are comprised of i and j contracted Gaussians (even though this example suggest two valence orbitals, it can be extended). The versions of these bases extended by polarization functions are designated by an asterisk or by the angular momentum of the functions being added, e.g., 6–31G* and 6–31G(d) refers to the same basis set [59]. For computational purposes, the Pople-style basis sets assume that the exponents of the $2s$ and $2p$ functions are the same, as the integrals can be computed faster this way. This is a restriction that other basis sets will not have.

Pople and co-workers eventually proposed a version of these basis sets that were correlated from the beginning via optimization of Gaussian exponents and contraction coefficients at the MP2 level with an unrestricted Hartree-Fock reference (UMP2) [60]. Very large versions of these bases can be found, with the largest being 6–311++G(3df, 3pd), where "++" means diffuse functions in all atoms, including hydrogen and helium, and heavy atoms count with three d and one f polarization functions and hydrogen and helium basis sets are increased by three p and one d polarization functions.

These basis sets are still used routinely in correlated calculations. Arguably, the main pitfall of the Pople-style basis sets is the lack of hierarchical design such that the progression to larger sets is closely connected to the asymptotic behavior to the CBS limit. On the other hand, two of the most famous composite schemes, the so-called Gaussian-n (Gn) theories of Pople et al. [61–64] and the CBS methods of Petersson and co-workers, [65–69] make use of these bases. The former is entirely dependent on the Pople-style basis sets, while the latter rely on the Pople-style basis sets in some of its several variants.

5.4.2 Atomic Natural Orbitals

There are several examples of the application of natural orbitals (NO) in the literature. The NOs are based on the insight that dates back to the seminal work of Löwdin [70]. As a first step toward the generation of a set of NOs, one computes the one-particle density matrix (1PDM) at a certain correlated level, simply referred to as density matrix hereafter, on a certain basis. This density matrix is diagonalized and a new MO basis can be obtained in which the eigenvectors can be viewed as coefficients of the expansion in terms of the original basis. The NOs have the property that, in the absence of using all MO functions, they effectively reduce the number of relevant orbitals for a system in question, improving convergence to the desired wave function. In other words, for an N-dimensional wave function, the K most occupied NOs provide the orbitals whose determinant best describes the 1PDM of the system, with $K < N$, and also form the K-dimensional determinant that most rapidly converge to the FCI expansion, limit which exactly solves the correlation problem, as exposed in Sect. 5.3.

Turning to atomic basis sets, if the basis representing the correlated 1PDM is a set of atomic primitive Gaussians, the corresponding eigenvectors, which are the atomic natural orbitals (ANO), provide coefficients through which the primitive set can be expanded, forming a new atomic basis set. For a primitive basis comprised of N elements, N ANOs are generated, and both bases span the exact same space, meaning that keeping the N ANOs leads to no error in contracting the primitive set. However, the primitive basis can be effectively contracted to K ANOs, $K < N$, reducing the associated computational cost relative to the primitive set. The corresponding eigenvalues are interpreted as the occupation number of a given ANO and usually guide the choice of the K most occupied ANOs to be retained in the ANO basis set.

An ANO basis set is naturally of the general contraction type, meaning that each primitive function contributes to all contractions. However, due to the orthogonality of the spherical harmonics, there is no mixing of functions associated to different ℓ numbers, meaning that the density matrix elements that arise from functions with different angular momentum quantum numbers are zero. For this reason, the ANOs automatically arrange themselves in the corresponding ℓ-shells. Below, we show some of the most notable cases of ANO basis sets.

5.4.2.1 NASA Ames

The first application of the ANO approach to the generation of atomic basis sets came from Almlöf and Taylor, labeled as NASA Ames ANO, with the focus on the valence correlation in atoms of the first row of the periodic table [71, 72]. The s and p shells are also spanned by the SCF-optimized Gaussians of van Duijneveldt, [73] and the polarization functions calculated from even-tempered expansions whose parameters were optimized at the CISD level of theory with predetermined factors between consecutive ℓ-shells. After the determination of the primitive set, the calculation of the 1PDMs of the corresponding atoms was also carried out at the CISD level, yielding the contraction coefficients.

These basis sets are available only in quadruple-zeta size, meaning that, while being relatively large, they do not follow a hierarchical scheme that allows for extrapolations to the CBS limit. Despite this shortcoming, the use of these basis sets has been reported to provide accurate results in several applications, notably in spectroscopic calculations, yielding results with significant accuracy and in cases where members of the correlation consistent family fail [74, 75]. Atomic and molecular properties which depend considerably on the outermost regions of space require further flexibility, which can be achieved by having the primitives with the smallest exponents be uncontracted, or even the addition of extra diffuse functions [76].

5.4.2.2 Roos' ANO

In the pursuit of a general purpose basis set, one can naively keep increasing the size of the basis set, guided by the hope that an infinite basis is exact and that a realistic basis large enough will be successful for the problem at hand. Given the slow convergence of the correlation energy with the size of the basis set, this is not an advisable strategy. Atomic basis sets tend to be obtained by some sort of optimization of an atomic property, usually an energetic quantity, and for this reason are inherently biased toward this quantity. It was seen above Sect. 5.4.2.1 that Almlöf and Taylor prescribed modifications in their basis set to enhance the performance in the calculation of certain properties. Roos and co-workers, on the other hand, devised a set of ANOs in which the effect of certain features was believed to be already incorporated into the contraction coefficients [7].

As in the case of the NASA Ames Sect. 5.4.2.1 bases, these basis sets also start with the SCF-optimized primitives of van Duijneveldt, [73] which are further augmented by primitives derived from a variant of the even-tempered expansion that depends on three parameters of the following form:

$$\ln(\alpha_k) = c_{-1}/k + c_0 + c_1 k \quad , \quad k = 1, \ldots, N \tag{5.7}$$

where c_{-1}, c_0, and c_1 are the expansion coefficients to be determined.

To these primitives are also added a diffuse function in all ℓ-shells. Once the primitives are generated, the distinctive feature of these basis sets come into play. Instead of straightforward computation of the atomic density matrix and the subsequent gathering of the most occupied ANOs, the 1PDMs of several relevant atomic states, the corresponding anions and cations, and the atom in the presence of an electric field are averaged and subject to the diagonalization step. This process was intended to allow for a more balanced performance over a wide variety of chemical situations for atoms in the first and second row of the periodic table [7, 77].

This framework is slightly modified in regard to the 1PDM to be considered in the case of transition metal atoms. The three most important electronic configurations in these atoms, d^n, $d^{n-1}s^1$, and $d^{n-2}s^2$ are considered in the density matrix averaging, along with the anion and cation in an external electric field. These basis sets are reported to provide very small contraction errors for most of the properties for which they were tested, such as ionization potentials, dipole moments, and spectroscopic constants [78].

This prescription for basis set development has also been successfully developed in the context of capturing relativistic effects [79–83]. Further investigation regarding the specifics of these basis sets is beyond the scope of the present document.

5.4.2.3 ANO-pV*n*Z

The distinct characteristic of the ANO basis sets of Neese and Valeev in comparison to the bases previously discussed is that they are not dependent in any way on the CISD method. In their take on the ANO procedure, ano-pV*n*Z, the considerably large cc-pV6Z primitive sets serve as the basis for the calculation of the 1PDM at the Multireference Averaged Coupled Pair Functional (MR-ACPF) level of theory, [84] with a Complete Active Space SCF (CASSCF) reference [85]. Given the extensive size of the primitive set, basis sets of sizes up to hextuple zeta can be obtained from a single 1PDM computation.

In the same paper where their constructions are detailed, the ano-pV*n*Z sets are tested for a variety of situations and are found to perform better than the cc-pV*n*Z bases of the same size and are reported to be superior to basis sets of the def2 and pc-*n* families for SCF and CCSD(T) energies as well as in extrapolations to the CBS limit.

5.4.2.4 ANO-VT-*n*Z

The Coupled-Cluster method in which all single and double excitations are considered as well as the contribution from triple excitations as estimated from perturbation theoryr, CCSD(T), is nowadays referred to as the "gold standard" of Quantum Chemistry. Despite the constant enhancement in computational power, allowing CCSD(T) to be used in an ever-growing number of applications, there had been no basis set in the literature based on this method, or any version of CC, until recently. The MR-ACPF method used above and its updated Multireference Averaged Quadratic Coupled-Cluster (MR-AQCC) [86] version are actually multireference variants of CI that partly correct for its lack of size-extensivity.

The Atomic Natural Orbital—Virial Theorem—*n*-tuple Zeta sets, ANO-VT-*n*Z, with n = D, T, and Q, of Claudino et al. [87, 88] is an attempt to address this gap, providing a basis set with all the benefits of CC theory. The highlight of these ANO-type bases is that they are obtained from the 1PDM calculated at the CCSD(T) level of theory with the primitives found in the correlation consistent sets of the same size. Moreover, they satisfy the Virial theorem for the corresponding atoms via scaling the primitives exponents in the original primitive set by a uniform scale factor [89].

In comparison with the cc-pV*n*Z counterparts of the same size, the ANO-VT-*n*Z bases were shown to provide smaller errors in the contraction of the primitives for atomic and molecular energies and the density, as probed by the moments of r, e.g., $\langle r^n \rangle, -3 \leq n \leq 3$.

5.4.3 Segmented Basis Sets Based on Atomic Natural Orbitals

5.4.3.1 Natural Orbital-Based Segmented Contracted Gaussian Basis Sets

The ANO paradigm relies on the desired physical feature that natural orbitals offer the fastest orbital convergence to the target wave function, while these basis sets are a notorious example of the general contraction scheme. Despite the current technology having made possible efficient integral calculations on generally contracted basis sets, segmented contractions are naturally more computationally amenable. In order to take advantage of both of these properties, Noro and co-workers introduced the Natural Orbital-based Segmented Contracted Gaussian Basis Sets, or NOSeC, for short [90].

The construction of the NOSeC basis sets starts with the functions of Tatewaki and Koga (TK) optimized to minimize the error from Numerical Hartree-Fock (NHF) energies [91–94]. The polarization functions intended to recover correlation energy are not optimized with the goal of minimizing the CISD energy, but rather exponents and contraction coefficients are optimized in order to minimize the difference in the radial distribution from a predetermined subset of the highest occupied CISD NOs. In other words, the polarization functions are obtained by fitting to CISD ANOs in a least-squares fashion.

The different versions of the NOSeC basis sets are reported to cover all 103 atoms from hydrogen to lawrencium. Valence correlated NOSeC bases are available in double, triple, and quadruple sizes for atoms H through Xe, [90, 95–99] and core-valence functions for alkali and alkaline metals are also provided [100]. Atoms beyond Xe in the periodic table are very dependent of basis sets specifically design to account for the high relativistic character of such heavy atoms and have also been addressed by the NOSeC group, discussion of which can be found elsewhere [100–105].

5.4.3.2 Sapporo Basis Sets

The NOSeC and Sapporo families of basis sets are very closely related. Both share the same primitive functions, with slightly different contraction patterns in the inner shells, but have the same basis functions in higher ℓ-shells. This alternative contraction scheme in the shells of lower ℓ values in the Sapporo sets is able to provide moderate additional energy improvement over the NOSeC analogs. The polarization functions are the same as those found in the NOSeC basis sets.

The initial procedure employed in the generation of the basis sets above is refined to better accommodate core-valence correlation effects. The functions to populate the shells containing AOs to be occupied are initially the same ones used in the first record of the Sapporo family. Since these functions are now found in the angular momentum shells where core correlation takes place, they are recontracted and, in some cases, it is found that the addition of extra s, p functions to the TK sets is necessary. The polarization functions are also primarily those found in the NOSeC sets. With the exception of the atoms in the s block, in which the core-valence aspect was previously addressed, [100] all other atoms require polarization functions optimized for a proper description of core-valence correlation. In order to avoid linear dependencies, the extended TK set and the polarization functions are compared and similar functions are removed from the final basis. This protocol leads to the Sapporo-nZP-2012 family, available for basis set of n = D, T, and Q, that is, double, triple, and quadruple-zeta sizes [106].

The construction of the Sapporo-nZP basis sets is also accompanied by a test of the percentage of correlation energy recovery in the atoms for which they were generated and performance in the computational of spectroscopic constants r_e, ω_e, and D_e at the CCSD(T) level of theory. As expected, it is shown that the improvement is consistently achieved as the basis set becomes larger, but convergence may be slow, as hinted by the studies on the extrapolation to the CBS limit in the case of HF energy that shows a large deviation of the former in comparison to Numerical HF results for a variety of extrapolation schemes [107].

5.4.4 Correlation Consistent Basis Sets

The correlation consistent basis sets were first introduced by Dunning, [108] and are arguably the most popular and widely used bases in correlated calculations. Part of their success can be attributed to the wide range of flavors to accommodate different chemical demands and the convergence to the CBS limit that is incorporated by construction.

The starting point in the generation of the correlation consistent polarization valence n-tuple zeta, more readily identified by the acronym cc-pVnZ, is a set of primitive Gaussians with exponents optimized in order to provide the lowest SCF energy, obtained from the work of van Duijneveldt [73]. This primitive set is subsequently contracted using the SCF coefficients of the occupied MOs based on a restricted open-shell HF (ROHF) reference. This idea guarantees that the SCF energy of the contracted set reproduces that of the primitive set with no contraction error. The exponents of the polarization functions to be added are those which minimize the CI with single and double excitations (CISD), which constitutes a truncation of the FCI expansion for all atoms with more than two correlated electrons.

This general framework was initially applied in basis sets devised to provide a satisfactory description of valence correlation [108–113]. The name "correlation consistent" has its origins in the fact that functions are added on top of the contracted SCF primitive set following a very specific pattern. The addition of the first polarization function recovers approximately the same amount of correlation energy that one extra function in each of the angular momentum shells already populated by the occupied orbitals. The polarization function with angular momentum number incremented in 1 compared to the first polarization function recovers roughly the same amount of correlation energy that an extra function added in all lower angular momentum shells on top of the previously introduced primitives. Using the oxygen atom as an example to illustrate the process of construction of these sets, as also done by Dunning, a $(9s4p)$ primitive set optimized at the SCF level is contracted to $[2s1p]$ with the ROHF vectors as contraction coefficients, which is, in turn, an MBS for this atom. It is then found that one extra $1d$ function provides approximately the same CISD valence correlation energy recovery that additional $1s$ and $1p$ functions, becoming $[3s2p1d]$. Alternatively, it is also systematically labeled as cc-pVDZ. The next set on the series, cc-pVTZ, is comprised of extra an $(1d1f)$ set as well as one more function of s and p type, therefore determining the triple-zeta member of the family.

Once the main idea behind the construction of the correlation consistent basis sets was established, it was used as the scaffold for the construction of several other variations suited for different problems. The cc-pCVnZ was designed to treat core-valence correlation, [114, 115] which was later found to be biased toward core correlation and was corrected by a weight factor, introducing the cc-pwCVnZ series [116–118]. Concurrently, situations that demand diffuse functions can be treated by an augmented variant of the cc-pVnZ family, aug-cc-pVnZ [110, 112, 119]. These two distinct features are also combined in the aug-cc-p(w)CVnZ series, as in the case of transition metals [118].

The rationale underlying the construction of the correlation consistent basis set naturally leads them to a smooth extrapolation to the CBS limit by several of the formulas shown in Table 5.1. This feature is extensively explored by Martin and co-workers in their composite method known as Weizmann-n (Wn), [120–122] by the HEAT protocol targeting highly accurate thermochemistry, [123–125] and the correlation consistent's own composite scheme, ccCA [126].

The concept that forms the backbone of the correlation consistent family have also found application in basis sets that target relativistic effects (cc-pV*n*Z-DK), [117, 118, 127] usage in conjunction with pseudopotentials (cc-pV*n*Z-PP), [128–132] and also employed in explicit correlated calculations (cc-pV*n*Z-F12) [115, 133, 134]. Auxiliary basis sets, which are functions to be used in approximations to the two-electron, four-center integrals, such as resolution-of-the-identity and density-fitting, have also been reported to be constructed based on the concept of correlation consistent basis sets [135, 136]. These cases are beyond the scope of the present document.

5.4.5 *n-tuple-Zeta Augmented Polarized Basis Sets*

One common factor in the first basis sets proposed by Dunning 5.4.4, [108] Almlöf and Taylor Sect. 5.4.2.1, [71] and Roos et al. 5.4.2.2 [7] is that the SCF solution they provide originates from the SCF-optimized primitives of van Duijneveldt [73]. Full optimization of the primitives can be a lengthy and demanding process, and depending on the level of accuracy being sought, many primitives may be required, which adds to the difficulty of the problem. This obstacle can be partially overcome with a suitable expansion, like the even-tempered one discussed in Sect. 5.2.1. As the primitive grows, the exponents consistently deviate from the individually optimized Gaussians and also tend to be so closely spaced that numerical instability rises due to the presence of linear dependencies in the basis, meaning that two basis functions are considered numerically identical [137].

In order to generate large primitive sets that do not run into the linear dependency problem and do not demand the amount of effort associated with independent optimization, Petersson et al. developed a scheme whereby the exponents are given by an expansion in terms of Legendre polynomials, which can also be viewed as a generalization of the even-tempered expansion by the inclusion of more terms [8].

$$\ln\alpha_j = \sum_{k=0}^{k_{\max}} A_k P_k\left(\frac{2j-2}{N_{\text{prim}}-1} - 1\right) \tag{5.8}$$

where α_j is the exponent of the j-th primitive, $0 \leq j \leq N_{\text{prim}} - 1$, $k_{\max} = 1$ reduces to the even-tempered expansion, and $k_{\max} = N_{\text{prim}}$ is a fully optimized primitive set, A_k is the kth coefficient of the expansion of the Legendre polynomials P_k with argument $\frac{2j-2}{N_{\text{prim}}-1} - 1$.

This new alternative in the construction of the primitive set is the foundation of the n-tuple-Zeta augmented Polarized basis sets, nZaP. A hierarchy of basis sets, ranging from 2ZaP to 6ZaP, have been reported, with valence primitive sets of varying sizes obtained from the Legendre polynomial expansion and contracted according to the segmented scheme. The criterion in determining the size of these primitive sets is based on thresholds in how much the SCF energy per electron deviates from that obtained by Numerical Hartree-Fock. This new requirement allows for sounder basis sets by accommodating the increasing complexity of atoms in a given row of the periodic table, as opposed to the correlation consistent bases that exhibit the same number of functions across a given row of the periodic chart. The split-valence and polarization functions are subsequently added in the same manner as in the correlation consistent series, except that here these functions are derived from an even-tempered expansion optimized at the UMP2 level (MP2 based on an unrestricted Hartree-Fock reference) instead of CISD. Given that polarization sets tend to be restricted to relatively few functions, they are not usually plagued with linear dependencies and lesser expansions are satisfactory.

In a set of 72 atoms, ions, diatomics, and hydrides containing first- and second-row atoms, very accurate and reliable MP2 energies followed from extrapolation to the CBS limit using the *n*ZaP series when compared to high-level numerical atomic calculations and explicitly correlated (F12, R12-A, and R12-B) results [138]. Similar investigation of each component of the correlation energy, that is, the convergence of the MP2, CCSD and (T) energy terms in an extended set is reported with CCSD(T) energies displaying 0.1% root-mean-square deviation with respect to F12b energies [139]. Revamping the *n*ZaP basis sets to accommodate core-valence correlation led to the same level of accuracy (0.1%) for ionization potentials, electron affinities, and total atomization energies in an even larger test set [140].

5.4.6 Oliveira and Jorge's nZP

Another take on correlated basis sets that take advantage of the segmented contraction scheme was formulated by Oliveira, Jorge, and co-workers. In this approach, the primitive set responsible for the Hartree-Fock solution is rather small, and for this reason can be fully optimized in a computationally affordable manner. To this end, it is employed the improved generator coordinate Hartree-Fock technique, which very successfully approaches NHF results [141]. Once the inner-shell functions are determined and contracted, polarization functions are added by means of optimization with respect to the MP2 energy based on a ROHF reference (ROMP2).

Having laid out the basic framework for the construction of such basis sets, Jorge and co-workers have generated a comprehensive collection of basis sets that cover the vast majority of elements in the periodic table, from the hydrogen atom to the actinide series. These functions are available in valence, augmented, and relativistic versions, which are collectively labeled as (A)*n*ZP-(DKH), where "A' indicates the presence of diffuse functions and "DKH" means that the basis sets are recommended for inclusion of relativistic effects through the Douglas-Kroll-Hess Hamiltonian.

Currently, there are *n*ZP bases for the first- and second-row elements that range from double to hextuple quality [142–145]. Third-row atoms, e.g, K through Kr, are also covered by Jorge's *n*ZP basis sets, but instead limited to *n*= D, [146] T, [147] and Q [148]. For elements in the fourth row of the periodic table, the available *n*ZP sets are of double- [149] and triple-zeta qualities [150]. These are also the same sizes that are currently accessible for fifth-row atoms and the lanthanide and actinide series. These atoms fall into the category of "heavy atoms", which require basis sets appropriate for relativistic calculations and will not be discussed here.

One central concern that has motivated Jorge and co-workers in the development of the augmented versions of the valence *n*ZP bases, which led to the A*n*ZP series, was the proper description of electric properties. From the beginning, the DZP group was built including exponents optimized for the anions of corresponding atoms in order to include diffuse functions [142]. Following this idea, augmented variations of the initial scheme are also available for atoms H and Li through Ar, [151, 152] K through Kr, [153, 154], and Rb through Xe [154, 155].

The ability of the (A)*n*ZP in the evaluation of atomic and molecular properties was extensively probed and benchmarked, including the convergence to the CBS limit as the basis set is increased in size via increments in the *n* number. Jorge et al. have reported such studies for the convergence of dipole moments and polarizabilities in nucleic acid bases based on the B3LYP functional, for which the double-zeta set is believed to provided satisfactory results and [156] nuclear magnetic shielding constants were investigated at the HF and MP2 levels, requiring larger basis sets and CBS extrapolations to achieve reliable numbers [157]. The basis convergence toward the limit of infinite basis is also examined in the case of alkali-metal clusters with HF, B3LYP, and MP2 methods, and it is found that the A*n*ZP basis sets converge very slowly to experimental values in this context [158].

5.4.7 Polarization Consistent Basis Sets

So far, we have only dealt with basis sets grounded on wave function methods. In those cases, one starts with a reference function that (often) lacks correlation, most commonly HF, but from which a multitude of excited determinants can be accessed, which are in turn responsible for the inclusion of correlation effects, as in MBPT, CI, and CC theories. Due to the number of determinants from the reference function that are usually required to reach a reliable and accurate picture of the correlation in the system, the computational cost associated with these approaches is remarkably high, in the range of orders of magnitude higher than that to evaluate the reference function. Additionally, it is well known that the convergence towards the FCI limit with respect to the basis set is relatively slow within the wave function paradigm, currently limiting the scope of applicability of these methods to small-to-medium size cases.

On the other hand, it is well known that the basis set dependence for a one-particle operator is quite different from that for a many-particle one, converging rather quickly to the CBS limit. With that in mind, Jensen proposed a family of basis sets similar in spirit to the correlation consistent series, and whose distinguishing feature is being tailored to exploit the faster convergence of the energy of a single determinant with respect to the basis set size. In practical terms, it means that relatively small sets are able to successfully saturate the single determinant space, more rapidly approaching the CBS limit [159, 160]. This is closely related to the fact that these methods often emphasize the importance of lower angular momentum functions in comparison with those associated with larger ℓ values, which tend to be cumbersome computationally.

Despite the analogous single determinantal nature of HF and DFT, the exchange in the latter is replaced by an operator responsible for both exchange and correlation, V_{xc}. Its effect on the exponential convergence of DFT seems to be inconsequential for most functionals, as shown for the BLYP functional [161, 162]. Furthermore, the exponents optimized for DFT with the BLYP functional exhibit little deviation from those obtained with HF, even with the presence of electron correlation in the former, [163] and this is also observed for different choices of functional [164]. Situations where the accurate description of the wave function tail is required follows the common procedure of augmenting the basis set with diffuse (small exponent) functions, as in the case of anions and electric properties, giving rise to the aug-pc-n variant in close resemblance to the augmented correlation consistent series [165]. Additionally, the convergence profile of structural parameters, vibrational frequencies, and infrared intensities, within the pc-n family is investigated and shown to benefit from being designed for single determinant methods, as it more rapidly approaches the KS CBS limit than other functions of similar size [166].

Polarization consistent basis sets have been developed for all atoms H through Kr, in sizes that vary from $n = 0$, which has no polarization functions, to $n = 4$, comparable in size to cc-pV5Z, but demonstrated to be superior to cc-pV6Z in HF and DFT calculations. Further discussion on the particular considerations of each atom can be found in the later papers on the series [167–170]. These sets were at first generally contracted, and upon evaluation of an alternative segmented scheme, [171] the segmented variants pcS-n [172] and pcSseg-n [173] have also been proposed, with the latter aiming at nuclear magnetic shielding constants.

With the proven feature of faster convergence to the CBS limit, in conjunction with the ever-growing community of DFT users, this feature of the polarization consistent basis sets has been explored and benchmarked in a variety of different contexts, such as harmonic frequencies and anharmonic corrections [174–179], zero-point corrections to nuclear magnetic shielding constants and anisotropies, [180] spin-spin coupling constants, [181–184], molecular binding energies, [185, 186] and assessment of functionals [187, 188].

5.5 Conclusion

With the virtually infinite number of combinations of chemical species, correlated methods, and molecular properties, there is an overwhelmingly large variety of basis sets currently at one's disposal. Over the years, some of these bases have received more attention than others and the list above is by no means comprehensive of all basis sets currently available. The basis sets that we cover in this manuscript are

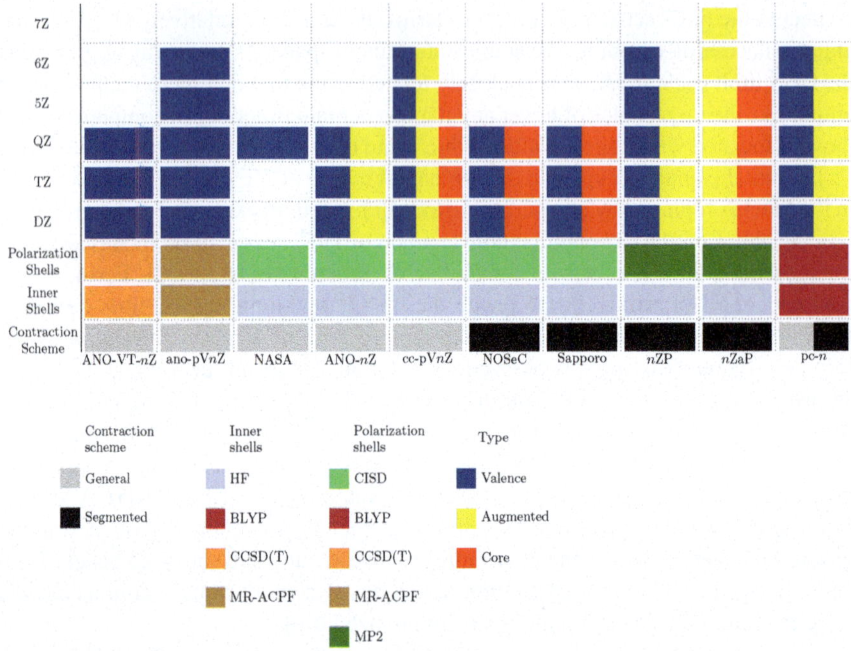

Fig. 5.4 Graphical summary of the discussed basis sets for correlated methods

those which we consider to be extensively used in correlated calculations and have also brought some novel aspects to basis set development.

Readers interested in other considerations that are important to the developments of basis sets or more detailed discussion on specifics of a particular type are referred to some reviews on these topics [72, 137, 189, 190]. Almost all basis sets presented above can be found on the EMSL Basis Set Exchange website (https://www.basissetexchange.org/) [191–193], with the exception of ano-pV*n*Z, *n*ZP, and NOSeC/Sapporo, for which links are provided in the respective references.

The different varieties and sizes of the basis sets discussed throughout this chapter are represented in the chart displayed in Fig. 5.4.

Quantum Chemistry has been based on the use of atomic basis sets centered on each atom in a molecule since its inception. Given that the alternative numerical solution for electron correlation has never been satisfactorily achieved (see Refs. [194–198] for some examples), it appears that basis sets of Gaussian type will be a part of most electronic structure calculations for the indefinite future. Nearly all useful variants have been considered, but introducing more accurate primitives, and improved contractions of those primitives, still can offer improvement. We will present such a new frozen atomic natural orbital (FANO) basis in a future paper [199]. It will be obtained from CCSD(T) density matrices with consideration of multi-atomic behavior.

Nonetheless, the use of atomic basis sets that can be extrapolated to the CBS limit along with CC theory has solved many problems that could not be solved even ten years ago. And such applications will continue to multiply in the foreseeable future.

References

1. Slater JC (1930) Phys Rev 36:57
2. Kato T (1957) Commun Pure Appl Math 10:151
3. Boys SF (1950) Proc Roy Soc [London] A 200:542
4. Stewart RF (1970) J Chem Phys 52:431
5. Ruedenberg K, Raffenetti RC, Bardo RD, In: Smith DW (ed) Proceedings of the 1972 boulder seminar research conference on theoretical chemistry
6. Huzinaga S, Klobukowski M (1985) Can J Chem 63:1812
7. Widmark PO, Malmqvist PA, Roos BO (1990) Theor Chim Acta 77:291
8. Petersson GA, Zhong S, Montgomery JA Jr, Frisch MJ (2003) J Chem Phys 118:1101
9. Feller D, Ruedenberg K (1979) Theor Chim Acta 52:231
10. Schmidt MW, Ruedenberg K (1979) J Chem Phys 71:3951
11. Dunning TH Jr (1970) J Chem Phys 53:2823
12. Raffenetti RC (1973) J Chem Phys 58:4452
13. Löwdin PO (1955) Phys Rev 97:1509
14. Shavitt I, Bartlett RJ (2009) Many-body perturbation theory and coupled-cluster methods in chemistry and physics. Cambridge University Press, Cambridge, MA
15. Møller C, Plesset MS (1934) Phys Rev 46:618
16. Bartlett RJ, Silver DM (1975) J Chem Phys 62:3258
17. Urban M, Noga J, Cole SJ, Bartlett RJ (1985) J Chem Phys 83:4041
18. Raghavachari K, Trucks GW, Pople JA, Head-Gordon M (1989) Chem Phys Lett 157:479
19. Watts JD, Gauss J, Bartlett RJ (1993) J Chem Phys 98:8718
20. Bartlett RJ (2010) Mol Phys 108:2905
21. Bartlett RJ, Musial M (2007) Rev Mod Phys 79:291
22. Hohenberg P, Kohn W (1964) Phys Rev 136:B864
23. Kohn W, Sham LJ (1965) Phys Rev 140:A1133
24. Zhao Y, Schultz NE, Truhlar DG (2005) J Chem Phys 123:161103
25. Zhao Y, Truhlar DG (2010) Theoretical and computational methods in mineral physics: geophysical applications. Mineralogical Society of America, Chantilly, VA; Reviews in Mineralogy and Geochemistry vol 71, Chap. The Minnesota Density Functionals and Their Applications to Problems in Mineralogy and Geochemistry, p 19
26. Becke AD (1997) J Chem Phys 107:8554
27. Chai JD, Head-Gordon M (2008) J Chem Phys 128:084106
28. Perdew JP, Burke K, Ernzerhof M (1996) Phys Rev Lett 77:3865
29. Perdew JP, Burke K, Ernzerhof M (1997) Phys Rev Lett 78:1396
30. Verma P, Bartlett RJ (2014) J Chem Phys 140:18A534
31. Jin Y, Bartlett RJ (2016) J Chem Phys 145:034107
32. Helgaker T, Jørgensen P, Olsen J (2000) Molecular electronic-structure theory. Wiley, Chichester
33. Schwartz C (1962) Phys Rev 126:1015
34. Hill RN (1985) J Chem Phys 83:1173
35. Kutzelnigg W (1985) Theor Chim Acta 68:445
36. Kutzelnigg W, Morgan JD III (1992) J Chem Phys 96:4484
37. Kutzelnigg W, Morgan JD III (1992) J Chem Phys 97:8821
38. Wang C (2013) Phys Rev A 88:032511
39. Wang C (2015) Phys Rev A 91:069904

40. Jensen F (1999) J Chem Phys 110:6601
41. Halkier A, Helgaker T, Jørgensen P, Klopper W, Olsen J (1999) Chem Phys Lett 302:437
42. Jensen F (2000) Theor Chim Acc 104:484
43. Christensen KA, Jensen F (2000) Chem Phys Lett 317:400
44. Helgaker T, Klopper W, Koch H, Noga J (1997) J Chem Phys 106:9639
45. Halkier A, Helgaker T, Jørgensen P, Klopper W, Koch H, Olsen J, Wilson AK (1998) Chem Phys Lett 286:243
46. Peterson KA, Woon DE, Dunning TH Jr (1994) J Chem Phys 100:7410
47. Martin JML, Lee TJ (1996) Chem Phys Lett 258:129
48. Martin JML (1996) Chem Phys Lett 259:669
49. Schwenke DW (2005) J Chem Phys 122:014107
50. Truhlar DG (1998) Chem Phys Lett 294:45
51. Varandas AJC (2000) J Chem Phys 113:8880
52. Huh SB, Lee JS (2003) J Chem Phys 118:3035
53. Bakowies D (2007) J Chem Phys 127:084105
54. Pansini FNN, Varandas AJC (2015) Chem Phys Lett 631–632:70
55. Zhong S, Barnes EC, Petersson GA (2008) J Chem Phys 129:184116
56. Feller D, Peterson KA, Hill JG (2011) J Chem Phys 135:044102
57. Feller D (2013) J Chem Phys 138:074103
58. Hehre WJ, Stewart RF, Pople JA (1969) J Chem Phys 51:2657
59. Ditchfield R, Hehre WJ, Pople JA (1971) J Chem Phys 54:724
60. Raghavachari K, Binkley JS, Seeger R, Pople JA (1980) J Chem Phys 72:650
61. Pople JA, Head-Gordon M, Fox DJ, Raghavachari K, Curtiss LA (1989) J Chem Phys 90:5622
62. Curtiss LA, Raghavachari K, Trucks GW, Pople JA (1991) J Chem Phys 94:7221
63. Curtiss LA, Raghavachari K, Redfern PC, Rassolov V, Pople JA (1998) J Chem Phys 109:7764
64. Curtiss LA, Redfern PC, Raghavachari K (2007) J Chem Phys 126:084108
65. Petersson GA, Tensfedlt TG, Montgomery JA Jr (1991) J Chem Phys 94:6091
66. Montgomery JA Jr, Ochterski JW, Petersson GA (1994) J Chem Phys 101:5900
67. Ochterski JW, Petersson GA, Montgomery JA Jr (1996) J Chem Phys 104:2598
68. Montgomery JA Jr, Frisch MJ, Ochterski JW, Petersson GA (1999) J Chem Phys 110:2822
69. Montgomery JA Jr, Frisch MJ, Ochterski JW, Petersson GA (2000) J Chem Phys 112:6532
70. Löwdin PO (1955) Phys Rev 97:1474
71. Almlöf J, Taylor PR (1987) J Chem Phys 86:4070
72. Almlöf J, Taylor PR (1991) Adv Quantum Chem 22:301
73. van Duijneveldt FB (1971) IBM Research Report RJ 945
74. Martin JML, Taylor PR, Lee TJ (1997) Chem Phys Lett 275:414
75. Martin JML, Lee TJ, Taylor PR (1998) J Chem Phys 108:676
76. Almlöf J, Taylor PR (1990) J Chem Phys 92:551
77. Widmark PO, Persson BJ, Roos BO (1991) Theor Chim Acta 79:419
78. Pou-Amérigo R, Merchán M, Nebot-Gil I, Widmark PO, Roos BO (1995) Theor Chim Acta 92:149
79. Roos BO, Veryazov V, Widmark PO (2004) Theor Chem Acc 345:111
80. Roos BO, Lindh R, Malmqvist PA, Veryazov V, Widmark PO (2004) J Phys Chem A 108:2851
81. Roos BO, Lindh R, Malmqvist PA, Veryazov V, Widmark PO (2005) J Phys Chem A 109:6575
82. Roos BO, Lindh R, Malmqvist PA, Veryazov V, Widmark PO (2005) Chem Phys Lett 409:295
83. Roos BO, Lindh R, Malmqvist PA, Veryazov V, Widmark PO, Borin AC (2008) J Phys Chem A 112:11431
84. Gdanitz RJ, Ahlrichs R (1988) Chem Phys Lett 143:413
85. Neese F, Valeev EF (2011) J Chem Theory Comput 7:33
86. Szalay PG, Bartlett RJ (1993) Chem Phys Lett 214:481
87. Claudino D, Gargano R, Bartlett R (2016) J Chem Phys 144:104106
88. Claudino D, Gargano R, Bartlett R (2016) J Chem Phys 145:019901
89. Löwdin PO (1959) J Mol Spec 3:46
90. Noro T, Sekiya M, Koga T (1997) Theor Chem Acc 98:25

91. Tatewaki H, Koga T (1996) J Chem Phys 104:8493
92. Tatewaki H, Koga T (1997) Theor Chem Acta 96:243
93. Koga T, Tatewaki H, Matsuyama H, Satoh Y (1999) Theor Chem Acc 102:105
94. Koga T, Yamamoto S, Shimazaki T, Tatewaki H (2002) Theor Chem Acc 108:41
95. Sekiya M, Noro T, Koga T, Matsuyama H (1998) J Mol Struct: Theochem 451:51
96. Noro T, Sekiya M, Koga T, Matsuyama H (2000) Theor Chem Acc 104:146
97. Sekiya M, Noro T, Osanai Y, Koga T (2001) Theor Chem Acc 106:297
98. Osanai Y, Sekiya M, Noro T, Koga T (2003) Mol Phys 101:65
99. Noro T, Sekiya M, Koga T (2003) Theor Chem Acc 109:85
100. Noro T, Sekiya M, Koga T (2008) Theor Chem Acc 121:289
101. Noro T, Sekiya M, Osanai Y, Miyoshi E, Koga T (2003) J Chem Phys 119:5142
102. Osanai Y, Noro T, Miyoshi E, Sekiya M, Koga T (2004) J Chem Phys 120:6408
103. Sekiya M, Noro T, Miyoshi E, Osanai Y, Koga T (2006) J Comput Chem 27:463
104. Noro T, Sekiya M, Osanai Y, Koga T, Matsuyama H (2007) J Comput Chem 28:2511
105. Sekiya M, Noro T, Koga T, Saito SL (2010) J Comput Chem 31:497
106. Noro T, Sekiya M, Koga T (2012) Theor Chem Acc 131:1124
107. Weber R, Hovda B, Schoendorff G, Wilson AK (2015) Chem Phys Lett 637:120
108. Dunning TH Jr (1989) J Chem Phys 90:1007
109. Wilson AK, van Mourik T, Dunning TH Jr (1996) J Mol Struct: Theochem 388:339
110. Woon DE, Dunning TH Jr (1993) J Chem Phys 98:1358
111. van Mourik T, Dunning TH Jr (2000) Int J Quantum Chem 76:205
112. Wilson AK, Woon DE, Peterson KA, Dunning TH Jr (1999) J Chem Phys 110:7667
113. Prascher B, Woon DE, Peterson KA, Dunning TH Jr, Wilson A (2011) Theor Chem Acc 128:69
114. Woon DE, Dunning TH Jr (1995) J Chem Phys 103:4572
115. Hill JG, Mazumder S, Peterson KA (2010) J Chem Phys 132:054108
116. Peterson KA, Dunning TH Jr (2002) J Chem Phys 117:10548
117. DeYonker NJ, Peterson KA, Wilson AK (2007) J Phys Chem A 111:11383
118. Balabanov NB, Peterson KA (2005) J Chem Phys 123:064107
119. Kendall RA, Dunning TH Jr, Harrison RJ (1992) J Chem Phys 96:6796
120. Martin JML, de Oliveira G (1999) J Chem Phys 111:1843
121. Boese AD, Oren M, Atasoylu O, Martin JML, Kállay M, Gauss J (2004) J Chem Phys 120:4129
122. Karton A, Rabinovich E, Martin JML, Ruscic B (2006) J Chem Phys 125:144108
123. Tatji A, Szalay PG, Császár A, Kállay M, Gauss J, Valeev EF, Flowers BA, Vázquez J, Stanton JF (2004) J Chem Phys 121:11599
124. Bomble YJ, Vázquez J, Kállay M, Michauk C, Szalay PG, Császár AG, Gauss J, Stanton JF (2006) J Chem Phys 125:064108
125. Harding ME, Vázquez J, Ruscic B, Wilson AK, Gauss J, Stanton JF (2008) J Chem Phys 128:114111
126. DeYonker NJ, Cundari TR, Wilson AK (2006) J Chem Phys 124:114104
127. de Jong WA, Harrison RJ, Dixon DA (2001) J Chem Phys 114:48
128. Peterson KA (2003) J Chem Phys 119:11099
129. Peterson KA, Puzzarini C (2005) Theor Chem Acc 114:283
130. Peterson KA, Figgen D, Dolg M, Stoll H (2007) J Chem Phys 126:124101
131. Figgen D, Peterson KA, Dolg M, Stoll H (2009) J Chem Phys 130:164108
132. Peterson KA, Yousaf KE (2010) J Chem Phys 133:174116
133. Peterson KA, Adler TB, Werner HJ (2008) J Chem Phys 128:084102
134. Hill JG, Peterson KA (2010) Phys Chem Chem Phys 12:10460
135. Weigend F (2002) Phys Chem Chem Phys 4:4285
136. Weigend F, Köhn A, Hättig C (2002) J Chem Phys 116:3175
137. Davidson ER, Feller D (1986) Chem Rev 86:681
138. Barnes EC, Petersson GA (2010) J Chem Phys 132:114111
139. Ranasinghe DS, Petersson GA (2013) J Chem Phys 138:144104
140. Ranasinghe DS, Frisch MJ, Petersson GA (2015) J Chem Phys 143:214110

141. Jorge FE, de Castro EVR (1999) Chem Phys Lett 302:454
142. Canal Neto A, Muniz EP, Centoducatte R, Jorge FE (2005) J Mol Struct: Theochem 718:219
143. Barbieri PL, Fantin PA, Jorge FE (2006) Mol Phys 104:2945
144. Jorge FE, Sagrillo PS, de Oliveira AR (2006) Chem Phys Lett 432:558
145. Campos CT, Ceolin GA, Canal Neto A, Jorge FE, Pansini FNN (2011) Chem Phys Lett 516:125
146. Camiletti GG, Machado SF, Jorge FE (2008) J Comput Chem 29:2434
147. Machado SF, Camiletti GG, Canal Neto A, Jorge FE, Jorge RS (2009) Mol Phys 107:1713
148. Ceolin G, de Berrêdo RC, Jorge FE (2013) Theor Chem Acc 132:1339
149. Barros CL, de Oliveira PJP, Jorge FE, Canal Neto A, Campos M (1965) Mol Phys 108:1965
150. Campos CT, Jorge FE (2013) Mol Phys 111:167
151. Fantin PA, Barbieri PL, Canal Neto A, Jorge FE (2007) J Mol Struct: Theochem 810:103
152. de Oliveira PJP, Jorge FE (2008) Chem Phys Lett 463:235
153. Camiletti GG, Canal Neto A, Jorge FE, Machado SF (2009) J Mol Struct: Theochem 910:122
154. Martins LSC, de Souza FAL, Ceolin GA, Jorge FE, de Berrêdo RC, Campos CT (2013) Comput Theor Chem 1013:62
155. de Oliveira PJP, Barros CL, Jorge FE, Canal Neto A, Campos M (2010) J Mol Struct: Theochem 948:43
156. Campos CT, Jorge FE (2009) Int J Quantum Chem 109:285
157. de Oliveira PJP, Jorge FE (2008) J Phys B: At Mol Opt Phys 41:145101
158. de Souza FAL, Jorge FE (2013) J Braz Chem Soc. 24:1357
159. Jensen F (2001) J Chem Phys 115:9113
160. Jensen F (2002) J Chem Phys 116:3502
161. Lee C, Yang W, Parr RG (1988) Phys Rev B 37:785
162. Becke AD (1988) Phys Rev A 38:3098
163. Jensen F (2002) J Chem Phys 116:7372
164. Jensen F (2005) Chem Phys Lett 402:510
165. Jensen F (2002) J Chem Phys 117:9234
166. Jensen F (2003) J Chem Phys 118:2459
167. Jensen F, Helgaker T (2004) J Chem Phys 121:3463
168. Jensen F (2007) J Phys Chem A 111:11198
169. Jensen F (2012) J Chem Phys 136:114107
170. Jensen F (2013) J Chem Phys 138:014107
171. Jensen F (2005) J Chem Phys 122:074111
172. Jensen F (2014) J Chem Theory Comput 10:1074
173. Jensen F (2015) J Chem Theory Comput 11:132
174. Buczek A, Kupka T, Broda MA (2011) J Mol Model 17:2029
175. Buczek A, Kupka T, Broda MA (2011) J Mol Model 17:2265
176. Broda MA, Buczek A, Kupka T, Kaminsky J (2012) Vib Spectrosc 63:432
177. Buczek A, Kupka T, Broda MA, Żiła A (2016) J Mol Model 22:42
178. Boese AD, Klopper W, Martin JML (2005) Int J Quantum Chem 104:830
179. Levi C, Martin JML, Bar I (2005) Int J Quantum Chem 104:830
180. Faber R, Buczek A, Kupka T, Sauer SPA (2017) Mol Phys 115:144
181. Jensen F (2009) Theor Chem Acc 126:371
182. Jensen F (2008) J Chem Theory Comput 4:719
183. Kupka T, Nieradka M, Stachów M, Pluta T, Nowak P, Kæjr H, Kongsted J, Kaminsky J (2012) J Phys Chem A 116:3728
184. Rusakov YY, Krivdin LB (2013) Russ Chem Rev 82:99
185. Witte J, Neaton JB, Head-Gordon M (2016) J Chem Phys 144:194306
186. ElSohly AM, Tschumper GS (2009) Int J Quantum Chem 109:91
187. Suponitsky KY, Tafur S, Masurov AE (2008) J Chem Phys 129:044109
188. Wang NX, Wilson AK (2005) Mol Phys 103:345
189. Shavitt I (1993) Isr J Chem 33:357
190. Jensen F (2013) WIREs Comput Mol Sci. 3:273

191. Feller D (1996) J Comput Chem 17(13):1571
192. Schuchardt KL, Didier BT, Elsethagen T, Sun L, Gurumoorthi V, Chase J, Li J, Windus TL (2007) J. Chem Inf Model 47:1045
193. Pritchard BP, Altarawy D, Didier B, Gibson TD, Windus TL (2019) J Chem Inf Model 59(11):4814–4820. https://doi.org/10.1021/acs.jcim.9b00725
194. Adamowicz L, Bartlett RJ, McCullough EA Jr (1985) Phys Rev Lett 54:426
195. Becke AD (1988) J Chem Phys 88:2547
196. Shiozaki T, Hirata S (2007) Phys Rev A 76:040503
197. Hirata S, Shiozaki T, Johnson CM, Talman JD (2017) Mol Phys 115:510
198. Harrison RJ, Beylkin G, Bischoff FA, Calvin JA, Fann GI, Fosso-Tande J, Galindo D, Hammond JR, Hartman-Baker R, Hill JC, Jia J, Kottmann JS, Yvonne Ou M, Pei J, Ratcliff LE, Reuter MG, Richie-Halford AC, Romero NA, Sekino H, Shelton WA, Sundahl BE, Thornton WS, Valeev EF, Vázquez-Mayagoitia A, Vence N, Yanai T, Yokoi Y (2017) SIAM J Sci Comput 38:S123
199. Claudino D, Bartlett RJ (2018) J. Chem. Phys. 149:064105. https://doi.org/10.1063/1.5039741

Chapter 6
Gaussian Basis Sets for Solid State Calculations

Klaus Doll

6.1 Introduction

The central problem in electronic structure theory is the solution of the Schrödinger equation. It is a many-body problem, and—with few exceptions such as the hydrogen atom—numerical solutions have to be computed. For this purpose, the wave function is expanded in terms of a basis set. In molecular quantum chemistry, a natural starting point is to use basis sets made from atom-centred orbitals. The approximate solution of the Schrödinger equation is then expressed as a linear combination of atomic orbitals (LCAO).

In calculations for periodic systems such as bulk and surfaces, the solutions must fulfill the Bloch condition. This has made plane waves very popular as basis functions for these systems. Nevertheless, also local basis functions are frequently used. In this latter case, Bloch functions have to be constructed from the local basis functions.

After a brief overview on the basis set types in Sect. 6.2, this chapter on basis sets for solid-state calculations will have its main focus on using Gaussian basis sets in Sect. 6.3. The choice of basis sets will be discussed in Sect. 6.4. While many basis sets are already available in databases, it is still an important issue on how to generate new basis sets. This is shown in Sect. 6.5, and should be useful not only if there is no reasonable basis set available. This section also aims at an understanding of where basis set deficiencies come from, and how they can be identified e.g. by computing band structures.

K. Doll (✉)
Institute of Theoretical Chemistry, Pfaffenwaldring 55, 70569 Stuttgart, Germany
e-mail: doll@theochem.uni-stuttgart.de

E. Perlt (ed.), *Basis Sets in Computational Chemistry*, Lecture Notes in Chemistry 107,
https://doi.org/10.1007/978-3-030-67262-1_6

6.2 Basis Set Types

In molecular quantum chemistry, local basis sets are usually used. Most programs employ Gaussian basis functions, and the topic of Gaussians for molecular calculations will be discussed in other chapters of this book. Local basis sets can be applied to periodic systems as well—one has to construct Bloch functions, as will be shown in Sect. 6.3. Gaussian basis functions have been employed for periodic systems such as surfaces and bulk systems by various groups: e.g. by Callaway and co-workers [8, 9], Feibelman [19], Boettger and Trickey [5], Harmon and co-workers [26], and is suggested in the recent periodic implementation in the Gaussian code, see e.g. [32]. Also, systems with one-dimensional periodicity have been studied with codes based on Gaussian basis sets [2, 43, 45]. The CRYSTAL code [18, 39] has been developed since the mid-1970s, and is based on Gaussian-type orbitals. This chapter will be mainly based on the CRYSTAL code, though the concepts are general.

Besides Gaussian-type functions, also Slater functions may be used, as is done for example in the ADF code and its periodic extension BAND [1].

A different approach is to use plane waves, and variants such as augmented plane waves. Plane waves are a natural approach to periodic boundary conditions, in order to fulfill the Bloch condition. The Bloch condition states that in a periodic potential, with $V(\mathbf{r}) = V(\mathbf{r}+\mathbf{R})$, the Schrödinger equation can be solved by a basis of solutions which must be of the form

$$\Psi(\mathbf{r}) = \exp(\mathrm{i}\mathbf{k}\mathbf{r})\, \mathrm{u}(\mathbf{r}) \tag{6.1}$$

with $u(\mathbf{r}) = u(\mathbf{r}+\mathbf{R})$ having the same periodicity as the potential (see, e.g. [3]). $\mathbf{R}$ is a lattice vector of the direct lattice. In the numerical solution, a supercell is usually generated, with lattice vectors which are multiples of the original primitive lattice vectors. The corresponding reciprocal lattice with vectors $\mathbf{k}$ is referred to as Pack-Monkhorst [35] net (the Monkhorst-Pack [34] is closely related).

In a plane wave basis set, the crystalline orbitals can be expressed as

$$\Phi(\mathbf{r},\mathbf{k}) = \sum_{\mathbf{K}} c_{\mathbf{k},\mathbf{K}} \exp(-\mathrm{i}(\mathbf{k}+\mathbf{K})\mathbf{r}) \tag{6.2}$$

where the sum over reciprocal lattice vectors $\mathbf{K}$ is truncated according to the condition

$$\frac{\hbar^2}{2m_{el}}(\mathbf{k}+\mathbf{K})^2 \leq E_{\text{cutoff}} \tag{6.3}$$

with a cutoff energy E_{cutoff}. In the following, $\hbar$ and electron mass m_{el} are assumed to be 1, and atomic units are used.

Plane waves are a general basis set independent of the atom to be described. However, to describe the core region, an enormous number of plane waves would be required. Thus, plane waves are frequently used in combination with a pseudopotential. Another approach is to combine atom-centred local basis sets in the region near the nuclei (with this region being defined by muffin-tin radii), and plane waves in the interstitial region, as is done in, e.g. the augmented plane wave method.

Widely used codes based on plane waves are CASTEP [36] or VASP [31], or WIEN [4] in the case of linear augmented plane waves.

Mixed basis sets with combinations of plane waves and local basis sets have also been suggested, e.g. [33].

Another different approach is the use of numerical basis sets, where the basis functions are given numerically on a grid, which was pioneered in DMol [11] and its periodic variant DMol^3 [12].

With a local basis set, any periodicity can be implemented: no periodicity (like in a molecule), periodicity in one dimension (like a polymer with infinite length), periodicity in two dimensions (surfaces) or periodicity in three dimensions (bulk). In the case of plane waves, the system is always treated periodic in three dimensions. For example, surfaces are then treated as periodic in three dimensions with a vacuum region.

6.3 Gaussian Basis Sets

Gaussian basis sets were introduced in quantum chemistry with the work of Boys [6]. Gaussian functions are no solutions of the hydrogen atom. Their derivatives exist everywhere, also at the position of the nucleus, and they do not satisfy the cusp condition at the nucleus—conditions which the exact solution must fulfill (see, e.g. Sect. 7.2 in [28]). Nevertheless, they have become very popular because integrals over Gaussians can be efficiently solved, restricting the numerical evaluation to the calculation of the error function, see e.g. [40, 44].

The mathematical structure is

$$\phi_{\text{primitive}}(\mathbf{r}-\mathbf{A}) = N(x-A_x)^a(y-A_y)^b(z-A_z)^c\exp(-\zeta(\mathbf{r}-\mathbf{A})^2) \qquad (6.4)$$

Exponents are traditionally labelled with the letter ζ, and N is a normalisation. Here, $\mathbf{A} = (A_x, A_y, A_z)$ is the point where the Gaussian is centred, usually at the nuclear position $\mathbf{A}$. There are however cases when they are at general places where no nucleus is situated. For two-dimensional periodic systems, an important case is to describe the charge density in a region far away from the surface, where additional Gaussians in the vacuum region are required to converge the work function [15, 20, 23].

Normally, the exponents ζ are given in atomic units, i.e. in $1/a_0^2$ with the Bohr radius $a_0 \approx 0.5291772$ Å. Large exponents are referred to as 'tight' (important for the core region), smaller ones (i.e. roughly the ones below 1), as 'diffuse'. Diffuse functions are important for the valence region and the chemical bond, and so care has to be taken that the diffuse functions are appropriate for the system considered.

The Gaussian in Eq. 6.4 is also called a Gaussian primitive. Usually, the core region does not change very much when a chemical bond is formed. Therefore, a linear combination of Gaussians is made, which describes the core region well; this is called a contraction:

$$\phi(\mathbf{r}-\mathbf{A}) = N \sum_{i=1}^{n} d_i (x-A_x)^a (y-A_y)^b (z-A_z)^c \exp(-\zeta_i (\mathbf{r}-\mathbf{A})^2) \tag{6.5}$$

The contraction coefficients d_i can be obtained, for example, by performing a calculation for an atom, and then extracting them from the computed Hartree-Fock (HF) orbitals: the coefficients which are obtained due to the Hartree-Fock minimisation, and which describe how the primitive Gaussians have to be combined to obtain the orbitals, may then be used as contraction coefficients.

By solving the Hartree-Fock or Kohn-Sham equations in the self-consistent field procedure (SCF), the coefficients $a_{i\mu}$ are determined. The molecular orbitals are then obtained as linear combinations of the basis functions ϕ_μ:

$$\Psi_i(\mathbf{r}) = \sum_{\mu} a_{i\mu} \phi_\mu(\mathbf{r}) \tag{6.6}$$

For periodic systems, one constructs basis functions $\Phi_\mu(\mathbf{r}, \mathbf{k})$ satisfying the Bloch condition:

$$\Phi_\mu(\mathbf{r}, \mathbf{k}) = \sum_{\mathbf{g}} \phi_\mu(\mathbf{r} - \mathbf{A}_\mu - \mathbf{g}) \exp(i\mathbf{k}\mathbf{g}) \tag{6.7}$$

where the summation is over the direct lattice vectors $\mathbf{g}$, and the basis functions now depend on $\mathbf{k}$. The crystalline orbitals are then obtained as linear combinations:

$$\Psi_i(\mathbf{r}, \mathbf{k}) = \sum_{\mu} a_{i\mu}(\mathbf{k}) \Phi_\mu(\mathbf{r}, \mathbf{k}) \tag{6.8}$$

In Hartree-Fock or density functional calculations, the orbitals are obtained at the end of the self-consistent field procedure. All basis functions contribute to each orbital, unless there is a symmetry that makes the contribution of certain basis functions to an orbital vanish.

6.3.1 Angular Part

The angular part is obtained by multiplying a spherical harmonic $Y_{lm}(\vartheta, \varphi)$ with a factor r^l. This results in the solid harmonics:

$$r^l Y_{lm}(\vartheta, \varphi) \tag{6.9}$$

These functions are in general complex. However, only in the case of atoms and linear molecules, L_z is a good quantum number. Therefore, real solid harmonics are usually used (see, e.g. Sect. 6.4.2 in [28]).

The combinations are as follows, when the Condon-Shortley convention is used [10] (ignoring normalisation):

$$l = 0: \quad Y_{00} \Rightarrow \quad \sim 1 \tag{6.10}$$
$$l = 1: Y_{11} - Y_{1-1} \Rightarrow \quad \sim x \tag{6.11}$$
$$Y_{11} + Y_{1-1} \Rightarrow \quad \sim y \tag{6.12}$$
$$Y_{10} \Rightarrow \quad \sim z \tag{6.13}$$
$$l = 2: \quad Y_{20} \Rightarrow \quad 3z^2 - r^2 = 2z^2 - x^2 - y^2 \tag{6.14}$$
$$Y_{21} - Y_{2-1} \Rightarrow \quad xz \tag{6.15}$$
$$Y_{21} + Y_{2-1} \Rightarrow \quad yz \tag{6.16}$$
$$Y_{22} + Y_{2-2} \Rightarrow \quad x^2 - y^2 \tag{6.17}$$
$$Y_{22} - Y_{2-2} \Rightarrow \quad xy \tag{6.18}$$

The order depends on how it is implemented in the code. The order is important to properly interpret data from population analyses and projected densities of states (i.e. a plot of the density of states, with the density of states projected on one or a set of basis functions).

In the context of solids, an important application of population analysis is to understand the d-occupancy in the case of metal ions in a crystal field. Typical cases are ions with partially filled d-shells in a tetrahedral or octahedral surrounding (Fig. 6.1). Usually, in textbooks, the atoms are arranged along the axes in this case. In a real solid, this may not be the case, and then the energetically favoured orbitals are not the same as in a textbook. With CRYSTAL, the keyword ROTREF may be used to rotate the eigenvectors and density matrix, and thus to rotate the Cartesian reference frame [18], and subsequently compute the population in this rotated coordinate system.

6.3.2 *Exponents: Even Tempered Basis Sets*

The possibly easiest choice is to use even tempered exponents (Sect. 8.2.3 in [28], [41] and references therein). A number k of exponents is obtained according to the formula

$$\alpha\beta^k \tag{6.19}$$

with the parameters α and β to be optimised. This has been done in [41] for the atoms in the first three periods. An excerpt is shown in Fig. 6.2. We can see that the energy converges well, and for hydrogen the limit of $-0.5\ E_h$ is practically reached. From the various values for β, we can derive that they are roughly in the range from

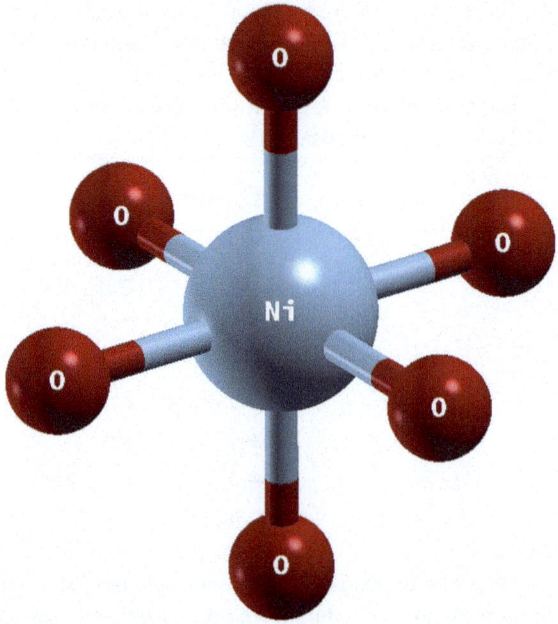

Fig. 6.1 A nickel ion in an octahedral surrounding, as in the case of NiO

2–4. We could thus generate an even tempered basis set by using a factor of 3, and a suitable value for α and k. This will actually be done in Sect. 6.5.5, to obtain a basis set for carbon and then compute the band structure of diamond.

6.4 Choice of Basis Sets

6.4.1 Exponents: Choice From Databases

For molecular calculations, large databases exist, especially:

The basis set database of the Environmental Molecular Sciences Laboratory (EMSL) employing the basis set exchange software [21, 42] at https://www.basissetexchange.org and the basis set library of the Molpro code at http://www.molpro.net/info/basis.php?portal=user&choice=Basis+library. For periodic systems, the basis set library of the CRYSTAL code is at http://www.crystal.unito.it/basis-sets.php

The web addresses were checked on 17 June 2020. Obviously, they may change, and it may be necessary to use a search engine to find them. For periodic systems, the corresponding section in the manual should be consulted (chapter 'Basis set' in the CRYSTAL manual [18]).

Hydrogen

N_s	α_s	β_s
3	0.0334000	4.88072
4	0.0330878	4.07414
5	0.0328187	3.58217
6	0.0319909	3.25769
7	0.0312701	3.02136
8	0.0304322	2.84371
9	0.0296728	2.70257
10	0.0289053	2.58878

	Hydrogen	
3S	−0.495842	−0.4952865
4S	−0.498751	−0.4986892
5S	−0.499562	−0.4995583
6S	−0.499840	−0.4998402
7S	−0.499937	−0.4999372
8S	−0.499974	−0.4999743
9S	−0.499988	−0.4999890
10S	−0.499995	−0.4999951

Carbon

N_s	α_s	β_s
6	0.063039	4.34055
7	0.057012	3.79734
8	0.050895	3.50889
9	0.050187	3.29366
10	0.049887	3.11117
11	0.047979	2.90316
12	0.044950	2.76467
14	0.043391	2.57899

N_p	α_p	β_p
3	0.045502	4.45043
4	0.041681	3.79202
5	0.038056	3.35026
6	0.035226	3.03818
7	0.032290	2.81511

	Carbon	
6S/3P	−37.5833929	−37.5828006
8S/4P	−37.6681259	−37.6679073
10S/5P	−37.6840452	−37.6840093
12S/6P	−37.6874371	−37.6874299
14S/7P	−37.6883035	−37.6883035
16S/8P		−37.6885182
18S/9P		−37.6885862
20S/10P		−37.6886074
22S/11P		−37.6886143
24S/12P		−37.6886170
26S/13P		−37.6886181

Fig. 6.2 Optimised values for α and β for hydrogen and carbon, and the energy obtained, in atomic units. Reprinted with permission from: Schmidt, M. W. and Ruedenberg, K., J. Chem. Phys, **71**, 3951–3962 (1979) [41], Copyright 1979 AIP Publishing LLC

6.4.1.1 Notes on the Basis Sets From the CRYSTAL Web Page

In the case of the CRYSTAL web page, it is recommended to check the basis sets:

Was the basis set calibrated for the neutral atom, or the negatively/positively charged atom? What was the coordination of the atom, when the basis set was optimised in a calculation on the bulk?

If the basis set will be used for an atom which will carry a positive charge, are there diffuse functions which might or should be omitted? This would apply, e.g. to the case of a basis set that was calibrated for copper metal, and is intended to be used in a system where Cu is expected to be essentially in a Cu^{+2} state. One should however not, in general, remove diffuse exponents. For heavy atoms with large ionic radius, diffuse functions may still be necessary. For example, for BaS, a diffuse $d = 0.15$ for Ba was found to be important to obtain the proper band structure [49].

If the basis set will be used for an atom that will carry a negative charge, are there enough diffuse exponents? Usually, *sp* shells in the range of 0.1 ... 0.15 should be present. Otherwise, it should be tested whether diffuse functions can be added and the calculation remains numerically stable. The population of an extra diffuse shell can be inspected, and the energy change when adding such a diffuse function.

If the basis set will be used for a metal, also the diffuse exponents should be inspected. Usually, *sp* shells in the range of 0.1 ... 0.15 should be present (see also the overview in [16]).

Note that the density matrix of a metallic system decays only algebraically in real space. If an exact exchange is used (or a hybrid functional which includes exact exchange), then this makes the summation of the exchange problematic, when it is done in real space as in the case of CRYSTAL. Thus, a finite temperature scheme should be used, which makes the decay exponential (see [24] and references therein).

When huge external pressure is applied, the basis functions may have to be chosen slightly less diffuse, to avoid numerical instabilities.

A large database is also available on Mike Towler's web site at https://vallico.net/mike_towler/crystal.html (accessed June 17, 2020), together with some very instructive and entertaining notes [46]. Also, basis sets for large parts of the periodic table have been published recently [37], but these basis sets should be carefully inspected before using them. Especially, in many cases, diffuse functions must be added.

6.4.2 Exponents: General Recommendations

The basis set choice depends on the type of calculation. For Hartree-Fock and density functional calculations, relatively small basis sets can be used, compared to basis sets required for wave function based correlation (e.g. Møller-Plesset perturbation theory such as MP2, coupled cluster, configuration interaction ...; and the corresponding multi-reference methods).

The basis set depends on the charge state of the atom in the periodic system. Negative ions have a larger radius than positive ions and thus require diffuse exponents (e.g. oxygen in NiO where O is roughly charged −2). Similarly, in the case of a positively charged ion, it may be possible to remove diffuse exponents (e.g. Ni in NiO where Ni is roughly charged +2). However, a metallic system will need diffuse exponents compared to an insulator (e.g. nickel metal compared to NiO—here we have the case of neutral Ni versus Ni^{+2}).

It is usually good to include polarisation functions: e.g. is *s* and *p* orbitals are occupied, a *d*-exponent may be added.

Basis sets with *sp* shells are computationally more efficient than basis sets where *s* and *p* exponents are different.

Effective core potentials are often advantageous from the first row of the transition metals onwards. They reduce the computational effort, and scalar-relativistic effects may be included. Nevertheless, if necessary, even for heavy atoms, all-electron calculations can be performed. This might be of interest, for example, to compute core-levels.

Adding more basis functions to a basis set will lower the energy because of the variational principle, if the basis set remains otherwise fixed (i.e. if we add a basis function, without changing anything else in the basis set). Usually, most care has to be taken about the diffuse exponents, and it is important that they are appropriate. Very large exponents may improve the core region, and thus also lead to a lower energy. However, this will have only little impact in the region where the chemical bond takes place. Therefore, usually the part of the basis set which describes the core region can be kept as is, when using a molecular basis set for solid-state calculations. Diffuse exponents often have to be altered: one may use an educated guess such as even tempered functions, with a minimum value which is safe enough to keep the calculations numerically stable (e.g. about 0.1 for *sp* shells), and possibly re-optimise these.

A reasonable first guess may be a basis set such as 6-31G for light atoms. For heavy atoms, one may use, e.g. the Stuttgart-Cologne pseudopotentials, and then re-optimise the diffuse exponents.

6.4.3 Test of a Basis Set

To test a basis set, one may compare with an enhanced basis set. This can be done by adding, e.g. extra diffuse functions or polarisation functions. Then, one has to compare properties such as the total energy, band structure, equilibrium geometry. Also, a population analysis gives usually good hints on the basis set, e.g. how strongly the outermost diffuse exponents are occupied. Vice versa, if an outermost diffuse exponent shows only little occupancy, one may consider to omit it. The Mulliken population is computationally practically without cost, and may thus always be computed.

It would be very expensive to calibrate a basis set in a system with a large unit cell. In this case, one may use a basis set from a system with a similar charge state; or if necessary calibrate the basis set in a system with a similar charge state. For example, the nickel atom in $NiCl_2(C_4N_2H_4)_2$ [30] has a similar charge state as in NiO, and thus a Ni basis set from the latter system may be chosen.

6.5 Guidelines to Generate a New Basis Sets

In the following, a description will be given on how to generate a new basis set. In many cases, existing basis sets can be used and work reasonably well. However, there may be cases without good basis sets, e.g. because the existing ones were calibrated for a different charge state. In the following, the cases of generating an all-electron basis set and a basis set to be used with an effective core potential are discussed.

6.5.1 Generating An All-Electron Basis Set

As an example, the question of a basis set for diamond is considered. From a library with basis sets calibrated for molecules (EMSL, see Sect. 6.4.1), we may choose a 6–31G* basis set (6–31G basis from [27], outermost *sp*-shell rescaled with 1.04^2 from [13], polarisation function from [25]). The basis set as obtained from EMSL is displayed in Table 6.1.

When transferring to a basis set for the bulk, the diffuse exponents have to be inspected. The outermost *sp* exponent is 0.1687144, and one may try to use it unchanged, which means, the whole basis set is left as is. Vice versa, if the most diffuse exponent had been below 0.1, one might have to consider omitting it or using a less diffuse exponent. When computing the band structure with this basis set on the level of the local density approximation (LDA), Fig. 6.3 is obtained. This agrees well with the literature, e.g. [7]. This basis set may thus be used for neutral carbon, or slightly (positively or negatively) charged carbon.

Deviations in the band structure will always appear with a local basis set for very high bands (i.e. high above the Fermi energy for metals, or high above the top of valence bands in the case of insulators). With a local basis set, the number of bands is equal to the number of basis functions. The basis functions are calibrated in such

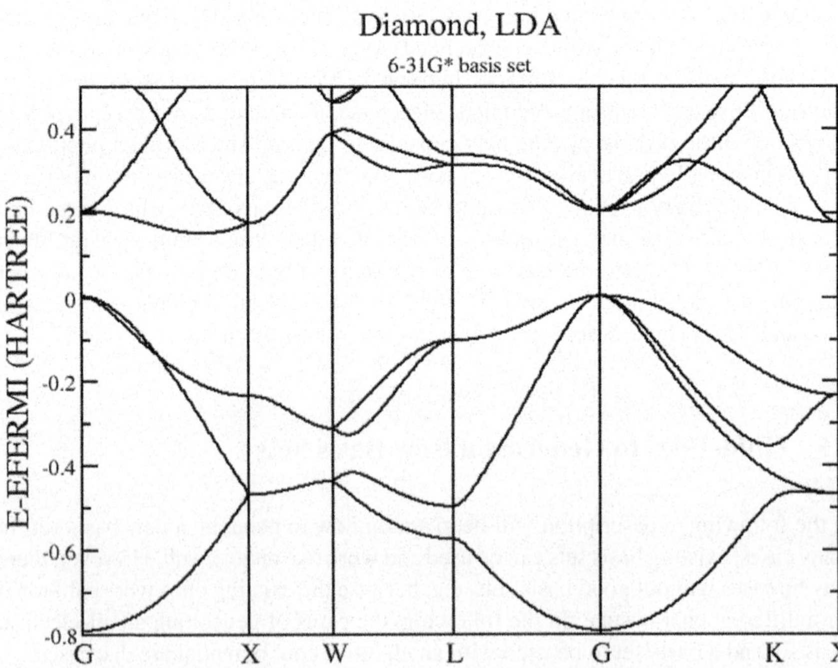

Fig. 6.3 Diamond band structure, LDA, with the basis set as in Table 6.1

Table 6.1 6–31G* basis set for carbon

Shell type	Exponent	Contraction coefficient	Contraction coefficient
s	3047.5249000	0.0018347	
	457.3695100	0.0140373	
	103.9486900	0.0688426	
	29.2101550	0.2321844	
	9.2866630	0.4679413	
	3.1639270	0.3623120	
sp	7.8682724	−0.1193324	0.0689991
	1.8812885	−0.1608542	0.3164240
	0.5442493	1.1434564	0.7443083
sp	0.1687144	1.0000000	1.0000000
d	0.8000000	1.0000000	

a way that the occupied states are well described, and thus high bands are less well described, or completely missing.

This is the same as in the molecular case, where the number of orbitals obtained in the calculation is identical to the number of basis functions. As a trivial example, one might choose a large basis set consisting of only s-exponents for hydrogen. Such a basis set might give a nearly perfect description of the ground state and its energy, but would not be able to describe a hydrogen atom with an electron in the $2p$ orbital. In general, a low total energy is no guarantee for a good basis set.

A good test is to compute the band structure, and to compare with band structures from plane-wave calculations. Especially interesting is the region around the Gamma point (i.e. $\mathbf{k} = \mathbf{0}$). The region with a small value of $\mathbf{k}$ corresponds to large spatial dimensions. If the band structure does not agree with the literature or e.g. if there are unusual oscillations, then this may hint at problems with the basis set: typically, when there are not enough diffuse exponents, this results in problems in the long range (long range in *direct* space), and thus deficiencies for $\mathbf{k} = \mathbf{0}$ (short range in *reciprocal* space).

Usually, with a good band structure, other properties such as the equilibrium geometry also agree reasonably well with, e.g. data from plane-wave calculations. In the case of diamond, the optimised lattice constant is 3.54 Å, which compares well with the one from [7] (3.52 Å).

As stated, this basis set can be used as is within the LDA. When using the same basis set within the HF approximation, and standard computational parameters, the calculation becomes numerically unstable. As will be discussed again in Sect. 6.5.3, this is in the present case due to a poor accuracy in the treatment of the exchange series. A remedy is to use stricter tolerances for the integral selection. Tests have been done for this system and using the same basis set as in the LDA. It is found that increasing the tolerances ITOL3, ITOL4, ITOL5 (for an explanation of these parameters, see [18]) for the summation of the exchange series from the default values (6, 6, 12) to stricter values (7, 7, 18) is necessary to achieve stability and convergence. The Coulomb series controlled by the first two parameters could be evaluated with default tolerances (6, 6), the same ones also used in the LDA. The instability is thus due to the exchange series. This is also shown in [46] (Sect. 4c).

6.5.2 Generating a Basis Set to be Used With an Effective Core Potential

The idea of an effective core potential is to reduce the computational effort with a potential that describes the nucleus and the inner electrons. Instead of, e.g. a Ca nucleus with a charge of +20, an effective core potential can be used which models the Ne-like core, and only the remaining charge of +10 is applied on the remaining 10 electrons. Projectors are used to approximate the potential of the Ne-like core. Besides reducing the computational effort, this also gives the possibility to include relativistic effects. One main variant is shape-consistent potentials, which are calibrated to fit the shape of the valence orbitals to the shape of the all-electron orbitals. Another variant is energy-consistent potentials, where the pseudopotentials are fitted to a set of energies of electronic states of an atom or ion. The following discussion will be based on the Stuttgart-Cologne energy consistent pseudopotentials by Stoll and co-workers (e.g. [22] or for a review, see [14]). They are referenced as ECPnXY, where ECP is for effective core potential, and n is the number of core electrons replaced with the ECP. X is for S or M, where S means a single-valence-electron ion fit was used—this should only be used for one- or two-valence electron atoms. M is for a multi-valence-electron fit, where states of the neutral atom and of ions with low charge are used for the fit. M should be used if available. Y is for the type of energy calculation: HF for Hartree-Fock, DF for fully relativistic Dirac-Fock, and WB for Wood-Boring, a simplification of Dirac-Fock. If available, DF or WB should be used. For example, ECP10MWB means that a pseudopotential is used which replaces the $1s^2 2s^2 2p^6$ electrons and the corresponding part of the nuclear charge, i.e. a neon core is used. The fit is of the type multi-valence-electron, with energies computed with the Wood-Boring Hamiltonian.

6.5.2.1 An Example: SrO

In the following, SrO is considered as an example of how to optimise a basis set to be used in combination with an effective core potential. SrO crystallises in a rock salt structure. For oxygen, the basis set from NiO is used [47], where oxygen is expected to be similarly charged as in SrO, (−2) due to the six-fold coordination. For Sr, the pseudopotential labelled with ECP28MWB is used, together with the basis set labelled with ECP28MWB [29]. The basis set is displayed in Table 6.2.

Table 6.2 Basis set ECP28MWB for Sr [29]

Shell type	Exponent	Contraction coefficient
s	5.8791570	0.1967090
	3.0924820	−0.6258980
	0.6446670	0.7357230
s	0.2988760	1.0
s	0.0572760	1.0
s	0.0238700	1.0
p	2.4324720	−0.3748990
	1.6642340	0.3876150
	0.5699890	0.6558380
p	0.2207180	1.0
p	0.0676290	1.0
p	0.0267270	1.0
d	3.6180810	−0.0075010
	0.9966560	0.1080980
	0.3907350	0.2785400
d	0.1227700	0.4773180
	0.0366550	0.4481830

A first inspection of the basis set shows that some of the exponents are very diffuse, e.g. two s exponents are below 0.1. The CRYSTAL code issues a warning when exponents below 0.06 are used (see Sect. 6.5.3). We may thus remove all exponents below 0.1 to be sure.

This would leave us with outermost diffuse exponents of 0.2988760 (s), 0.2207180 (p) and 0.3907350 (d). These are probably not diffuse enough, and we have to augment with additional exponents. An educated guess may be to use 0.1 for s, p and d, in accordance with the rule from Sect. 6.3.2, where it was stated that exponents of an even tempered basis set should be separated by factors in the range from 2–4.

This leaves us with the basis set in Table 6.3. To test this, we may perform a band structure calculation for SrO and compare with the literature. Of course, it should be the same level of theory, e.g. in the present case an LDA band structure. The result of such a band structure calculation is displayed in Fig. 6.4. It agrees well with the literature [38], and we can thus be confident to have a reasonably chosen basis set.

We can now check the importance of diffuse exponents. For this purpose, we can omit selected exponents. Figure 6.5 shows the band structure, when the diffuse s function on Sr is omitted. The green arrows indicate regions where the band structure particularly deviates, here at the Γ point.

Figure 6.6 shows the band structure, when the diffuse p function on Sr is omitted. The bands deviate only little from the original ones in Fig. 6.4. Mainly higher bands are affected. One might thus consider to omit the diffuse p function, if computing resources are limited, and to speed up calculations.

Table 6.3 Basis set for Sr of the size [3*s*3*p*2*d*], modified for calculations on bulk SrO, starting from ECP28MWB in Table 6.2

Shell type	Exponent	Contraction coefficient	Changes compared to type the original basis set
s	5.8791570	0.1967090	
	3.0924820	−0.6258980	
	0.6446670	0.7357230	
s	0.2988760	1.0	
s	0.1	1.0	← Diffuse *s* exponents below 0.1 replaced with this
p	2.4324720	−0.3748990	
	1.6642340	0.3876150	
	0.5699890	0.6558380	
p	0.2207180	1.0	
p	0.1	1.0	← Diffuse *p* exponents below 0.1 replaced with this
d	3.6180810	−0.0075010	
	0.9966560	0.1080980	
	0.3907350	0.2785400	
d	0.1	1.0	← Diffuse *d* contraction containing an exponent below 0.1 replaced with this

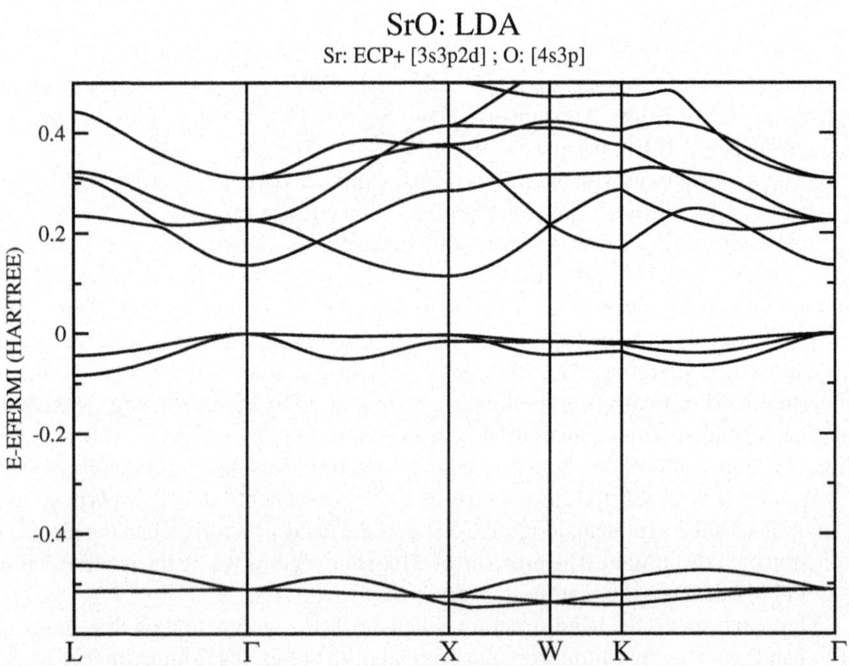

Fig. 6.4 SrO band structure, LDA, with the basis set as in Table 6.3

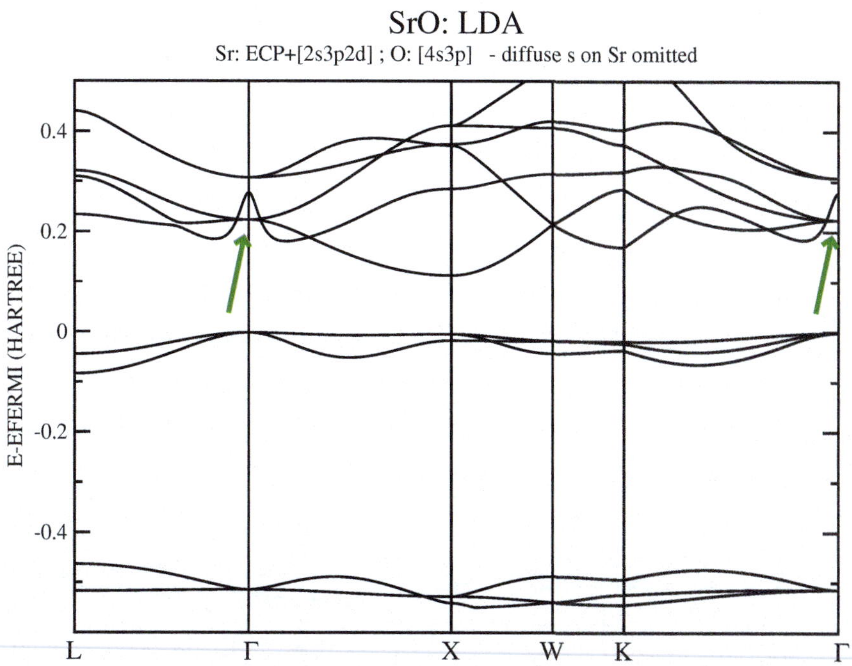

Fig. 6.5 SrO band structure, LDA, with the basis set as in Table 6.3, but the diffuse *s* exponent with value 0.1 is omitted. The arrows indicate where this leads to obvious errors in the band structure

Figure 6.7 shows the band structure, when the diffuse *d* function on Sr is omitted. The bands strongly deviate from the original ones in Fig. 6.4, and thus the diffuse *d* is mandatory. Especially, the bandgap is wrong: Fig. 6.4 shows an indirect gap with a maximum of the occupied bands at the Γ point, and the minimum of the unoccupied bands is at the X point. But when omitting the diffuse *d* function, the minimum of the conduction band is at the Γ point, and a direct gap would be obtained.

When looking at the periodic table and the position of Sr, these findings can be easily explained: in the case of SrO, we have Sr^{+2}, and thus the lowest unoccupied states are expected to be 5*s* and 4*d*. We must thus particularly take into account, that a diffuse *s* and diffuse *d* for Sr will be necessary.

6.5.3 Improve An Existing Basis Set: Re-optimise Exponents

Often, a basis set is already available, but for a system where the respective element is in a different charge state. An example would be a basis set for Cu in $KCuF_3$ [48], which describes Cu^{+2}. This basis set is displayed in Table 6.4. If such a basis set is supposed to be used for neutral Cu metal, then the outermost diffuse exponents with

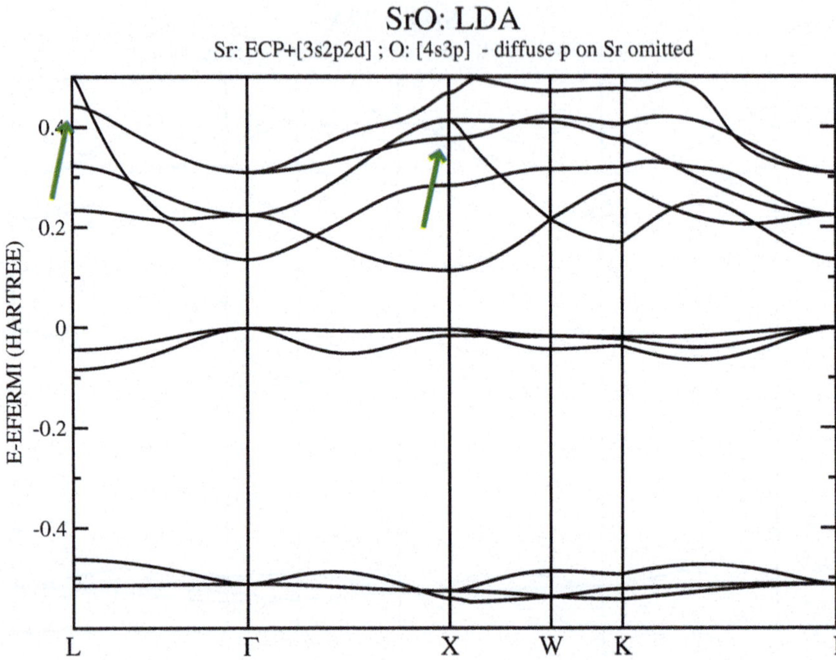

Fig. 6.6 SrO band structure, LDA, with the basis set as in Table 6.3, but the diffuse p exponent with value 0.1 is omitted. The arrows indicate where this leads to obvious errors in the band structure

$sp = 0.559$ is not sufficiently diffuse. Therefore, the basis set must be improved, and diffuse shells need to be added.

A simple strategy is to keep the inner exponents fixed, and add new diffuse exponents, and re-optimise these. In the present case, one may keep all exponents larger than 1 fixed, and remove the more diffuse ones. Then one can start with an educated guess, again based on the rule from Sect. 6.3.2, and thus separate the exponents by factors in the range from 2–4. As the outermost diffuse sp exponent is 1.582, one may, e.g. use 0.6 and 0.2 as an initial guess, and start with 0.43 for the d exponent. The optimisation is then done in an iterative way: one varies the larger of the two sp-shells, and optimises with respect to the total energy. Then, the diffuse sp-shell is optimised, while keeping all other exponents fixed. Then, the d-exponent is optimised. After this first round of optimisations, the next round can be done, starting again with the larger of the two sp-shells. This iterative procedure usually converges quickly. This strategy is similar to a geometry optimisation, where a geometrical minimum can be determined in the same way, by iteratively optimising a set of geometrical parameters. For a small set of parameters, such an approach is still reasonably efficient.

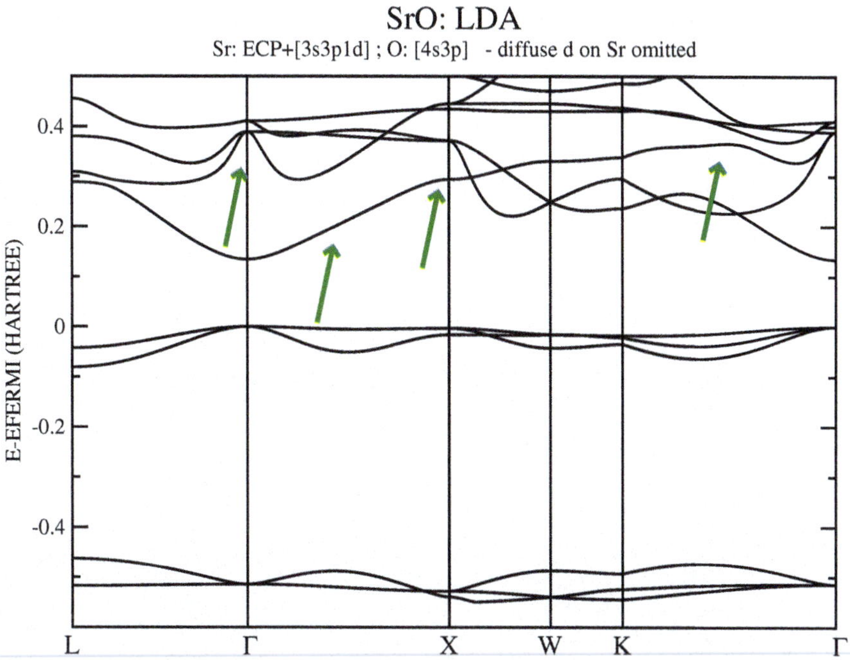

Fig. 6.7 SrO band structure, LDA, with the basis set as in Table 6.3, but the diffuse *d* exponent with value 0.1 is omitted. The arrows indicate where this leads to obvious errors in the band structure

The finally obtained basis set is displayed in Table 6.5. The equilibrium lattice constant and band structure as displayed in Fig. 6.8 agree well with the literature, see the discussion in [17].

As explained in the manual ([18], Sect. 12), a numerical instability and catastrophic behaviour can occur for very diffuse exponents. This is argued to be due to poor accuracy in the treatment of the Coulomb and exchange series. An attempt for a remedy could be to choose very strict cutoff parameters for the selection of the integrals; together with techniques used in molecular quantum chemistry to avoid linear dependency of basis functions by using a truncated transformation matrix during the SCF cycles (Sect. 3.4.5 in [44]). A more pragmatic way is to put lower bounds to the exponents, and not allow lower values, even though an optimisation would favour this.

The level of optimisation is recommended to be the same as the level of calculation for the system to be considered, i.e. the same density functional should be chosen, or in case of Hartree-Fock, also an optimisation on the Hartree-Fock level is preferable. The differences may, however, not be huge. The most important point is to have enough diffuse functions. This can be tested by adding another diffuse function, and trying to perform an optimisation.

Table 6.4 Basis set for Cu in $KCuF_3$ [48], where the Cu charge is essentially Cu^{+2}

Shell type	Exponent	Contraction coefficient	Contraction coefficient
s	398000.0	0.000227	
	56670.0	0.001929	
	12010.0	0.01114	
	3139.0	0.05013	
	947.2	0.17031	
	327.68	0.3693	
	128.39	0.4030	
	53.63	0.1437	
sp	1022.0	−0.00487	0.00850
	238.9	−0.0674	0.06063
	80.00	−0.1242	0.2118
	31.86	0.2466	0.3907
	13.33	0.672	0.3964
	4.442	0.289	0.261
sp	54.7	0.0119	−0.0288
	23.26	−0.146	−0.0741
	9.92	−0.750	0.182
	4.013	1.031	1.280
sp	1.582	1.0	1.0
sp	0.559	1.0	1.0
d	48.54	0.031	
	13.55	0.162	
	4.52	0.378	
	1.47	0.459	
d	0.430	1.0	

6.5.4 A Basis Set for the Free Atom

In order to compute binding energies, the energies of the free atoms may be required. However, the basis set for an atom within the bulk is usually not sufficient to describe a free atom. One possible remedy is to build a cluster where the atom is surrounded by the same neighbours as in the bulk. The neighbouring atoms then have the same basis functions as in the bulk, but no nuclear charge and no electrons, i.e. similar to the case of a counterpoise correction calculation. The downside is that this may be demanding, as many neighbours may be required, if one tries to converge the energy.

An alternative approach is to take a single free atom and add more diffuse functions to the bulk basis set, as the problems concerning numerical instability mentioned in Sect. 6.5.3 are much less severe for a single atom. The inner exponents must be the same as for the atom in the bulk, in order to be able to compare the energies. For

Table 6.5 Basis set for Cu, modified for metallic Cu bulk [17], starting from the basis set in Table 6.4

Shell type	Exponent	Contraction coefficient	Contraction coefficient	Changes compared to the original basis set in Table 6.4
s	398000.0	0.000227		
	56670.0	0.001929		
	12010.0	0.01114		
	3139.0	0.05013		
	947.2	0.17031		
	327.68	0.3693		
	128.39	0.4030		
	53.63	0.1437		
sp	1022.0	−0.00487	0.00850	
	238.9	−0.0674	0.06063	
	80.00	−0.1242	0.2118	
	31.86	0.2466	0.3907	
	13.33	0.672	0.3964	
	4.442	0.289	0.261	
sp	54.7	0.0119	−0.0288	
	23.26	−0.146	−0.0741	
	9.92	−0.750	0.182	
	4.013	1.031	1.280	
sp	1.582	1.0	1.0	
sp	0.596	1.0	1.0	← Re-optimised
sp	0.150	1.0	1.0	← Diffuse *sp* shell added
d	48.54	0.031		
	13.55	0.162		
	4.52	0.378		
	1.47	0.459		
d	0.392	1.0		← Re-optimised

the outer exponents, e.g. exponents below 1.0, we may again employ even tempered exponents. An example of a basis set suitable for a free Cu atom is displayed in Table 6.6. Here, even tempered basis functions separated by a factor of 2.5 have been added to describe the free atom. Though the basis set is now much larger, it is computationally inexpensive, as it is used only for the free atom.

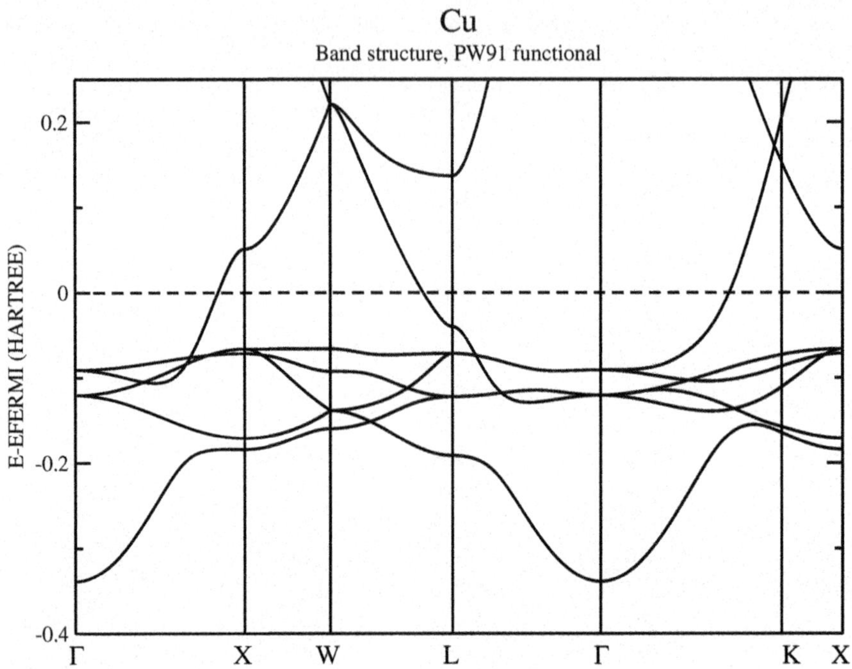

Fig. 6.8 Cu band structure, PW91 functional, with the basis set as in Table 6.5

6.5.5 A Simple Basis Set From Scratch: Even Tempered Basis Sets for Solids

If a completely new basis set shall be generated, then a very simple approach is to use even tempered basis functions. One has to choose parameters α, β and k as described in Sect. 6.3.2. An example is shown in Table 6.7 where a factor of roughly 3 has been chosen between the exponents. s-exponents from 100000 to 0.2 are used, and p-exponents from 1000 to 0.2. The band structure computed with this basis set for diamond agrees well with that from the 6–31G* basis set in Fig. 6.3, see Fig. 6.9.

6.6 Conclusion

The field of electronic structure calculations has been growing enormously in the past decades. Numerical solutions of the Schrödinger equation are required, and this implies that an expansion of the solution in terms of basis functions is required. There are nowadays many powerful tools available, which complement each other. Plane waves may be more black-box like, as they are determined by cutoff energy.

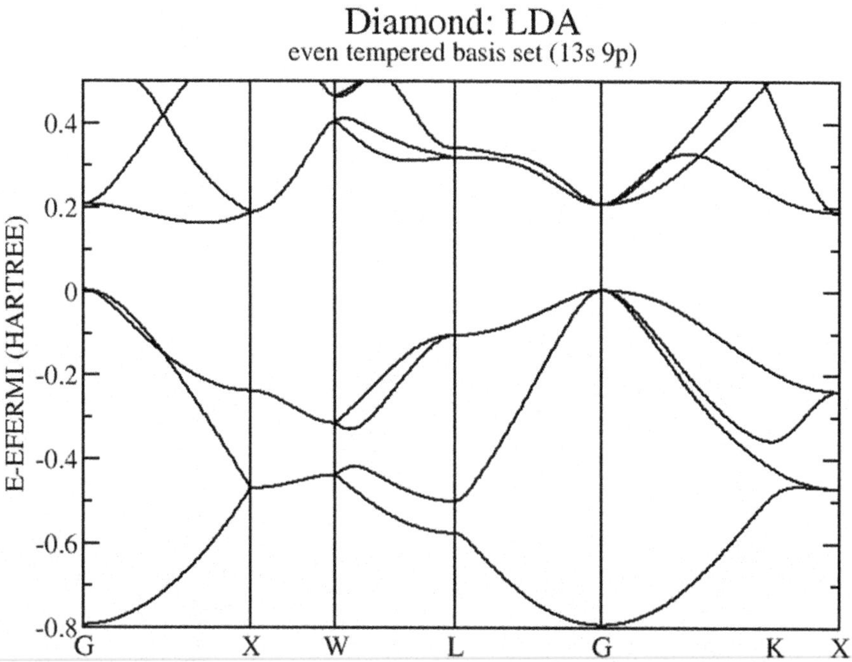

Fig. 6.9 Diamond band structure, LDA, with the even tempered basis set as in Table 6.7

The core region has to be described with, e.g. a pseudopotential. Local basis sets on the other hand have the advantage that the core region is efficiently described with few basis functions. The basis functions for the valence region have to be carefully chosen, which poses some more effort to the user and makes it less black-box like. However, with a careful choice, the results of both approaches converge well and

Table 6.6 Basis set for a free Cu atom, with inner exponents as in bulk Cu. Even tempered basis functions separated by a factor of 2.5 have been added to describe the free atom

Shell type	Exponent	Contraction coefficient	Contraction coefficient	Changes compared to the original basis set in Table 6.5
s	398000.0	0.000227		
	56670.0	0.001929		
	12010.0	0.01114		
	3139.0	0.05013		
	947.2	0.17031		
	327.68	0.3693		
	128.39	0.4030		
	53.63	0.1437		

(continued)

Table 6.6 (continued)

Shell type	Exponent	Contraction coefficient	Contraction coefficient	Changes compared to the original basis set in Table 6.5
sp	1022.0	−0.00487	0.00850	
	238.9	−0.0674	0.06063	
	80.00	−0.1242	0.2118	
	31.86	0.2466	0.3907	
	13.33	0.672	0.3964	
	4.442	0.289	0.261	
sp	54.7	0.0119	−0.0288	
	23.26	−0.146	−0.0741	
	9.92	−0.750	0.182	
	4.013	1.031	1.280	
sp	1.582	1.0	1.0	
sp	0.596	1.0	1.0	
sp	0.250	1.0	1.0	← New diffuse function, replaces 0.150
sp	0.100	1.0	1.0	← New diffuse function
sp	0.040	1.0	1.0	← New diffuse function
sp	0.016	1.0	1.0	← New diffuse function
sp	0.0064	1.0	1.0	← New diffuse function
d	48.54	0.031		
	13.55	0.162		
	4.52	0.378		
	1.47	0.459		
d	0.6	1.0		← New diffuse function, replaces 0.392
d	0.25	1.0		← New diffuse function
d	0.10	1.0		← New diffuse function
d	0.04	1.0		← New diffuse function
d	0.016	1.0		← New diffuse function

can be compared for benchmark purposes. As a whole, the choice and calibration of basis sets for solid-state calculations have become more like routine issues.

Table 6.7 An even tempered ($13s9p$) basis set to be used, e.g. for bulk diamond. The basis set is fully uncontracted

Shell type	Exponent	Contraction coefficient
s	100000	1.
s	30000	1.
s	10000	1.
s	3000	1.
s	1000	1.
s	300	1.
s	100	1.
s	30	1.
s	10	1.
s	3	1.
s	1	1.
s	0.4	1.
s	0.2	1.
p	1000.	1.
p	300.	1.
p	100.	1.
p	30.	1.
p	10.	1.
p	3.	1.
p	1.	1.
p	0.4	1.
p	0.2	1.

References

1. ADF2017, SCM, Theoretical Chemistry, Vrije Universiteit, Amsterdam, The Netherlands
2. André JM (1969) Self-consistent field theory for the electronic structure of polymers. J Chem Phys 50:1536–1542
3. Ashcroft NW, Mermin ND (1976) Solid state physics. Saunders, Philadelphia
4. Blaha P, Schwarz K, Madsen GKH, Kvasnicka D, Luitz J (2001) WIEN2k, An augmented plane wave + Local orbitals program for calculating crystal properties. Karlheinz Schwarz, Technische Universität Wien, Austria
5. Boettger JC, Trickey SB (1984) Total energy and pressure in the Gaussian-orbitals technique. I. Methodology with application to the high-pressure equation of state of neon. Phys Rev B 29:6425–6433
6. Boys SF (1950) I. A general method of calculation for the stationary states of any molecular system. Proc R Soc A 200:542–554
7. Bross H, Bader R (1995) Calculation of the Ground State Properties of Diamond and Cubic Boron Nitride. Phys Stat Sol (b) 191:369–385
8. Callaway J, Zou X, Bagayoko D (1983) Total energy of metallic lithium. Phys Rev B 27:631–635

9. Ching WY, Callaway J (1974) Band structure, cohesive energy, optical conductivity, and Compton profile of lithium. Phys Rev B 9:5115–5121
10. Condon EU, Shortley G (1951) The theory of atomic spectra. Cambridge University Press, Cambridge
11. Delley B (1990) An all-electron numerical method for solving the local density functional for polyatomic molecules. J Chem Phys 92:508–517
12. Delley B (2000) From molecules to solids with the DMol3 approach. J Chem Phys 113:7756–7764
13. Ditchfield R, Hehre WJ, Pople JA (1971) Self-consistent molecular-orbital methods. IX. An extended gaussian-type basis for molecular-orbital studies of organic molecules. J Chem Phys 54:724–728
14. Dolg M, Xiaoyan Cao X (2012) Relativistic pseudopotentials: their development and scope of applications. Chem Rev 112:403–480
15. Doll K (2006) Calculation of the work function with a local basis set. Surf Sci 600:L321–L325
16. Doll K (2009) Ab initio calculations with a Gaussian basis set for metallic surfaces and the adsorption thereon. In: Basiuk V, Ugliengo P (eds) Quantum chemical calculations of surfaces and interfaces of materials. American Scientific Publishers, pp 41–53
17. Doll K, Harrison NM (2000) Chlorine adsorption on the Cu(111) surface Chem Phys Lett 317:282–289
18. Dovesi R, Saunders VR, Roetti C, Orlando R, Zicovich-Wilson CM, Pascale F, Civalleri B, Doll K, Harrison NM, Bush IJ, D'Arco Ph, Llunell M, Causà M, Noël Y (2014) CRYSTAL14. University of Torino, Torino
19. Feibelman PJ (1988) Local-orbital basis for defect electronic structure calculations of an Al(100) film. Phys Rev B 38:1849–1855
20. Feibelman PJ (1995) Anisotropy of the stress on fcc(110) surfaces. Phys Rev B 51:17867–17875
21. Feller D (1996) The role of databases in support of computational chemistry calculations. J Comp Chem 17:1571–1586
22. Fuentealba P, Szentpály LV, Preuß H, Stoll H (1985) Pseudopotential calculations for alkaline-earth atoms. J Phys B 18:1287–1296
23. García-Gil S, García A, Lorente N, Ordejón P (2009) Optimal strictly localized basis sets for noble metal surfaces. Phys Rev B 79:075441
24. Goedecker S (1999) Linear scaling electronic structure methods. Rev Mod Phys 71:1085–1123
25. Hariharan PC, Pople JA (1973) The influence of polarization functions on molecular orbital hydrogenation energies. Theor chim Acc 28:213–222
26. Harmon BN, Weber W, Hamann DR (1982) Total-energy calculations for Si with a first-principles linear-combination-of-atomic-orbitals method. Phys Rev B 25:1109–1115
27. Hehre WJ, Ditchfield R, Pople JA (1972) Self-consistent molecular orbital methods. xii. further extensions of gaussian-type basis sets for use in molecular orbital studies of organic molecules. J Chem Phys 56:2257–2261
28. Helgaker T, Jørgensen P, Olsen J (2000) Molecular electronic-structure theory. Wiley, Chichester
29. Kaupp M, Schleyer PVR, Stoll H, Preuss H (1991) Pseudopotential approaches to Ca, Sr, and Ba hydrides. Why are some alkaline earth MX_2 compounds bent? J Chem Phys **94**, 1360–1366
30. Kreitlow J, Menzel D, Wolter AUB, Schoenes J, Süllow S, Feyerherm R, Doll K (2005) Pressure dependence of $C_4N_2H_4$-mediated superexchange in $XCl_2(C_4N_2H_4)_2(X = \mathrm{Fe, Co, Ni})$. Phys Rev B 72:134418
31. Kresse G, Furthmüller J (1996) Efficiency of ab-initio total energy calculations for metals and semiconductors using a plane-wave basis set. Comput Mat Sci 6:15–50
32. Kudin KN, Scuseria GE, Martin RL (2002) Hybrid density-functional theory and the insulating gap of UO_2. Phys Rev Lett 89:266402
33. Louie SG, Ho K-M, Cohen ML (1979) Self-consistent mixed-basis approach to the electronic structure of solids. Phys Rev B 19:1774–1782

34. Monkhorst HJ, Pack JD (1976) Special points for Brillouin-zone integrations. Phys Rev B 13:5188–5192
35. Pack JD, Monkhorst HJ (1977) Special points for Brillouin-zone integrations—a reply. Phys Rev B 16:1748–1749
36. Payne MC, Teter MP, Allan DC, Arias TA, Joannopoulos JD (1992) Iterative minimization techniques for ab initio total-energy calculations: molecular dynamics and conjugate gradients. Rev Mod Phys 64:1045–1097
37. Peintinger MF, Oliveira DV, Bredow T (2013) Consistent Gaussian basis sets of triple-zeta valence with polarization quality for solid-state calculations. J Comput Chem 34:451–459
38. Pan B, Wang N-P, Rohlfing M (2015) Electron-hole excitations and optical spectra of bulk SrO from many-body perturbation theory. Appl Phys A 120:587–593
39. Pisani C, Dovesi R, Roetti C (1988) Hartree-Fock Ab initio treatment of crystalline systems. Lecture notes in Chemistry, vol 48. Springer, Berlin
40. Saunders VR (1983) Molecular integrals for Gaussian type functions. In: Diercksen GHF, Wilson S (eds) Methods in computational molecular physics, pp 1–36. Reidel, Dordrecht, Netherlands
41. Schmidt MW, Ruedenberg K (1979) Effective convergence to complete orbital bases and to the atomic Hartree-Fock limit through systematic sequences of Gaussian primitives. J Chem Phys 71:3951–3962
42. Schuchardt KL, Didier BT, Elsethagen T, Sun L, Gurumoorthi V, Chase J, Li J, Windus TL (2007) Basis set exchange: a community database for computational sciences. J Chem Inf Model 47:1045–1052
43. Suhai S (1974) The electronic structure of periodic protein models. Theor Chim Acta 34:157–163
44. Szabo A, Ostlund NS (1989) Modern quantum chemistry. McGraw-Hill, New York
45. Teramae H, Yamabe T, Imamura A (1983) Ab initio effective core potential studies on polymers. Theor Chim Acta 64:1–12
46. Towler M (2000) An introductory guide to Gaussian basis sets in solid-state electronic structure calculations. https://vallico.net/mike_towler/basis_sets/basis_sets_2000.ps. Accessed 17 June 2020
47. Towler MD, Allan NL, Harrison NM, Saunders VR, Mackrodt WC, Aprà E (1994) Ab initio study of MnO and NiO. Phys Rev B 50:5041–5054
48. Towler MD, Dovesi R, Saunders VR (1995) Magnetic interactions and the cooperative Jahn-Teller effect in $KCuF_3$. Phys Rev B 52:10150–10159
49. Zagorac D, Doll K, Schön JC, Jansen M (2012) Sterically active electron pairs in lead sulfide? an investigation of the electronic and vibrational properties of PbS in the transition region between the rock salt and the α-GeTe-type modifications. Chem Eur J 18:10929–10936

Chapter 7
Basis Sets for Heavy Atoms

Diego Fernando da Silva Paschoal, Mariana da Silva Gomes, Larissa Pereira Nogueira Machado, and Hélio Ferreira Dos Santos

7.1 Motivation

The chemistry of heavy atoms plays important roles in various branches of science. Here, we named as heavy atoms the second- and third-row of the transition elements (transition d-metal), the lanthanides, and the actinides. The chemistry of transition d-metal atoms is very rich, with a broad application such as catalysts [1], luminescent [2], and in medicinal chemistry [3–6]. Lanthanides elements ($_{57}$La – $_{71}$Lu) have attracted interest because of their unique optical and magnetic properties. The focus is to find out lanthanide complexes with application in lasers technology, optical fibers, and in the biomedical field, being used as therapeutic and diagnostic agents [7–11]. The actinide elements ($_{89}$Ac – $_{103}$Lr) are mostly synthesized and radioactive and have a short half-life, consequently, their use is very limited. Among the actinides, we may highlight uranium and plutonium elements that have been used in nuclear weapons and nuclear power generation [12, 13]. In the present contribution, we will focus on the applications of d-transition metals in Medicinal Inorganic Chemistry, mainly the platinum complexes.

Medicinal Inorganic Chemistry has made great advances since the discovery of antitumor activity of *cis*-diamminedichloroplatinum(II), cisplatin, in 1965 [14]. Since then, there has been a growing interest in the search for transition metal complexes for the treatment and diagnosis of various diseases [15–17]. When compared with organic molecules, transition metal complexes offer several therapeutic advantages.

D. F. S. Paschoal (✉) · M. S. Gomes · L. P. N. Machado
NQTCM: Núcleo de Química Teórica e Computacional de Macaé, Polo Ajuda, Universidade Federal do Rio de Janeiro, Campus UFRJ-Macaé, 27.971-525, Macaé, RJ, Brasil
e-mail: diegofspaschoal@macae.ufrj.br; diegopaschoal01@gmail.com

H. F. Dos Santos
NEQC: Núcleo de Estudos em Química Computacional, Departamento de Química – ICE, Universidade Federal de Juiz de Fora, Campus Universitário, 36.036-900, Juiz de Fora, MG, Brasil

E. Perlt (ed.), *Basis Sets in Computational Chemistry*, Lecture Notes in Chemistry 107,
https://doi.org/10.1007/978-3-030-67262-1_7

The coordinated ligands can be modified to tune the steric and electronic properties, allowing greater selectivity and reactivity of the metal against a specific target. In addition, metal ions with distinct oxidation states can coordinate with ligands in different geometries and participate in various biological redox reactions. Transition metal complexes with labile ligands may also coordinate with potential targets through the ligand-exchange reaction [2, 18]. Such variety and complexity are an inexhaustible source of resources with constant challenges to be solved [19].

Metallodrugs can provide unique mechanisms of action based on the choice of metal, its oxidation state, types and number of coordinated ligands, and coordination geometry. There are transition metal complexes as agents used in therapy and diagnostics [16, 17, 19]. In addition to the success of cisplatin, which is used in approximately 50% of the cancer chemotherapy treatments [6, 19], we can highlight others successful cases in Medicinal Inorganic Chemistry: the use of gold(I) complexes as anti-arthritic agents [15], gadolinium(III) complexes as MRI contrast agent [6, 15, 19, 20] and the Tc-99m radionuclide as an emitter of gamma radiation in nuclear medicine [6, 15, 19–21]. In Fig. 7.1, we highlight the platinum(II) complexes in clinical use for cancer therapy.

Lastly, in recent years, it has become common to use quantum chemistry calculations to assist the experimentalists to interpret data and to provide information that cannot be obtained experimentally. However, when the interest is on heavy atoms, a physical problem arises due to the inner shell electrons that must be represented by relativistic theories. Moreover, based on the standard Hartree–Fock–Roothaan (HFR) approach, commonly used in quantum chemistry calculations for molecules, the atomic basis sets must also play a primary role in the description of the electronic structure of heavy metals compounds. In the current literature, we find effective core potentials (ECPs), non-relativistic and relativistic basis sets, Slater- and Gaussian-Type Orbitals (STOs and GTOs, respectively). The basis sets for heavy atoms and their applications are the central points addressed in this chapter.

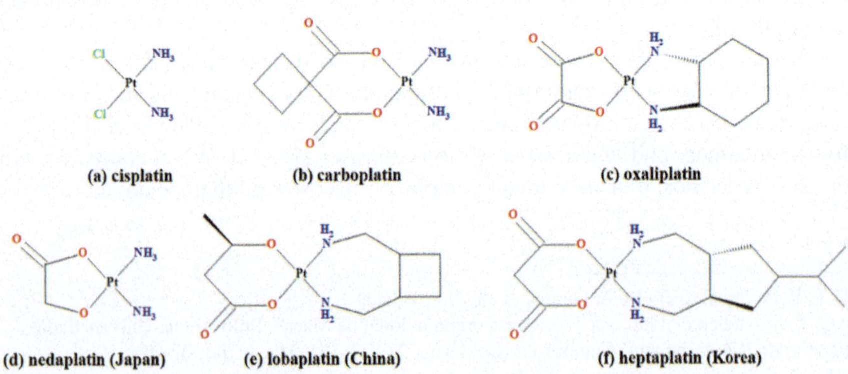

Fig. 7.1 Pt(II) complexes approved for clinical use in the cancer treatment. The structures were obtained with the BIOVIA Draw program [151]

7.2 The Atomic Basis Sets

In quantum mechanics, we are interested in solving the Schrödinger equation. However, for practically all chemical systems of interest, an exact solution of the Schrödinger equation cannot be found, so it is necessary to use approximations to find their solutions. One of the approximations that are characteristic of basically all ab initio methods is the HFR method where the introduction of an atomic basis set is needed [22, 23].

The basis set is a set of mathematical functions from which the wave function is constructed. The choice of basis set for a quantum chemistry calculation is a key step, which fundamentally determines the quality of the wave function and properties derived from it. The molecular orbitals (MOs) are constructed by linear expansion in a set of basis functions. We expand the MOs $\phi_p(\mathbf{r})$ in a set of simple analytical one-electron functions [22, 23]:

$$\phi_p(r) = \sum_{\mu} C_{\mu p} \chi_\mu(r) \tag{7.1}$$

The major problem is to define which atomic functions $\chi_\mu(\mathbf{r})$ are suitable for orbital expansions, once the series in Eq. (7.1) is finite (the C_p are the variational coefficients). Some requirements are important to be considered when choosing a basis set:

(a) The basis set should allow a rapid convergence of the system under study;
(b) The resulting integrals should be easy to calculate.

We have two distinct types of basis functions commonly used in quantum chemistry calculations: Slater-Type Orbitals (STO) and Gaussian-Type Orbitals (GTO). The STO function has the form given in Eq. (7.2) in spherical polar coordinates [24]:

$$\chi_{\zeta,\mathrm{n,l,m}}(\mathrm{r}, \theta, \varphi) = \mathrm{N}\mathrm{Y}_{\mathrm{l,m}}(\theta, \varphi)\mathrm{r}^{n-1}\mathrm{e}^{-\zeta r} \tag{7.2}$$

Where N is a normalization constant and $Y_{l,m}$ are spherical harmonic functions. The exponential dependence on the distance between the nucleus and electron mirrors the exact orbitals for the hydrogen atom.

The GTO function has the form given in Eq. (7.3) in spherical polar (7.3a) [25] and cartesian coordinates (7.3b) [26]:

$$\chi_{\zeta,\mathrm{n,l,m}}(\mathrm{r}, \theta, \varphi) = \mathrm{N}\mathrm{Y}_{\mathrm{l,m}}(\theta, \varphi)\mathrm{r}^{2\mathrm{n}-2-\mathrm{l}}\mathrm{e}^{-\zeta r^2} \tag{7.3a}$$

$$\chi_{\zeta,\mathrm{n,l,m}}(\mathrm{r}, \theta, \varphi) = \mathrm{N}\mathrm{x}^{\mathrm{l_x}}\mathrm{y}^{\mathrm{l_y}}\mathrm{z}^{\mathrm{l_z}}\mathrm{e}^{-\zeta r^2} \tag{7.3b}$$

The sum of l_x, l_y, and l_z determines the type of orbital (for example, $l_x + l_y + l_z = 2$ is a d-orbital). Despite the apparent similarity, there is a difference between GTO in the two sets of coordinates. A GTO written in spherical coordinates has five components of d-type and seven components of f-type, while a GTO written in Cartesian coordinates has six and ten components of d- and f-types, respectively [25].

Analyzing the forms of STO and GTO functions, we observe that the r^2 dependence in the exponential makes the GTOs inferior to the STOs in two respects. The GTO decay very rapidly far from the nucleus, consequently the "tail" of the wave function is poorly represented and a GTO does not adequately represent the behavior of atomic orbital near the nucleus. This is because a GTO has a zero slope at the nucleus, while an STO has a "cusp" (discontinuous derivative) [22, 23, 25, 27].

The GTOs do not reproduce the behavior of the orbitals as well as the STOs, so a greater number of GTOs are required for achieving the accuracy compared with STOs. However, the difference in the exponential term gives specific properties for the GTOs. The product of two or more Gaussians will also be a Gaussian function, which significantly simplifies the three- and four-center two electrons integrals. This is the main reason why GTOs are used in electronic structure calculations, since the greater speed and simplicity to solve the integrals compensates for the high number of integrals that arise when using Gaussian functions [22, 23, 25, 27].

Another crucial factor to take into account in molecular calculations is the number of functions to be used. The simplest basis set is called a minimal basis set. A minimal basis set has only one set of basis functions for each atomic orbital occupied with distinct n and l quantum numbers. In the chemical context, an atomic basis set is not flexible enough to describe the distortions that the orbitals undergo upon the formation of chemical bonds or as a result of some external perturbation (electric field, solvent, etc.). For such purposes, larger and more flexible basis sets are needed; thus, functions of the same type as those already present can be added to improve the radial flexibility of the wave function. Thereby, a double-zeta (DZ) basis set has two sets of GTOs for each occupied AO [22, 23, 25, 27].

For molecular calculations, functions of symmetry different from those of the occupied orbitals can also contribute to the wave function and molecular properties. These functions of higher angular momentum are called polarization functions and the introduction of polarization functions improves the flexibility of angular part of the wave function. The polarization functions describe the polarization of the charge distribution in a molecular environment [22, 23, 25, 27].

Over the years, many different Gaussian basis sets have been proposed for molecular calculations and the GTOs have been used in most computational packages such as GAUSSIAN 09 [28], ORCA 4.2.1 [29], and NWCHEM 6.6 [30]. However, currently, STOs have also been used in softwares such as the Amsterdam Density Functional (ADF) [31–33]. Although many basis sets are presented as suitable in the description of the most different systems and properties, it is very difficult to construct a universal basis set that is good to be applied under any circumstance. Jorge and Castro [34] generated an accurate Universal Gaussian Basis Set (UGBS) for all atoms of the Periodic Table. However, the use of UGBS for molecular calculations is impracticable due to the size of the basis set, which generates a very high computational cost.

In the main basis sets database, the Basis Set Exchange [35–37], we found approximately 611 basis sets available. The users must be very careful when choosing a suitable basis set for their calculations and are strongly recommended to look in the literature for reviews' work or to test the chosen basis set on the system of interest or for smaller models. Moreover, when working with heavy atoms, greater care must be taken. This is mainly due to the relativistic effects that must be included for heavy metals [38–40]. Thus, before dealing with the basis sets for heavy elements, a brief introduction about the relativistic effects will be presented.

7.2.1 *Relativistic Effects*

The non-relativistic binding energy for a hydrogen-like system (any atomic nucleus with one electron) is $E_n = -\frac{Z^2}{2n^2}$. For the 1s-electron in a hydrogen-like system the energy is given as Eq. (7.4) [41, 42]:

$$E = -\frac{Z^2}{2} \tag{7.4}$$

Where Z is the atomic number. According to Bohr´s model, the energy is equal to minus the kinetic energy, owing to the Virial theorem, Eq. (7.5) [41, 42].

$$E = -T = -\frac{1}{2}mv^2 \tag{7.5}$$

In atomic units m = 1 and, using Eqs. (7.4) and (7.5), we have the classical velocity of a 1s-electron, Eq. (7.6) [41, 42].

$$v_{1s-\text{electron}} = Z \tag{7.6}$$

One of the consequences of the constant speed of light is that the mass of a particle, which moves at a substantial fraction of c, increases over the rest mass m_0, Eq. (7.7) [41, 42].

$$m = m_0\left(\sqrt{1-\frac{v^2}{c^2}}\right)^{-1} \tag{7.7}$$

As a consequence, the effective Bohr radius (a_0) decreases, Eq. (7.8) [41, 42].

$$a_0 = \frac{\hbar^2}{mc^2} \tag{7.8}$$

The speed of light in atomic units is 137.036. So, it is clear that relativistic effects cannot be neglected in heavy nuclei. For atoms with large Z, the first and direct consequence is a relativistic contraction and stabilization of all s and most p orbitals. This provides a more effective shielding of the nuclear charge for the higher angular momentum orbitals, which consequently increases in size. So, the second and indirect effect is the relativistic expansion and destabilization of d- and f-orbitals, which become larger and more diffuse. The third relativistic effect is the spin–orbit interaction, important for levels where $l > 0$ that split into $j = l \pm 1/2$ [41–43].

Then, as the relativistic scalar effects lead to a contraction of orbitals with smaller angular momentum and expansion of orbitals with larger angular momentum, we can use scalar relativistic Hamiltonians to deal with these effects properly, such as the Douglas–Kroll–Hess (DKH) [44–48] and the zeroth-order regular approximation (ZORA) [49–51] approximations. In the next session, the basis sets effects will be discussed for heavy atoms.

7.2.2 *Basis Sets for Heavy Atoms*

7.2.2.1 Effective Core Potentials (ECPs)

In both atomic and molecular systems, the occupied orbitals can be divided into two classes: the core orbitals and the valence orbitals. The core orbitals are the innermost orbitals, located close to the atomic nuclei and the valence orbitals are the outermost orbitals, spatially more extended than the core orbitals. For heavy atoms, there are many core electrons. These electrons are not generally important for describing chemical properties; however, to represent them, many basis functions are required. Besides, for these heavier elements the relativistic effects become important. One way to solve both problems simultaneously is to fit an effective core potential (ECP) to describe the core electrons using relativistic calculations, reducing the computational problem by treating only the valence electrons explicitly [22, 25, 27]. Most ECP calculations give results comparable to those obtained using an all-electron basis set with a lower computational cost. Added to this, a fraction of relativistic effects (scalar effects) can also be recovered, without the need to perform a full relativistic calculation [24].

We find in the literature a very large number of effective core potentials. Hay and Wadt [52–54] have developed ECPs for transition metals incorporating the mass–velocity and one-electron Darwin corrections into their potentials for the heavier elements ($Z > 36$). Weigend and Ahlrichs [55] have developed Gaussian basis sets (def2-basis set) of split valence, triple-zeta valence, and quadruple-zeta valence quality to be used with ECPs for H to Rn, excluding the lanthanides. Peterson et al. [56–59] have developed correlation consistent ECPs. Besides, full series of correlation consistent basis sets from double to quintuple-zeta have also been developed for use with the ECPs. We show in Table 7.1 some ECPs with their corresponding valence basis sets available in Basis Set Exchange [35–37] for the third-row transition metal atoms, ${}_{72}$Hf to ${}_{80}$Hg.

Table 7.1 Some ECPs and valence basis sets available in EMSL Basis Set Exchange for the third-row transition metal atoms

ECPs	Contraction scheme	CGTO	ECPs	Contraction scheme	CGTO
cc-pVDZ-PP	[4s4p3d1f]	38	LANL2DZ	[3s3p2d]	22
cc-pVTZ-PP	[5s5p4d2f1g]	63	LANL2TZ	[5s5p3d]	35
cc-pVQZ-PP	[6s6p5d3f2g1h]	99	SBKJC-VDZ	[4s4p3d]	31
cc-pV5Z-PP	[7s7p6d4f3g2h1i]	148	Stuttgart-RSC	[6s5p3d]	36
cc-pwCVDZ-PP	[5s5p4d2f]	54	CRENBS[a]	[1s1p1d]	9
cc-pwCVTZ-PP	[7s7p6d3f2g]	88	CRENBL	[5s5p4d]	40
cc-pwCVQZ-PP	[8s8p7d4f3g2h]	135	def2-SV	[6s3p2d]	25
cc-pwCV5Z-PP	[9s9p8d5f4g3h2i]	206	def2-SVP	[6s3p2d1f]	32
aug-cc-pVDZ-PP	[5s5p4d2f]	54	def2-SVPD	[6s4p2d1f]	35
aug-cc-pVTZ-PP	[6s6p5d3f2g]	88	def2-TZVP	[6s4p3d1f]	40
aug-cc-pVQZ-PP	[7s7p6d4f3g2h]	135	def2-TZVPD	[6s5p3d1f]	43
aug-cc-pV5Z-PP	[8s8p7d5f4g3h2i]	197	def2-TZVPP	[6s4p3d2f1g]	56
dhf-SV	[6s3p2d]	25	def2-TZVPPD	[6s5p3d2f1g]	59
dhf-SVP	[6s3p2d1f]	32	def2-QZVP	[7s5p4d3f1g]	72
dhf-TZVP	[6s5p3d1f]	43	def2-QZVPD	[7s5p4d3f1g]	72
dhf-TZVPP	[6s5p3d2f1g]	59	def2-QZVPP	[7s5p4d4f2g]	88
dhf-QZVP	[7s6p4d3f1g]	75	def2-QZVPPD	[7s5p4d4f2g]	88
dhf-QZVPP	[7s6p4d3f2g]	84			

[a] All ECPs describe 60 electrons, except the CRENBS ECP which describes 68 electrons. All these ECPs can be found in https://www.basissetexchange.org/; All contraction schemes and number of CGTO are relative to the Pt atom

In general, the various papers found in the literature show that ECPs provide reliable results mainly for geometries and relative energies [60]. However, limitations tend to appear for properties in which the core electrons are important. In these cases, it is necessary to use an all-electron basis set and to include relativistic effects explicitly in the Hamiltonian.

7.2.2.2 All-electron basis sets (ABS)

The number of all-electron basis sets (ABS) available for the heavy atoms is sparse when compared to the quantity available for the light atoms. However, in recent years, due to the increase of computational resources and availability of scalar relativistic Hamiltonians implemented in quantum chemistry packages, an increasing number of ABS appeared in the literature. The use of ABS added to the inclusion of relativistic effects on the Hamiltonian can provide more accurate results than the use of ECPs, mainly when we are interested in properties where the core electrons are important, such as chemical shifts and coupling constants in nuclear magnetic resonance (NMR) spectroscopy.

We found in the literature non-relativistic and relativistic ABS. Jorge et al. [61–68] developed non-relativistic and DKH relativistic double-zeta quality all-electron Gaussian basis set for $_{1}$H to $_{103}$Lr atoms, named as DZP and DZP-DKH, respectively. Besides, triple-zeta quality basis set can be found for H to Rn atoms [62, 69–73]. Despite the importance of including relativistic effects when dealing with heavy atoms, good results can be achieved with non-relativistic ABS for geometries and relative energies. Successful results were obtained by Paschoal et al. [74] based on the non-relativistic Jorge's DZP basis set for platinum atom. It was proposed a modified basis set with the addition of a d-diffuse function, named as mDZP [74]. Besides, Paschoal et al. [75, 76] developed a set of Gaussian basis sets contracted with DKH2 approximation to predict NMR properties of the H-Ar, K, Ca, Ga-Kr, Rb, Sr, In-Xe, and Pt atoms. These basis sets were called as NMR-DKH and present a triple-zeta character with two sets of polarization functions (TZ2P).

Neese et al. [77–80] have developed a segmented all-electron relativistically contracted (SARC) basis set for third-row transition metal atoms ($_{72}$Hf – $_{80}$Hg), lanthanides ($_{57}$La – $_{71}$Lu), actinides ($_{89}$Ac – $_{103}$Lr), and 6p elements ($_{81}$Tl – $_{86}$Rn). The SARC basis sets can be found contracted for the second-order Douglas–Kroll–Hess (SARC-DKH) and the zeroth-order regular approximation (SARC-ZORA) scalar relativistic Hamiltonians. The SARC basis set is well-balanced and has an affordable size for routine density functional theory (DFT) calculations. It is important to mention that the contraction coefficients of the ZORA and DKH basis sets are determined separately. This happens because in these two approaches the behavior close to the nucleus is different. The authors recommended the use of SARC basis sets for heavier elements than the Xe and for the light atoms they recommended the use of re-contracted def-TZVP basis set for DKH2 or ZORA calculations. All these basis sets are part of the freely available ORCA program [29].

Relativistic segmented contracted basis set for DKH calculations were reported by Koga et al. [81–83] for $_{1}$H to $_{88}$Rn atoms. Roos et al. [84–86] have been developed new relativistic atomic natural orbital (ANO) basis sets for the first, second, and third-row transition metal atoms, lanthanides, and actinides. Watanabe et al. [87] presented relativistic Gaussian basis sets for molecular calculations for $_{1}$H up to $_{80}$Hg atoms. Highly accurate relativistic Gaussian basis sets for DKH calculations were developed for $_{1}$H up to $_{103}$Lr atoms by Hirao et al. [88, 89].

Relativistic double-, triple- and quadruple-zeta basis sets have also been developed by Dyall et al. [90–97] and are available from the DIRAC program [98]. It is worth mentioning the basis sets of Dyall, are constructed for four-component calculations. Recently, Pollack and Weigend [99] proposed segmented contracted Gaussian basis sets for elements up to Rn, for calculation within the exact two-component decoupling, X2C.

It is important to mention the basis sets used in the Amsterdam Density Functional (ADF) software [31–33]. These are commonly known as Slater-Type Orbitals (STOs), and are different from most other quantum chemistry programs, which usually employ Gaussian-Type Orbitals (GTOs). We found distinct STOs for all

Table 7.2 Some non-relativistic and relativistic all-electron Gaussian basis sets found in the literature for the third-row transition metal atoms

Basis set	Contraction scheme	CGTO	Basis set	Contraction scheme	CGTO
Non-relativistic					
jorge-DZP	[8s7p4d2f]	63	jorge-ADZP	[9s8p5d3f]	79
jorge-TZP	[9s5p5d3f1g]	79	jorge-ATZP	[9s6p6d4f2g]	103
mDZP[c]	[8s7p5d2f]	68	UGBS	[32s23p18d13f]	282
Relativistic					
jorge-DZP-DKH	[8s7p4d2f]	63	Sapporo-DKH3-DZP-2012[a]	[8s6p5d2f]	65
jorge-TZP-DKH	[9s5p5d3f1g]	79	Sapporo-DKH3-TZP-2012[a]	[10s8p6d3f1g]	94
SARC-DKH	[17s11p8d2f]	104	Sapporo-DKH3-QZP-2012[a]	[11s9p7d4f2g1h]	130
SARC-ZORA	[17s11p8d3f]	111	cc-pVTZ-DK[b]	[9s8p6d3f1g]	93
NMR-DKH[d]	[18s12p9d3f2g]	136	cc-pwCVTZ-DK[b]	[11s10p8d4f2g]	127

All these ABS can be found in https://www.basissetexchange.org/, except [a]Sapporo basis sets that can be found in https://sapporo.center.ims.ac.jp:8080/sapporo/Welcome.do and [b]Correlation consistent basis sets that can be found in https://www.grant-hill.group.shef.ac.uk/ccrepo/. [c]Basis set available only for Pt atom. [d]Basis sets available for H-Ar, K, Ca, Ga-Kr, Rb, Sr, In-Xe, and Pt atoms. All contraction schemes and number of CGTO are relative to the Pt atom

elements ($Z = 1 - 118$) ranging from a valence double-zeta to a valence quadruple-zeta quality. The exponents of these STOs were optimized for the use in the ZORA approach [100].

Overall, there are available in the literature both non-relativistic and relativistic atomic basis sets with the latter recommended for heavy elements. Thus, since the relativistic effects are concentrated largely in the core electrons, good results can also be obtained with the use of non-relativistic basis sets when valence molecular properties are concerning. In Table 7.2, we listed some ABS available in the literature for the third-row transition metal atoms. In short, there is a wide variety of ECPs and ABS for the heavy atoms. In the next section, we will present some case studies using these basis sets in the calculation of atomic and molecular properties involving the platinum atom ($Z = 78$).

7.3 Applications of Basis Sets for Heavy Atoms

Barnett Rosenberg et al. [14] discovered the antitumor properties of cisplatin in 1965. Since then, cisplatin has been approved for the treatment of some tumors and it has become one of the most widely used antitumor drugs in the world. However, the occurrence of side effects motivated researchers to search for new Pt(II) complexes that exhibited the same activity, but with less side effects [101–103]. Thus, with the success of cisplatin, new complexes have been synthesized and two important steps of these studies are the characterization and the reactivity of the Pt(II) complexes.

So, in this section, we present the importance of the platinum basis set in the study of the structure, reactivity, and prediction of NMR parameters of Pt(II) complexes with potential antitumor activity.

7.3.1 Structure and Reactivity of Cisplatin

Cisplatin, *cis*-$[Pt(NH_3)_2Cl_2]$ is a relatively non-reactive molecule, that is, it does not react directly with any molecules that are present in the biological environment. But, in aqueous solution, the chloride ligands are replaced by water, forming the aqua complexes, *cis*-$[Pt(NH_3)_2(H_2O)Cl]^+$ and *cis*-$[Pt(NH_3)_2(H_2O)_2]^{2+}$. The Pt-$OH_2$ bond is more reactive than the Pt-Cl bond. Therefore, the aqua complexes may react rapidly with N-donor ligands, such as the nitrogenous bases of DNA. Therefore, the aquation reaction of cisplatin is considered a necessary activation step for its biological response [101–104]. Besides, we found in the literature some studies that correlate the rate constant of the aquation reaction with the biological response of cisplatin derivatives [105]. Within this context, electronic structures and geometries, as well as the activation barriers for the aquation reaction, are important molecular descriptors to be considered theoretically in the search for new platinum complexes.

The theoretical study of platinum complexes is not so simple due to the presence of the platinum atom ($Z=78$), which is a heavy element that in computational chemistry presents difficulties due to the incompleteness of the basis set and the need to include relativistic effects. Many theoretical studies are available in the literature addressing the structure and reactivity of cisplatin [74, 106–110]. Among some other studies [106, 107], Paschoal et al. [74] present a comprehensive analysis of the platinum basis sets in the prediction of structure and reactivity of cisplatin. An extensive analysis of the role of the basis set and the level of theory in the prediction of structure and reactivity of cisplatin was accomplished. All calculations were performed using the GAUSSIAN 09 program [28] and the authors concluded that the effective core potentials (ECPs) for Pt atom are good alternatives to describe the cisplatin structure. On the other hand, it was shown that to represent satisfactorily the aquation reaction of cisplatin, the non-relativistic mDZP basis set for Pt atom lead to the best result in comparison with the experimental data, even though some ECP also produced good results. A more complete study involving the use of an all-electron Pt basis set and the influence of the inclusion of relativistic effects on these properties has not yet been reported. Herein we expand our previous work [74] and present a detailed study of the role of Pt basis set and the relativistic effects on the prediction of the geometry and reactivity of cisplatin. The next sections include a brief description of the computational protocols and a discussion of the results.

7.3.1.1 Theoretical Methodology

Figure 7.2 represents the reaction process analyzed in the present work. The geometry of cisplatin and all reactive species were fully optimized in the gas phase and characterized as a stationary point on the potential energy surface (PES) through harmonic frequency calculation. For the aquation reaction, an initial guess for transition state (TS) structure was proposed with the entering (H_2O) and leaving (Cl^-) groups located on the equatorial plane of a distorted trigonal bipyramidal geometry. The TS structure was further optimized and characterized as saddle point (1st order) on the PES by having only one imaginary vibrational frequency. From the TS structure, the intermediates were found using the intrinsic reaction coordinate (IRC) method. The final points were then optimized and characterized as true minima on the PES at the same level of theory used for TS structure. The solvent effect was taken into account using the Conductor-like Polarizable Continuum Model (C-PCM) approximation [111].

The activation Gibbs free energy ($\Delta G^{\ddagger}$) for the aquation reaction was calculated as in Eq. (7.9) [74–100]:

$$\Delta G^{\ddagger} = G_{TS} - G_{I1} \tag{7.9}$$

Where the TS is the transition state structure and I1 is the intermediate of the reaction, Fig. 7.3. The second-order rate constant was obtained from the Eyring-Polanyi, Eq. (7.10) [74], under normal conditions, namely $T = 298,15$ K, $p = 1$ atm and $c^0 = 1$ mol L^{-1}.

$$k_2 = \frac{k_B T}{hc^0} \exp\left(-\frac{\Delta G^{\ddagger}}{RT}\right) \tag{7.10}$$

Where k_B, h, and R are the Boltzmann, Planck, and ideal gas constants, respectively.

The calculations were performed at the Density Functional Theory (DFT) level with the B3LYP [112] functional. Non-relativistic and Douglas–Kroll–Hess with second-order scalar relativistic correction (DKH2) [44–47, 113–115] Hamiltonians

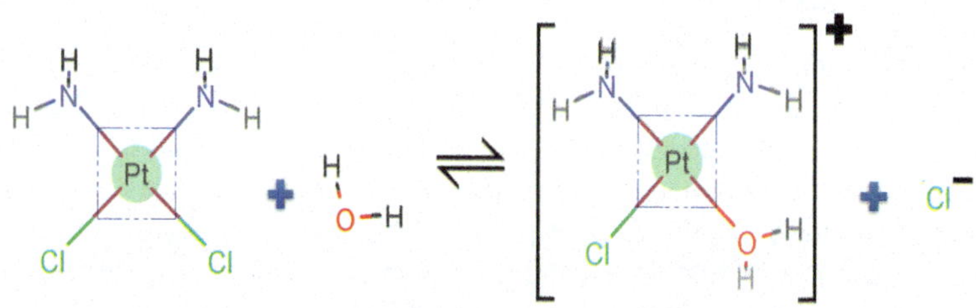

Fig. 7.2 First aquation process of cisplatin. The structures were obtained with the Marvin program [152]

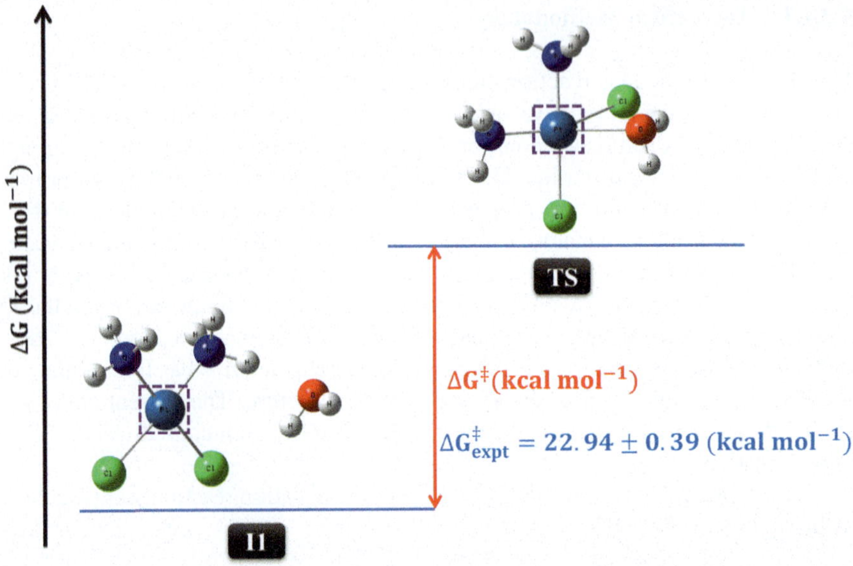

Fig. 7.3 Rate-limiting step, $I1 \rightarrow TS$, for aquation reaction of cisplatin. The intermediate I1 is a stable molecular complex. The structures were obtained with the Avogadro 1.2.0 program [148, 149]

were used. In order to assess the role of the Pt basis set, effective core potential (ECP), non-relativistic all-electron basis set, and relativistic all-electron basis set listed in Table 7.3 were used. Non-relativistic Hamiltonian was used with ECPs and non-relativistic ABS, and DKH2 Hamiltonian was used with relativistic ABS. For ligands' atoms, the jorge-DZP and the jorge-DZP-DKH basis set were considered for non-relativistic and DKH2 relativistic calculations, respectively. All calculations were performed in ORCA 4.2.1 software [29].

7.3.1.2 Results and Discussion

Cisplatin structure

The structure of the cisplatin is discussed first. The structural parameters calculated for cisplatin are presented in Table 7.4 and compared with the experimental data available in the solid state [116]. It is important to mention that the comparison must be made with caution since the matrix effect in the solid state can modify the structure. The quality of the distinct platinum basis set was assessed through the relative deviation ($\delta_{i,j}$), Eq. (7.11), and mean relative deviation (MRD), Eq. (7.12), of structural parameters including Pt – N, Pt – Cl bonds and N – Pt – N, Cl – Pt – Cl, N – Pt – Cl angles.

Table 7.3 Pt basis set used in the prediction of structure and reactivity of cisplatin

Pt basis set	Contraction scheme	CGTO
ECP		
def2-SVP	[6s3p2d1f]	32
def2-TZVP	[6s4p3d1f]	40
def2-TZVPP	[6s4p3d2f1g]	56
Stuttgart-RSC	[6s5p3d]	36
LANL2DZ	[3s3p2d]	22
LANL2TZ	[5s5p3d]	35
LANL2TZ(f)	[5s5p3d1f]	42
CRENBS	[1s1p1d]	9
CRENBL	[5s5p4d]	40
Non-relativistic all-electron		
jorge-DZP	[8s7p4d2f]	63
mDZP	[8s7p5d2f]	68
jorge-TZP	[9s5p5d3f1g]	79
Relativistic all-electron		
jorge-DZP-DKH	[8s7p4d2f]	63
jorge-TZP-DKH	[9s5p5d3f1g]	79
Sapporo-DKH3-DZP	[8s6p5d2f]	65
Sapporo-DKH3-TZP	[10s8p6d3f1g]	94
Sapporo-DKH3-QZP	[11s9p7d4f2g1h]	130
SARC-DKH	[17s11p8d2f]	104

$$\delta_{i,j} = \frac{s_j^{expt} - s_{i,j}^{calc}}{s_i^{expt}} \times 100 \tag{7.11}$$

$$MRD = \frac{1}{n_k} \sum_{k=1}^{n_k} \left| \delta_{i,j} \right| \tag{7.12}$$

Where i corresponds to the structural parameter of interest and j corresponds to the platinum basis set used. For example, $s_{Pt-N,jorge-DZP}$ corresponds to the Pt – N bond using the non-relativistic all-electron jorge-DZP basis set.

The calculated structural parameters do not change much with the Pt basis set. The calculated values with the ECPs and non-relativistic ABS presented a good agreement with the calculated values with the relativistic ABS. The MRD were less than 8% for the calculations in gas phase and less than 4% for the calculation with the C-PCM model. The most pronounced effect in predicting the structure was the solvent effect. As expected, the bond angles are more sensitive, thus, the largest errors were found for the calculated angles in gas phase. From the calculated deviations in Fig. 7.4, we note that the N – Pt – N and Cl – Pt – Cl angles are overestimated, whereas

Table 7.4 Calculated structural parameters for cisplatin at B3LYP/Pt basis set/jorge-DZP level for non-relativistic calculations and B3LYP-DKH2/Pt basis set/jorge-DZP-DKH level for relativistic calculations

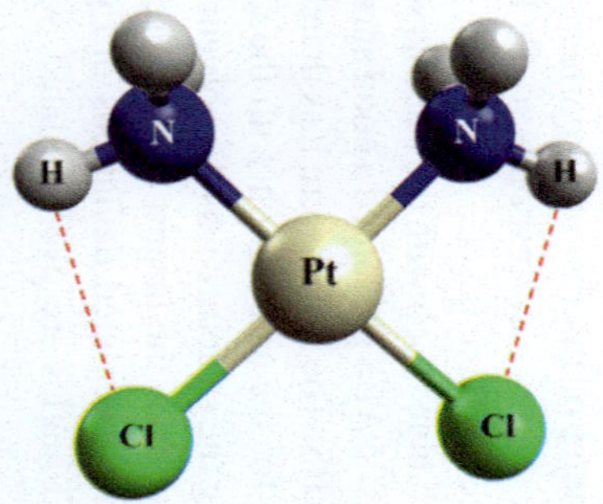

Pt basis set	Pt–N		Pt-Cl		N-Pt–N		Cl–Pt–Cl		N-Pt-Cl		(N-)H…Cl		$\delta_{i,j}$(%)	
ECP basis set														
def2-SVP	2.12	(2.07)	2.33	(2.37)	98.4	(90.7)	95.5	(93.1)	83.1	(88.1)	2.43	(2.74)	6.1	(2.6)
def2-TZVP	2.11	(2.07)	2.32	(2.36)	98.2	(90.6)	95.5	(93.1)	83.2	(88.1)	2.42	(2.73)	6.0	(2.4)
def2-TZVPP	2.10	(2.06)	2.32	(2.36)	98.2	(90.6)	95.6	(93.2)	83.1	(88.1)	2.42	(2.72)	6.0	(2.4)
Stuttgart-RSC	2.11	(2.07)	2.34	(2.38)	98.2	(90.6)	95.3	(92.9)	83.2	(88.2)	2.45	(2.75)	6.0	(2.5)
LANL2DZ	2.12	(2.07)	2.35	(2.39)	98.5	(90.7)	95.5	(93.1)	83.0	(88.1)	2.44	(2.75)	6.3	(2.8)
LANL2TZ	2.11	(2.07)	2.33	(2.37)	98.2	(90.6)	95.3	(93.1)	83.2	(88.2)	2.44	(2.75)	5.9	(2.5)
LANL2TZ(f)	2.10	(2.06)	2.32	(2.36)	98.1	(90.6)	95.3	(93.1)	83.3	(88.2)	2.43	(2.73)	5.8	(2.3)
CRENBS	2.17	(2.11)	2.35	(2.40)	98.3	(90.5)	94.9	(93.1)	83.4	(88.2)	2.43	(2.76)	6.5	(3.1)
CRENBL	2.10	(2.07)	2.33	(2.37)	98.4	(90.8)	95.7	(93.1)	83.0	(88.1)	2.41	(2.73)	6.0	(2.5)
Non-relativistic ABS														

(continued)

Table 7.4 (continued)

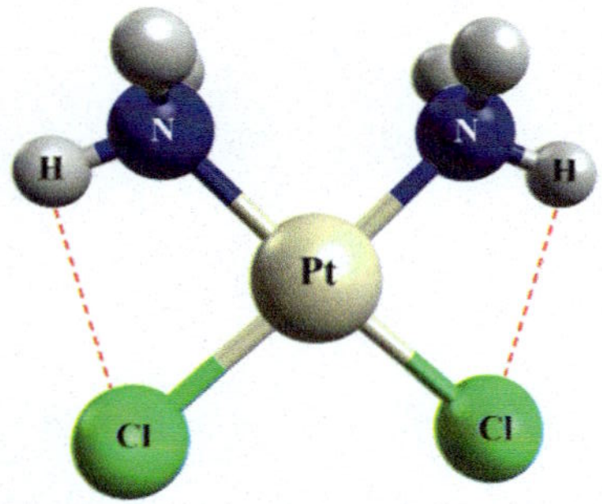

Pt basis set	Pt–N		Pt-Cl		N-Pt–N		Cl–Pt–Cl		N-Pt-Cl		(N-)H…Cl		$\delta_{i,j}$(%)	
jorge-DZP	2.06	(2.04)	2.32	(2.37)	95.7	(90.5)	97.6	(94.9)	83.3	(87.3)	2.45	(2.72)	5.3	(2.7)
mDZP	2.13	(2.10)	2.36	(2.39)	98.9	(90.6)	95.6	(93.1)	82.7	(88.1)	2.45	(2.78)	6.7	(3.0)
jorge-TZP	2.21	(2.16)	2.38	(2.41)	97.9	(90.8)	96.5	(93.3)	82.8	(87.9)	2.47	(2.82)	7.6	(3.9)
Relativistic ABS														
jorge-DZP-DKH	1.97	(1.96)	2.26	(2.30)	95.1	(90.8)	96.6	(94.6)	84.1	(87.3)	2.41	(2.63)	5.2	(2.9)
jorge-TZP-DKH	2.13	(2.08)	2.34	(2.37)	98.0	(90.6)	96.3	(93.3)	82.9	(88.0)	2.42	(2.75)	6.4	(2.7)
Sapporo-DKH3-DZP-2012	2.09	(2.05)	2.33	(2.37)	98.3	(90.9)	95.8	(93.1)	82.9	(88.0)	2.40	(2.71)	5.9	(2.4)
Sapporo-DKH3-TZP-2012	2.08	(2.05)	2.32	(2.36)	98.0	(90.8)	95.9	(93.2)	83.0	(88.0)	2.40	(2.70)	5.8	(2.3)
Sapporo-DKH3-QZP-2012	2.08	(2.04)	2.32	(2.36)	98.1	(90.8)	95.9	(93.2)	83.0	(88.0)	2.40	(2.70)	5.8	(2.3)
SARC-DKH	2.08	(2.05)	2.32	(2.36)	98.0	(90.7)	95.6	(93.1)	83.2	(88.1)	2.42	(2.72)	5.7	(2.2)

(continued)

Table 7.4 (continued)

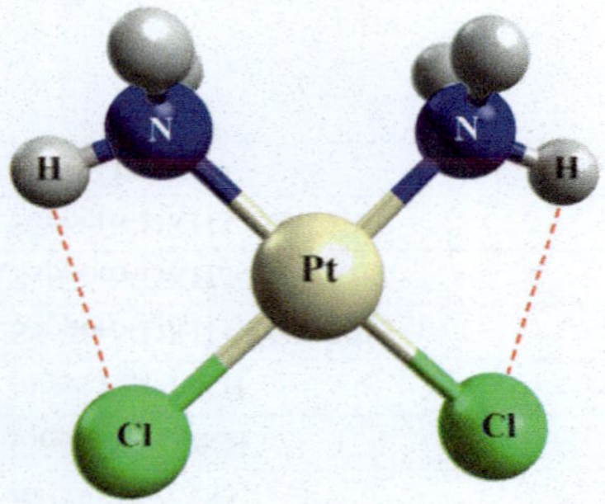

Pt basis set	Pt–N	Pt-Cl	N-Pt–N	Cl–Pt–Cl	N-Pt-Cl	(N-)H…Cl	$\delta_{i,j}$(%)
Experimental[a]	**2.01**	**2.33**	**87.0**	**91.9**	**90.3**	-	-

The first entries correspond to the values obtained in gas phase and the entries in parentheses correspond to the values obtained with the solvent effect by means of the C-PCM(water) approach. [a]Experimental values obtained by Milburn and Truter [116]. Bond lengths in Å and bond angles in degree. The structure was obtained with the Avogadro 1.2.0 program [148, 149]

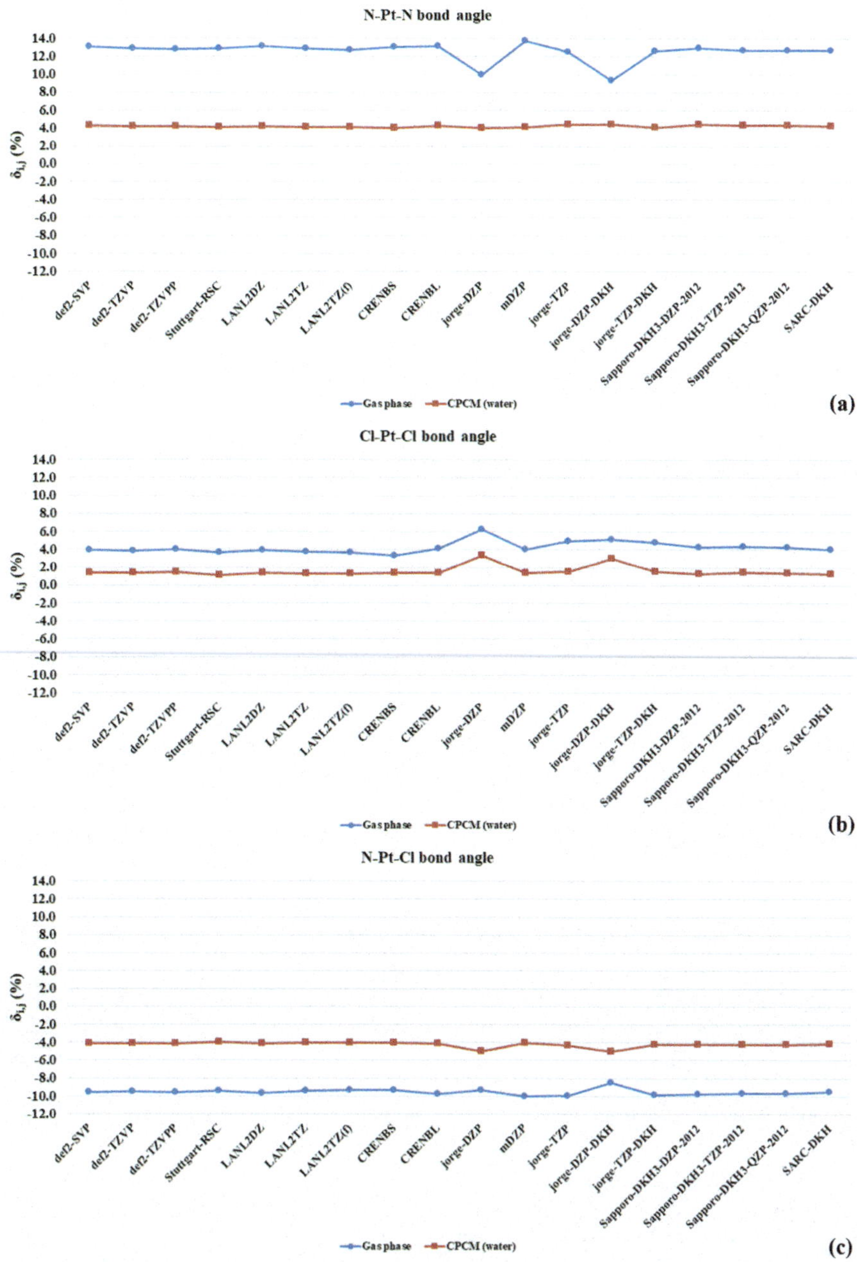

Fig. 7.4 Calculated $\delta_{i,j}(\%)$ for N-Pt–N (**a**), Cl–Pt–Cl (**b**) and N-Pt-Cl (**c**) bond angles with distinct Pt basis sets

N – Pt – Cl angle is underestimated compared to the experimental. This result is due to the N – H · · · – Cl intramolecular hydrogen bond, which is not present in the solid state. The calculated values in the gas phase for this intramolecular hydrogen bond ranged from 2.40 to 2.47 Å. When the solvent effect is included, the N – H · · · – Cl intramolecular hydrogen bond increased, with values ranged from 2.63 to 2.82 Å. As a consequence, the calculated errors in the angles decreased, improving the agreement with the experimental data (Fig. 7.4).

For the bond lengths, we note that the relativistic and solvent effects shorten the Pt – N bond distance, Fig. 7.5, and the solvent effect lengthen the Pt – Cl bond distance. Considering only the calculated values with the C-PCM model, we observe that the calculated Pt – N and Pt – Cl distances were systematically overestimated at all levels, except with the jorge-DZP-DKH Pt basis set that underestimated the values with relative deviations of −2.5 and −1.2% for Pt – N and Pt – Cl, respectively. The largest deviations were found with the non-relativistic jorge-TZP ABS with a relative deviation of 7.3% (Pt – N) and 3.5% (Pt – Cl).

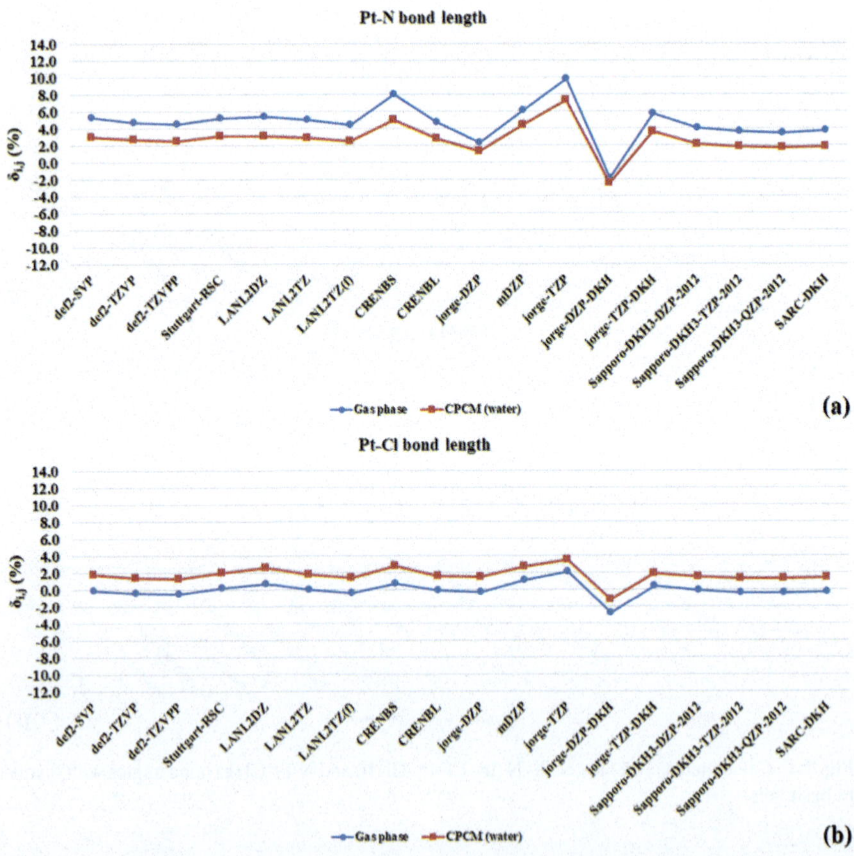

Fig. 7.5 Calculated $\delta_{i,j}(\%)$ for Pt–N (**a**) and Pt-Cl (**b**) bond lengths with distinct Pt basis sets

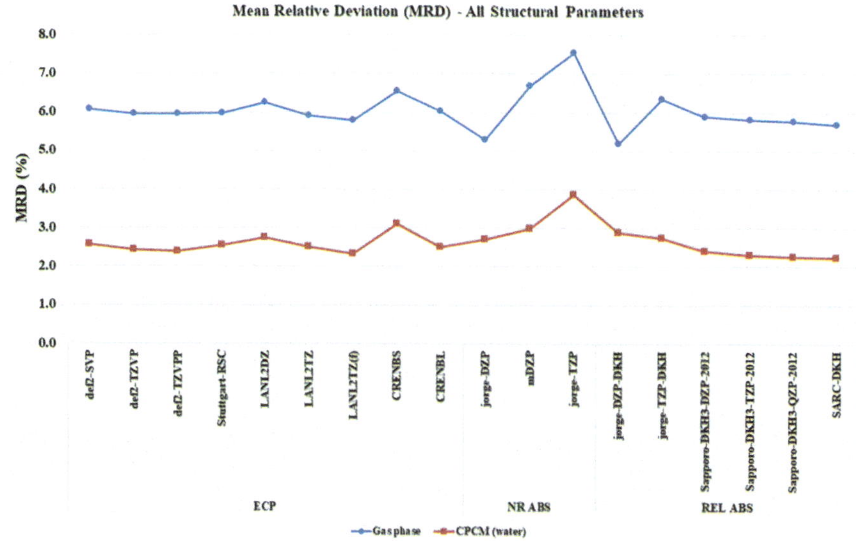

Fig. 7.6 Mean relative deviation (MRD) considering all five structural parameters calculated with distinct Pt basis sets

When we look at all the calculated values for the structural parameters of the cisplatin (Fig. 7.6), we conclude that the quality of the ECPs is comparable to the relativistic ABS, indicating that the scalar relativistic corrections included in the ECPs are sufficient to represent the geometry. Moreover, we also note similar results for double- and triple-zeta basis set for Pt atom.

In short, despite the calculated values with the relativistic ABS are found in better agreement with the experimental data, MRD of 2.2, 2.3, and 2.3% with the SARC-DKH, Sapporo-DKH3-TZP-2012, and Sapporo-DKH3-QZP-2012 basis sets, respectively, the ECPs and the non-relativistic ABS are also considered alternative basis sets to predict structural parameters, with the advantage of the lower computational cost of ECPs.

Aquation reaction

The rate-limiting step of the aquation reaction of cisplatin is shown in Fig. 7.2, represented by the I1 $\rightarrow$ TS process in Fig. 7.3. Herein the activation Gibbs free energy ($\Delta G^{\ddagger}$, Eq. 7.9) and the second-order rate constant (k_2, Eq. 7.10) for the aquation process is discussed. The experimental values for $\Delta G^{\ddagger}$ range from 22.55 kcal mol^{-1} [117] to 23.32 kcal mol^{-1} [118]. The average value of 22.94 $\pm$ 0.39 kcal mol^{-1} will be used in the present work as a reference for comparison with the calculated values. Table 7.5 presents the calculated values using a distinct Pt basis set.

Figure 7.7 shows the calculated values for $\Delta G^{\ddagger}$ with distinct Pt basis set in gas phase and including the solvent effect (C-PCM model). We can observe that the relativistic effect increases the energy barrier, when compared with the values obtained with non-relativistic ABS, and the solvent effect decreases. For the calculations in

Table 7.5 Activation Gibbs free energy ($\Delta G^{\ddagger}$, kcal mol^{-1}) and rate constant (k_2, M^{-1} s^{-1}) for aquation process of cisplatin at B3LYP/Pt basis set/jorge-DZP level for non-relativistic calculations and B3LYP-DKH2/Pt basis set/jorge-DZP-DKH level for relativistic calculations

Pt basis set	$\Delta G^{\ddagger}_{gas}$	$\Delta G^{\ddagger}_{CPCM}$	k_{2-gas}	k_{2-CPCM}
ECP basis set				
def2-SVP	26.20	21.89	3.88×10^{-7}	5.60×10^{-4}
def2-TZVP	26.69	21.97	1.68×10^{-7}	4.88×10^{-4}
def2-TZVPP	26.51	22.22	2.31×10^{-7}	3.23×10^{-4}
Stuttgart-RSC	24.94	21.21	3.25×10^{-6}	1.76×10^{-3}
LANL2DZ	24.57	20.96	6.03×10^{-6}	2.71×10^{-3}
LANL2TZ	25.76	22.24	8.21×10^{-7}	3.09×10^{-4}
LANL2TZ(f)	26.13	22.34	4.34×10^{-7}	2.62×10^{-4}
CRENBS	18.49	15.66	1.75×10^{-1}	2.06×10^{1}
CRENBL	27.02	23.15	9.72×10^{-8}	6.70×10^{-5}
Non-relativistic ABS				
jorge-DZP	18.01	14.25	3.88×10^{-1}	2.22×10^{2}
mDZP	21.78	17.75	6.72×10^{-4}	6.10×10^{-1}
jorge-TZP	12.72	9.88	2.95×10^{3}	3.55×10^{5}
Relativistic ABS				
jorge-DZP-DKH	26.16	21.90	4.17×10^{-7}	5.51×10^{-4}
jorge-TZP-DKH	20.70	16.63	4.19×10^{-3}	4.00×10^{0}
Sapporo-DKH3-DZP-2012	26.53	23.49	2.22×10^{-7}	3.75×10^{-5}
Sapporo-DKH3-TZP-2012	26.14	22.10	4.30×10^{-7}	3.90×10^{-4}
Sapporo-DKH3-QZP-2012	25.91	22.54	6.32×10^{-7}	1.87×10^{-4}
SARC-DKH	26.02	22.06	5.23×10^{-7}	4.19×10^{-4}
Experimental[a]	22.55 – 23.32		$1.02 \times 10^{-4} – 5.20 \times 10^{-5}$	

[a]Experimental values obtained from Bancroft et al. [117] ($\Delta G^{\ddagger} = 22.55$ kcal mol^{-1} and $k_2 = 1.02 \times 10^{-4}$ M^{-1} s^{-1}) and Hindmarsh et al. [118] ($\Delta G^{\ddagger} = 23.32$ kcal mol^{-1} and $k_2 = 5.20 \times 10^{-5}$ M^{-1} s^{-1})

gas phase using the ECPs, the barriers are overestimated by 7.1% (Stuttgart-RSC) and 17.8% (CRENBL), except the value with CRENBS, which is underestimated by 19.4%. However, when the solvent effect is included, the calculated values with the ECPs presented an excellent agreement with the experimental values, with deviations between 0.9% (CRENBL) and 8.7% (LANL2DZ), except again the CRENBS value, underestimated by 31.7%. This behavior presented by ECP CRENBS, different from that presented by the other ECPs, can be explained by its worse description of the Pt valence, since it is the only ECP of 68 electrons, presenting a very short valence shell. The importance of the valence basis set can also be noted by the decrease in the relative error from 4.6% (def2-SVP) to 4.2% (def2-TZVP) and from 8.7% (LANL2DZ) to 3.0% (LANL2TZ), which shows the importance of using a triple-zeta basis set

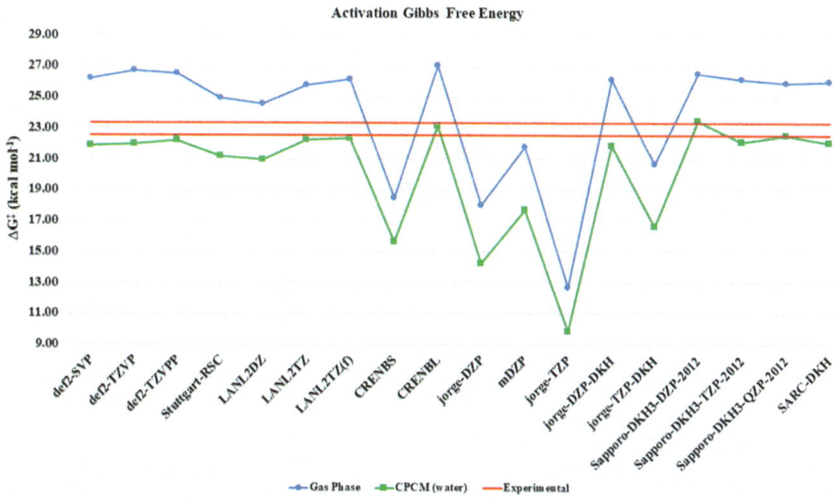

Fig. 7.7 Activation Gibbs free energy ($\Delta G^{\ddagger}$) for aquation reaction of cisplatin calculated using distinct Pt basis sets. The experimental limits are indicated as solid lines

for the valence of heavy atoms. Besides, the decrease from 4.2% (def2-TZVP) to 3.2% (def2-TZVPP) and from 3.0% (LANL2TZ) to 2.6% (LANL2TZ(f)) shows the importance of adding more polarization functions to heavy atoms.

Considering the all-electron non-relativistic basis sets, the jorge-DZP gave an energy barrier of 18.01 kcal mol^{-1} in gas phase (error of 21.5%). Surprisingly, the addition of a set of d-diffuse functions (mDZP basis set) leads to a satisfactory agreement with experimental data, for which the energy barrier in gas phase was 21.78 kcal mol^{-1} (error of 5.0%). This result suggests that the jorge-DZP basis set is not well-balanced and shows the importance of the d-diffuse function in the mDZP basis set, these same behaviour between jorge-DZP and mDZP Pt ABS was also observed in ref. [74]. The calculated barrier in solution (C-PCM) for all non-relativistic ABS decrease and the relative errors were 37.9% and 56.9% for the jorge-DZP and jorge-TZP basis set, respectively. The calculated value with the mDZP basis set using the C-PCM model also decreases and the relative error was 22.6%. The importance of d-diffuse functions can be understood based on the fact that platinum is a d-transition metal and the reaction analyzed involves different structures with distinct coordination shells that require a great flexibility of the valence shell of metal atom, therefore the d-functions play a key role in the description of reactivity of Pt complexes.

When the relativistic all-electron basis sets are used, Fig. 7.7 shows they give similar barriers as the ECPs. The calculated values with these ABS in gas phase are overestimated and, when the solvent effect is included, the calculated values present an excellent agreement with the experimental value for most of them. The calculated values with the relativistic ABS in gas phase presented errors of ~ 14%, except when the jorge-TZP-DKH basis set is used, error of 9.8%. Nevertheless, when the solvent effect is included, except with the jorge-TZP-DKH basis set, all other relativistic ABS presented an excellent agreement with the experimental values, with errors less than 5%. Regarding the calculations with the C-PCM model, it is interesting to note the difference presented when the jorge-DZP and jorge-DZP-DKH basis sets are used. The energy barrier with the jorge-DZP basis set, 14.25 kcal mol^{-1} (error of 37.9%), increased to 21.90 kcal mol^{-1} with the jorge-DZP-DKH, with a relative error of only 4.5%, stressing the importance of the relativistic effects in the basis set. Besides, the size of the basis set also plays a role for the activation energy, as noted for the calculations with the Sapporo-DKH3-QZP-2012 basis set, which relative deviation was only 1.7%. Then, for standard calculations in the gas phase and without the inclusion of relativistic effects, the mDZP basis set for Pt is a good choice, given an error of 5.0%. For calculations with the inclusion of solvent effect, the ECPs lead to satisfactory results, with errors of 0.9% and 2.6% with the CRENBL and LANL2TZ(f), respectively. As expected, for the most complete calculations, considering solvent and relativistic effects, an excellent agreement with the experimental value was obtained using the DKH basis sets jorge-DZP-DKH, Sapporo-DKH3-DZP-2012, Sapporo-DKH3-TZP-2012 and Sapporo-DKH3-QZP-2012, for which the relative deviations were 4.5%, 2.4%, 3.6%, and 1.7%, respectively.

7.3.2 NMR Spectroscopy for Pt(II) Complexes

The Nuclear Magnetic Resonance (NMR) spectroscopy is the main technique for structural characterization of molecules in solution. Moreover, with the advances in recent years, the NMR has played a primary role in drug discovery and design [119]. The nucleus of Pt-195 is the only NMR active isotope of platinum [119] and it has the necessary properties for use in NMR spectroscopy (spin quantum number, $I = 1/2$; natural abundance of 33.8%; relative sensitivity of 9.94×10^{-3} (^{1}H: 1.00); gyromagnetic ratio $\gamma = 5.768 \times 10^{7} rad s^{-1} T^{-1}$; Larmor frequency of 64.5 MHz at 7.05 T) [120–123]. The NMR parameters, chemical shift (δ) and coupling constants, are particularly sensitive to changes in the metal oxidation state (the most common oxidation states of Pt are + 2 and + 4), the type of the ligands, and the coordination mode [123, 124]. The NMR chemical shift of Pt-195 nucleus can be found in a wide range of values from + 8,000 to −7,000 ppm, relative to $[PtCl_6]^{-2}$ in D_2O [120, 124]. The Pt-195 NMR is used in many applications, such as structural elucidations, kinetics, and mechanistic studies [117, 125].

From a theoretical point of view, the computational prediction of Pt-195 NMR parameters is still a challenging task. There are several aspects that should be taken into account for a good calculation, such as the relativistic [126] and the solvent effects [126], the electronic correlation [127], and the basis sets [75]. Regarding the basis set, the platinum basis set plays a key role since a good description of the NMR parameters requires a good description of both core and valence electrons. Thus, an extended and little contracted basis set with sets of polarization functions is required. In this context, the use of ECP to describe the Pt atom is not recommended.

We have found in the literature several papers involving theoretical studies of Pt-195 NMR spectroscopy [75, 76, 127–141]. Recently, Paschoal et al. [75] presented a complete study concerning the prediction of NMR chemical shift of Pt-195 for Pt(II) complexes. The authors showed the importance to include the relativistic effects in predicting properties such as NMR parameters involving heavy nuclei. The authors studied the *cis*-$[Pt(NH_3)_2Cl_2]$, *cis*-$[Pt(NH_3)_2Br_2]$ and *cis*-$[Pt(NH_3)_2I_2]$. The non-relativistic results for the Pt-195 NMR chemical shift showed deviations greater than 1,300 ppm relative to the experimental values for all complexes. For the relativistic Spin–Orbit ZORA approach the largest deviation was 117 ppm for *cis*-$[Pt(NH_3)_2I_2]$. Besides, Paschoal et al. proposed the NMR-DKH basis sets for the H–He, Li-Ne, Na-Ar, K-Ca, Ga-Kr, Rb–Sr, In-Xe, and Pt atoms. Lastly, from an empirical model, the authors calculated the Pt-195 NMR chemical shift for a set of 258 Pt(II) complexes with a mean relative deviation (MRD) of ~ 6% at PBEPBE/NMR-DKH/IEF-PCM(UFF)// B3LYP/LANL2DZ/def2-SVP/IEF-PCM(UFF) level. These same protocols were further applied to predict $^1J(^{195}Pt\text{-}^{15}N)$ coupling constant in Pt(II) complexes with an average error of 10.4%, considering 98 coupling constants [76]. In general, the papers found in the literature do not emphasize the role of the platinum basis set (heavy element) in the Pt-195 NMR calculations. Therefore, in this section, we investigate the influence of Pt basis set in the computation of Pt-195 NMR chemical shift. Relativistic calculations using the Douglas–Kroll–Hess (DKH2) approximation are employed using a distinct relativistic contracted DKH all-electron platinum basis set.

7.3.2.1 Theoretical Methodology

DFT non-relativistic calculations were used to predict the structures of the Pt(II) complexes. Three simple Pt(II) models with available experimental Pt-195 NMR chemical shift were chosen: *cis*-$[Pt(NH_3)_2Cl_2]$ [142], *cis*-$[Pt(NH_3)_2Br_2]$ [143] and *cis*-$[Pt(NH_3)_2I_2]$ [144]. The geometries were optimized and characterized as a minimum point on the PES (all vibrational frequencies found real) at B3LYP level [112] with the def2-SVP effective core potential (ECP) for Pt atom and def2-SVP basis set for ligands atoms [35–37]. The solvent effect was accounted for in the structure and NMR chemical shift calculations, using the Solvation Model Based on Density (SMD) [145]. The solvent considered in the optimization of structure and NMR calculations is the same as that used in the experiments.

The gauge-independent atomic orbitals (GIAO) method [146, 147] was used for calculation of magnetic shielding constant (σ) and the NMR chemical shift was predicted through Eq. (7.12) [75], in which σ_{ref} is the shielding constant for the internal reference, $[PtCl_6]^{-2}$ in D_2O.

$$\delta_{\text{calc}} = \sigma_{\text{ref}} - \sigma_{calc} \tag{7.12}$$

The NMR calculations were carried at the B3LYP level, using the NMR-DKH basis set for ligands atoms [75]. The relativistic effect was considered using the Douglas–Kroll–Hess (DKH) approach [44–47, 113–115] within the DKFULL approximation. In order to assess the role of the Pt basis set in the NMR chemical shift, the relativistic ABS presented in Table 7.3 was considered. In addition, the NMR-DKH basis set for Pt atom [75] found in Table 7.2 was also used. All calculations were performed in NWCHEM 6.6 program [30] and compared with the experimental data [142–144].

7.3.2.2 Results and Discussion

The NMR chemical shifts were calculated using Eq. (7.12) at B3LYP-DKFULL/Pt basis set/NMR-DKH//B3LYP/def2-SVP/def2-SVP. The solvent effect was considered using the SMD method and the relativistic effects by DKFULL approximation as implemented in NWCHEM 6.6 program. As previously discussed, the relativistic effects are of fundamental importance in the prediction of NMR properties of heavy nuclei, such as Pt nucleus, so only relativistic calculations were performed. The results are shown in Table 7.6.

From the results, it is clear that the size of the Pt basis set is very important. When the results obtained with the jorge-DZP-DKH and jorge-TZP-DKH basis set are compared, the mean relative deviation decreases from 73.0% (jorge-DZP-DKH) to 9.1% (jorge-TZP-DKH). The Sapporo and NMR-DKH basis sets lead to the smallest relative deviations, 3.2, 3.4, and 4.4% for Sapporo-DKH3-DZP-2012, Sapporo-DKH3-TZP-2012, and NMR-DKH, respectively. The calculations with the SARC-DKH basis set presented a mean relative deviation of 20%. When the individual errors are analyzed, we note that the error is of the same order for a specific basis set, regardless of the complex, suggesting that the ligand basis sets are somehow well-balanced.

7.4 Concluding Remarks

This chapter has provided a description of the basis sets for heavy atoms and their importance in the context of computational chemistry. Relevant issues that should be considered when dealing with systems containing heavy atoms were discussed. Moreover, unpublished results involving the role of the platinum basis set in describing the structure and reactivity of cisplatin and NMR chemical shift of *cis*-$[Pt(NH_3)_2Cl_2]$, *cis*-$[Pt(NH_3)_2Br_2]$ and *cis*-$[Pt(NH_3)_2I_2]$ complexes were presented.

From the results obtained for the structure and reactivity of cisplatin, we showed that the solvent and relativistic effects play a very important role in the description of the properties. Moreover, a well-balanced Pt basis set is required for a good description of the reactivity of cisplatin. The results suggest that for non-relativistic calculation in the gas phase the mDZP basis set can be considered a good choice. However, for a more accurate description, taking into account the relativistic and

Table 7.6 Pt-195 NMR chemical shifts (in ppm) calculated at B3LYP-DKFULL/Pt basis set/NMR-DKH/SMD level for the $[Pt(NH_3)_2X_2]$, with X = Cl, Br and I

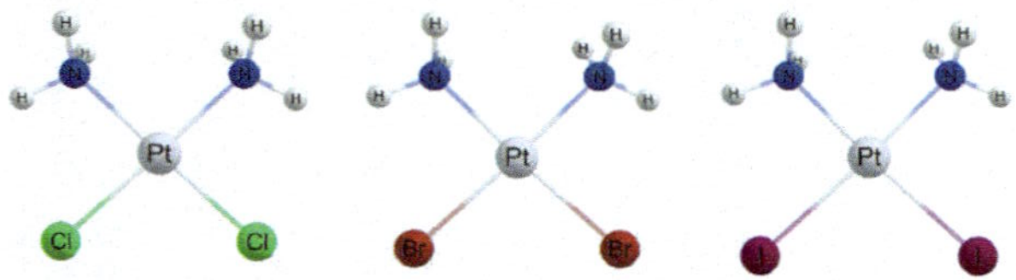

Pt basis set	Cl		Br		I		MRD
jorge-DZP-DKH	−535	(74.5%)	−640	(73.4%)	−939	(71.2%)	73.0%
SARC-DKH	−1715	(18.2%)	−3013	(25.0%)	−3815	(16.9%)	20.0%
jorge-TZP-DKH	−2297	(9.5%)	−2714	(12.6%)	−3436	(5.3%)	9.1%
NMR-DKH	−2123	(1.2%)	−2635	(9.3%)	−3346	(2.5%)	4.4%
Sapporo-DKH3-TZP-2012	−2110	(0.6%)	−2626	(9.0%)	−3284	(0.6%)	3.4%
Sapporo-DKH3-DZP-2012	−1954	(6.8%)	−2416	(0.3%)	−3184	(2.4%)	3.2%
Experimental	−2097[a]		−2410[b]		−3264[c]		

Experimental values obtained from: [a]Rochon and Morneau [142]; [b]Kerrison and Sadler [143]; [c]Rochon et al. [144]. The values in parentheses correspond to the relative deviation. MRD = mean relative deviation. The structures were obtained with the Chemcraft program [150]

solvent effects, the Sapporo-DKH3-XZP-2012 (X = D, T, or Q) basis sets are recommended. Besides, the calculated properties using ECPs presented a satisfactory agreement with the experimental data for both structure and reactivity of cisplatin, which are good choices for large molecules. Overall, the reactivity of cisplatin is much more sensitivity then the structure to the atomic basis set.

Considering the results obtained for NMR chemical shift of Pt-195, the calculated values suggest that a triple-zeta basis set with polarization function should be used, with the Sapporo-DKH3-TZP-2012 and NMR-DKH presenting good agreement with the experimental values.

Acknowledgements The authors would like to thank the PIBIC-CNPq and the Brazilian agency FAPERJ (E-26/200.934/2018 – BOLSA). HFDS also than CNPq and FAPEMIG for the continuing support to NEQC-UFJF laboratory. This study was financed in part by the Coordenação de Aperfeiçoamento de Pessoal de Nível Superior - Brasil (CAPES) - Finance Code 001.

References

1. Dedieu A (2000) Chem Rev 100:543–600. https://doi.org/10.1021/cr980407a
2. Ma DL, He HZ, Leung KH, Chan DSH, Leung CH (2013) Angew Chemie - Int Ed 52:7666–7682. https://doi.org/10.1002/anie.201208414
3. Schwietert CW, McCue JP (1999) Coord Chem Rev 184:67–89. https://doi.org/10.1016/S0010-8545(98)00205-7
4. Trudu F, Amato F, Vaňhara P, Pivetta T, Peña-Méndez EM, Havel J (2015) J Appl Biomed 13:79–103. https://doi.org/10.1016/j.jab.2015.03.003
5. Medici S, Peana M, Nurchi VM, Lachowicz JI, Crisponi G, Zoroddu MA (2015) Coord Chem Rev 284:329–350. https://doi.org/10.1016/j.ccr.2014.08.002
6. Barry NPE, Sadler PJ (2013) Chem Commun 49:5106. https://doi.org/10.1039/c3cc41143e
7. Wang X, Chang H, Xie J, Zhao B, Liu B, Xu S, Pei W, Ren N, Huang L, Huang W (2014) Coord Chem Rev 273–274:201–212. https://doi.org/10.1016/j.ccr.2014.02.001
8. Ma Y, Wang Y (2010) Coord Chem Rev 254:972–990. https://doi.org/10.1016/j.ccr.2010.02.013
9. Bünzli JCG (2015) Coord Chem Rev 293–294:19–47. https://doi.org/10.1016/j.ccr.2014.10.013
10. Armelao L, Quici S, Barigelletti F, Accorsi G, Bottaro G, Cavazzini M, Tondello E (2010) Coord Chem Rev 254:487–505. https://doi.org/10.1016/j.ccr.2009.07.025
11. de Sá G, Malta O, de Mello DC, Simas A, Longo R, Santa-Cruz P, da Silva E (2000) Coord Chem Rev 196:165–195. https://doi.org/10.1016/S0010-
12. Jones MB, Gaunt AJ (2013) Chem Rev 113:1137–1198. https://doi.org/10.1021/cr300198m
13. Kaltsoyannis N (2003) Chem Soc Rev 32:9–16. https://doi.org/10.1039/b204253n
14. Rosenberg B, VanCamp L, Trosko JE, Mansour VH (1969) Nature 222:385–386. https://doi.org/10.1038/222385a0
15. Wai-Yin Sun R, Ma D-L, Wong EL-M, Che C-M (2007) Dalt Trans 4884–4892. https://doi.org/10.1039/b705079h.
16. Ronconi L, Sadler PJ (2007) Coord Chem Rev 251:1633–1648. https://doi.org/10.1016/j.ccr.2006.11.017
17. Mjos KD, Orvig C (2014) Chem Rev 114:4540–4563. https://doi.org/10.1021/cr400460s
18. Van RSH, Sadler PJ (2009) Drug Discovery Today 14:1089–1097. https://doi.org/10.1016/j.drudis.2009.09.003

19. Barry NPE, Sadler PJ (2014) Pure Appl Chem 86:1897–1910. https://doi.org/10.1515/pac-2014-0504
20. Reichert DE, Lewis JS, Anderson CJ (1999) Coord Chem Rev 184:3–66. https://doi.org/10.1016/S0010-8545(98)00207-0
21. Thompson KH, Orvig C (2003) Science 300:936–939. https://doi.org/10.1126/science.1083004
22. Helgaker T, Jorgensen P, Olsen J (2000) In Molecular Eletronic-Structure Theory. Wiley, England
23. Levine IN (2009) In Quantum Chemistry, 6th edn. Pearson, United States of America
24. Davidson ER, Feller D (1986) Chem Rev 86:681–696. https://doi.org/10.1021/cr00074a002
25. Jensen F (2007) In Introduction to Computational Chemistry, 2nd edn. Wiley, England
26. Boys SF (1950) Proc R Soc A Math Phys. Eng Sci 200:542–554. https://doi.org/10.1098/rspa.1950.0036
27. Cramer CJ (2004) In Essentials of Computational Chemistry - Theories and Models, 2nd edn. Wiley, England
28. Frisch MJ, Trucks GW, Schlegel HB, Scuseria GE, Robb MA, Cheeseman JR, Scalmani G, Barone V, Petersson GA, Nakatsuji H, Li X, Caricato M, Marenich A, Bloino J, Janesko BG, Gomperts R, Mennucci B, Hratchian HP, Ortiz JV, Izmaylov AF, Sonnenberg JL, Williams-Young D, Ding F, Lipparini F, Egidi F, Goings J, Peng B, Sonnenberg JL, Williams-Young D, Ding F, Lipparini F, Egidi F, Goings J, Peng B, Petrone A, Henderson T, Ranasinghe D, Zakrzewski VG, Gao J, Rega N, Zheng G, Liang Petrone A, Henderson T, Ranasinghe D, Zakrzewski VG, Gao J, Rega N, Zheng G, Liang W, Hada M, Ehara M, Toyota K, Fukuda R, Hasegawa J, Ishida M, W, Hada M, Ehara M, Toyota K, Fukuda R, Hasegawa J, Ishida M, Nakajima T, Honda Y, Kitao O, Nakai H, Vreven T, Throssell K, Montgomery Jr JA, Peralta JE, Ogliaro F, Bearpark M, Heyd JJ, Brothers E, Kudin KN, Staroverov VN, Keith T, Kobayashi R, Normand J, Raghavachari K, Rendell A, Burant JC, Iyengar SS, Tomasi J, Cossi M, Millam JM, Klene M, Adamo C, Cammi R, Ochterski JW, Martin RL, Morokuma K, Farkas O, Foresman JB, Fox DJ (2016) In Gaussian 09, Revision D.01. Gaussian, Inc., Wallingford CT, 2009.
29. Neese F (2018) WIREs Comput Mol Sci 8:e1327. https://doi.org/10.1002/wcms.1327
30. Valiev M, Bylaska EJ, Govind N, Kowalski K, Straatsma TP, Van Dam HJJ, Wang D, Nieplocha J, Apra E, Windus TL, de Jong WA (2010) Comput Phys Commun 181:1477–1489. https://doi.org/10.1016/j.cpc.2010.04.018
31. Fonseca Guerra C, Snijders JG, te Velde G, Baerends EJ (1998) Theor Chem Accounts 99:391–403. https://doi.org/10.1007/s002140050353
32. te Velde G, Bickelhaupt FM, Baerends EJ, Fonseca Guerra C, van Gisbergen SJA, Snijders JG, Ziegler T (2001) J Comput Chem 22:931–967. https://doi.org/10.1002/jcc.1056
33. Baerends EJ, Ziegler T, Atkins AJ, Autschbach J, Bashford D, Baseggio O, Bérces A, Bickelhaupt FM, Bo C, Boerritger PM, Cavallo L, Daul C, Chong DP, Chulhai D V, Deng L, Dickson RM, Dieterich JM, Ellis DE, van Faassen M, Ghysels A, Giammona A, van Gisbergen SJA, Goez A, Götz AW, Gusarov S, Harris FE, van den Hoek P, Hu Z, Jacob CR, Jacobsen H, Jensen L, Joubert L, Kaminski JW, van Kessel G, König C, Kootstra F, Kovalenko A, Krykunov M, van Lenthe E, McCormack DA, Michalak A, Mitoraj M, Morton SM, Neugebauer J, Nicu VP, Noodleman L, Osinga VP, Patchkovskii S, Pavanello M, Peeples CA, Philipsen PHT, Post D, Pye CC, Ramanantoanina H, Ramos P, Ravenek W, Rodríguez JI, Ros P, Rüger R, Schipper PRT, Schlüns D, van Schoot H, Schreckenbach G, Seldenthuis JS, Seth M, Snijders JG, Solà M, M. S, Swart M, Swerhone D, te Velde G, Tognetti V, Vernooijs P, Versluis L, Visscher L, Visser O, Wang F, Wesolowski TA, van Wezenbeek EM, Wiesenekker G, Wolff SK, Woo TK, Yakovlev AL (2019) In ADF 2019.3, SCM, Theoretical Chemistry, Vrije Universiteit, Amsterdam, The Netherlands, https://www.scm.com.
34. de Castro EVR, Jorge FE (1998) J Chem Phys 108:5225–5229. https://doi.org/10.1063/1.475959
35. Feller D (1996) J Comput Chem 17:1571–1586. https://doi.org/10.1002/(SICI)1096-987X(199610)17:13%3c1571::AID-JCC9%3e3.0.CO;2-P

36. Schuchardt KL, Didier BT, Elsethagen T, Sun L, Gurumoorthi V, Chase J, Li J, Windus TL (2007) J Chem Inf Model 47:1045–1052. https://doi.org/10.1021/ci600510j
37. Pritchard BP, Altarawy D, Didier B, Gibson TD, Windus TL (2019) J Chem Inf Model 59:4814–4820. https://doi.org/10.1021/acs.jcim.9b00725
38. Pyykko P, Desclaux JP (1979) Acc Chem Res 12:276–281. https://doi.org/10.1021/ar50140a002
39. Pitzer KS (1979) Acc Chem Res 12:271–276. https://doi.org/10.1021/ar50140a001
40. Thayer JS (2010) Relativistic Effects and the Chemistry of the Heavier Main Group Elements. In Relativistic Methods for Chemists, Springer, The Netherlands
41. Pyykko P (1988) Chem Rev 88:563–594. https://doi.org/10.1021/cr00085a006
42. Pyykkö P (2012) Annu Rev Phys Chem 63:45–64. https://doi.org/10.1146/annurev-physchem-032511-143755
43. Rose SJ, Grant IP, Pyper NC (1978) J Phys B at Mol Phys 11:1171–1176. https://doi.org/10.1088/0022-3700/11/7/016
44. Jansen G, Hess BA (1989) Phys Rev a 39:6016–6017. https://doi.org/10.1103/PhysRevA.39.6016
45. Hess BA (1986) Phys Rev a 33:3742–3748. https://doi.org/10.1103/PhysRevA.33.3742
46. Hess BA (1985) Phys Rev a 32:756–763. https://doi.org/10.1103/PhysRevA.32.756
47. Douglas M, Kroll NM (1974) Ann Phys 82:89–155. https://doi.org/10.1016/0003-4916(74)90333-9
48. Wolf A, Reiher M, Hess BA (2002) J Chem Phys 117:9215–9226. https://doi.org/10.1063/1.1515314
49. Van Lenthe E, Snijders JG, Baerends EJ (1996) J. Chem. Phys. 105:6505–6516. https://doi.org/10.1063/1.472460
50. Van Lenthe E, Baerends EJ, Snijders JG (1994) J Chem Phys 101:9783–9792. https://doi.org/10.1063/1.467943
51. Van Wüllen C (1998) J Chem Phys 109:392–399. https://doi.org/10.1063/1.476576
52. Hay PJ, Wadt WR (1985) J Chem Phys 82:299–310. https://doi.org/10.1063/1.448975
53. Wadt WR, Hay PJ, Wadt WA, Hay PJ (1997) J Chem Phys 284:284–298. https://doi.org/10.1063/1.448800
54. Hay PJ, Wadt WR (1985) J Chem Phys 82:270–283. https://doi.org/10.1063/1.448799
55. Weigend F, Ahlrichs R (2005) Phys Chem Chem Phys 7:3297. https://doi.org/10.1039/b508541a
56. Peterson KA (2003) J Chem Phys 119:11099–11112. https://doi.org/10.1063/1.1622923
57. Peterson KA, Puzzarini C (2005) Theor Chem Acc 114:283–296. https://doi.org/10.1007/s00214-005-0681-9
58. Peterson KA, Figgen D, Dolg M, Stoll H (2007) J Chem Phys 126:124101. https://doi.org/10.1063/1.2647019
59. Figgen D, Peterson KA, Dolg M, Stoll H (2009) J Chem Phys 130:164108. https://doi.org/10.1063/1.3119665
60. Frenking G, Antes I, Böhme M, Dapprich S, Ehlers AW, Jonas V, Neuhaus A, Otto M, Stegmann R, Veldkamp A, Vyboishchikov SF (1996) Rev Comput Chem 8:63–144. https://doi.org/10.1002/9780470125854.ch2
61. Martins LSC, Jorge FE, Franco ML, Ferreira IB (2016) J Chem Phys 145:244113. https://doi.org/10.1063/1.4973377
62. Campos CT, de Oliveira AZ, Ferreira IB, Jorge FE, Martins LSC (2017) Chem Phys Lett 675:1–5. https://doi.org/10.1016/j.cplett.2017.02.080
63. Jorge FE, Martins LSC, Franco ML (2016) Chem Phys Lett 643:84–88. https://doi.org/10.1016/j.cplett.2015.11.013
64. Canal Neto A, Jorge FE (2013) Chem Phys Lett 582:158–162. https://doi.org/10.1016/j.cplett.2013.07.045
65. de Berrêdo RC, Jorge FE (2010) J Mol Struct: THEOCHEM 961:107–112. https://doi.org/10.1016/j.theochem.2010.09.007

66. Barros CL, De Oliveira PJP, Jorge FE, Canal Neto A, Campos M (2010) Mol Phys 108:1965–1972. https://doi.org/10.1080/00268976.2010.499377
67. Canal Neto A, Muniz EP, Centoducatte R, Jorge FE (2005) J Mol Struct: THEOCHEM 718:219–224. https://doi.org/10.1016/j.theochem.2004.11.037
68. Camiletti GG, Machado SF, Jorge FE (2008) J Comput Chem 29:2434–2444. https://doi.org/10.1002/jcc.20996
69. Campos CT, Jorge FE (2013) Mol Phys 111:165–171. https://doi.org/10.1080/00268976.2012.709282
70. Martins LSC, Jorge FE, Machado SF (2015) Mol Phys 113:3578–3586. https://doi.org/10.1080/00268976.2015.1040095
71. Machado SF, Camiletti GG, Neto AC, Jorge FE, Jorge RS (2009) Mol Phys 107:1713–1727. https://doi.org/10.1080/00268970903042258
72. Barbieri PL, Fantin PA, Jorge FE (2006) Mol Phys 104:2945–2954. https://doi.org/10.1080/00268970600899018
73. Jorge FE, Canal Neto A, Camiletti GG, Machado SF (2009) J Chem Phys 130:064108. https://doi.org/10.1063/1.3072360
74. Paschoal D, Marcial BL, Lopes JF, De Almeida WB, Dos Santos HF (2012) J Comput Chem 33:2292–2302. https://doi.org/10.1002/jcc.23061
75. Paschoal D, Guerra CF, de Oliveira MAL, Ramalho TC, Dos Santos HF (2016) J Comput Chem 37:2360–2373. https://doi.org/10.1002/jcc.24461
76. Carvalho J, Paschoal D, Fonseca Guerra C, Dos Santos HF (2020) Chem Phys Lett 745:137279. https://doi.org/10.1016/j.cplett.2020.137279
77. Pantazis DA, Neese F (2011) J Chem Theory Comput 7:677–684. https://doi.org/10.1021/ct100736b
78. Pantazis DA, Neese F (2012) Theor Chem Acc 131:1–7. https://doi.org/10.1007/s00214-012-1292-x
79. Pantazis DA, Neese F (2009) J Chem Theory Comput 5:2229–2238. https://doi.org/10.1021/ct100736b
80. Pantazis DA, Chen X-Y, Landis CR, Neese F (2008) J Chem Theory Comput 4:908–919. https://doi.org/10.1021/ct800047t
81. Noro T, Sekiya M, Koga T (2012) Theor Chem Acc 131:1124. https://doi.org/10.1007/s00214-012-1124-z
82. Noro T, Sekiya M, Koga T (2013) Theor Chem Acc 132:1363. https://doi.org/10.1007/s00214-013-1363-7
83. Sekiya M, Noro T, Koga T, Shimazaki T (2012) Theor Chem Acc 131:1247. https://doi.org/10.1007/s00214-012-1247-2
84. Roos BO, Lindh R, Malmqvist P-Å, Veryazov V, Widmark P-O (2005) Chem Phys Lett 409:295–299. https://doi.org/10.1016/j.cplett.2005.05.011
85. Roos BO, Lindh R, Malmqvist PÅ, Veryazov V, Widmark PO (2005) J Phys Chem a 109:6575–6579. https://doi.org/10.1021/jp0581126
86. Roos BO, Lindh R, Malmqvist PÅ, Veryazov V, Widmark PO, Borin AC (2008) J Phys Chem a 112:11431–11435. https://doi.org/10.1021/jp803213j
87. Watanabe Y, Tatewaki H, Koga T, Matsuoka O (2006) J Comput Chem 27:48–52. https://doi.org/10.1002/jcc.20313
88. Tsuchiya T, Abe M, Nakajima T, Hirao K (2001) J Chem Phys 115:4463–4472. https://doi.org/10.1063/1.1390515
89. Nakajima T, Hirao K (2002) J Chem Phys 116:8270–8275. https://doi.org/10.1063/1.1470496
90. Gomes ASP, Dyall KG, Visscher L (2010) Theor Chem Acc 127:369–381. https://doi.org/10.1007/s00214-009-0725-7
91. Dyall KG (2004) Theor Chem Acc 112:403–409. https://doi.org/10.1007/s00214-004-0607-y
92. Dyall KG (2007) Theor Chem Acc 117:483–489. https://doi.org/10.1007/s00214-006-0174-5
93. Dyall KG (2009) J Phys Chem a 113:12638–12644. https://doi.org/10.1021/jp905057q
94. Dyall KG (2011) Theor Chem Acc 129:603–613. https://doi.org/10.1007/s00214-011-0906-z
95. Dyall KG (2006) Theor Chem Acc 115:441–447. https://doi.org/10.1007/s00214-006-0126-0

96. Dyall KG, Gomes ASP (2009) Theor Chem Acc 125:97–100. https://doi.org/10.1007/s00214-009-0717-7
97. Dyall KG (2012) Theor Chem Acc 131:1–20. https://doi.org/10.1007/s00214-012-1172-4
98. Gomes ASP, Saue T, Visscher L, Jensen HJAa., Bast R, Aucar I A, Bakken V, Dyall KG, Dubillard S, Ekström U, Eliav E, Enevoldsen T, Faßhauer E, Fleig T, Fossgaard O, Halbert L, Hedegård ED, Helgaker T, Helmich-Paris B, Henriksson J, Iliaš M, Jacob ChR, Knecht S, Komorovský S, Kullie O, Lærdahl JK, Larsen CV, Lee YS, Nataraj HS, Nayak MK, Norman P, Olejniczak G, Olsen J, Olsen JMH, Park YC, Pedersen JK, Pernpointner M, Di Remigio R, Ruud K, Sałek P, Schimmelpfennig B, Senjean B, Shee A, Sikkema J, Thorvaldsen AJ, Thyssen J, van Stralen J, Vidal ML, Villaume S, Visser O, Winther T, Yamamoto S (2019) DIRAC, A relativistic ab initio electronic structure program, Release DIRAC19. doi: https://doi.org/10.5281/zenodo.3572669.
99. Pollak P, Weigend F (2017) J Chem Theory Comput 13:3696–3705. doi: https://doi.org/10.1021/acs.jctc.7b00593
100. Van Lenthe E, Baerends EJ (2003) J Comput Chem 24:1142–1156. https://doi.org/10.1002/jcc.10255
101. Apps M, Choi E, Wheate N (2015) Endocr Relat Cancer 22:R219–R233. https://doi.org/10.1530/ERC-15-0237
102. Fanelli M, Formica M, Fusi V, Giorgi L, Micheloni M, Paoli P (2016) Coord Chem Rev 310:41–79. https://doi.org/10.1016/j.ccr.2015.11.004
103. Dasari S, Bernard Tchounwou P (2014) Eur J Pharmacol 740:364–378. https://doi.org/10.1016/j.ejphar.2014.07.025
104. Johnstone TC, Suntharalingam K, Lippard SJ (2016) Chem Rev 116:3436–3486. https://doi.org/10.1021/acs.chemrev.5b00597
105. Costa LAS, Rocha WR, De Almeida WB, Dos Santos HF (2005) J Inorg Biochem 99:575–583. https://doi.org/10.1016/j.jinorgbio.2004.11.017
106. Pavankumar PNV, Seetharamulu P, Yao S, Saxe JD, Reddy DG, Hausheer FH (1999) J Comput Chem 20:365–382. https://doi.org/10.1002/(SICI)1096-987X(199902)20:3%3c365::AID-JCC8%3e3.0.CO;2-1
107. Amado AM, Fiuza SM, Marques MPM, De Carvalho LAEB (2007) J Chem Phys 127:1–10. https://doi.org/10.1063/1.2787528
108. Burda JV, Zeizinger M, Šponer J, Leszczynski J (2000) J Chem Phys 113:2224–2232. https://doi.org/10.1063/1.482036
109. Zhang Y, Guo Z, You XZ (2001) J Am Chem Soc 123:9378–9387. https://doi.org/10.1021/ja0023938
110. Burda JV, Zeizinger M, Leszczynski J (2004) J Chem Phys 120:1253–1262. https://doi.org/10.1063/1.1633757
111. Barone V, Cossi M (1998) J Phys Chem a 102:1995–2001. https://doi.org/10.1021/jp9716997
112. Stephens PJ, Devlin FJ, Chabalowski CF, Frisch MJ (1994) J Phys Chem 98:11623–11627. https://doi.org/10.1021/j100096a001
113. Visscher L, Dyall KG (1997) at Data Nucl Data Tables 67:207–224. https://doi.org/10.1006/adnd.1997.0751
114. de Jong WA, Harrison RJ, Dixon DA (2001) J Chem Phys 114:48. https://doi.org/10.1063/1.1329891
115. Barysz M, Sadlej AJ (2001) J Mol Struct: THEOCHEM 573:181–200. https://doi.org/10.1016/S0166-1280(01)00542-5
116. Milburn GHW, Truter MR (1966) J Chem Soc A Inorganic, Phys Theor 1609-1616. doi: https://doi.org/10.1039/j19660001609.
117. Bancroft DP, Lepre CA, Lippard SJ (1990) J Am Chem Soc 112:6860–6871. https://doi.org/10.1021/ja00175a020
118. Hindmarsch K, House DA, Turnbull MM (1997) Inorg Chim Acta 257:11–18. https://doi.org/10.1016/S0020-1693(96)05403-5
119. Vinje J, Sletten E (2007) Anticancer Agents Med Chem 7:35–54. https://doi.org/10.2174/187152007779313982

120. Pregosin PS (1986) Annu Reports NMR Spectrosc 17:285–349. https://doi.org/10.1016/S0066-4103(08)60238-0
121. Pesek JJ, Mason WR (1977) J Magn Reson 25:519–529. https://doi.org/10.1016/0022-2364(77)90217-7
122. Freeman W, Pregosin P, Sze S, Venanzi L (1976) J Magn Reson 22:473–478. https://doi.org/10.1016/0022-2364(76)90010-X
123. Still BM, Kumar PGA, Aldrich-Wright JR, Price WS (2007) Chem Soc Rev 36:665–686. https://doi.org/10.1039/B606190G
124. Pregosin PS (1982) Coord Chem Rev 44:247–291. https://doi.org/10.1016/S0010-8545(00)80523-8
125. Albrecht M, Rodríguez G, Schoenmaker J, van Koten G (2000) Org Lett 2:3461–3464. https://doi.org/10.1021/ol0064841
126. Bühl M, Kaupp M, Malkina OL, Malkin VG (1999) J Comput Chem 20:91–105. https://doi.org/10.1002/(SICI)1096-987X(19990115)20:1%3c91::AID-JCC10%3e3.0.CO;2-C
127. Pawlak T, Munzarová ML, Pazderski L, Marek R (2011) J Chem Theory Comput 7:3909–3923. https://doi.org/10.1021/ct200366n
128. Koch KR, Burger MR, Kramer J, Westra AN (2006) Dalt Trans 3277–3284. doi: https://doi.org/10.1039/B605182K.
129. Vícha J, Patzschke M, Marek R (2013) Phys Chem Chem Phys 15:7740. https://doi.org/10.1039/c3cp44440f
130. Sutter K, Autschbach J (2012) J Am Chem Soc 134:13374–13385. https://doi.org/10.1021/ja3040762
131. Sterzel M, Autschbach J (2006) Inorg Chem 45:3316–3324. https://doi.org/10.1021/ic052143y
132. Burger MR, Kramer J, Chermette H, Koch KR (2010) Magn Reson Chem 48:S38–S47. https://doi.org/10.1002/mrc.2607
133. Autschbach J, Zheng S (2008) Magn Reson Chem 46:S45–S55. https://doi.org/10.1002/mrc.2289
134. Autschbach J, Le Guennic B (2004) Chem - Eur J 10:2581–2589. https://doi.org/10.1002/chem.200305513
135. Tsipis AC, Karapetsas IN (2016) Magn Reson Chem 54:656–664. https://doi.org/10.1002/mrc.4426
136. Tsipis AC, Karapetsas IN (2015) J Coord Chem 68:3788–3804. https://doi.org/10.1080/00958972.2015.1083095
137. Gabano E, Marengo E, Bobba M, Robotti E, Cassino C, Botta M, Osella D (2006) Coord Chem Rev 250:2158–2174. https://doi.org/10.1016/j.ccr.2006.02.011
138. Gilbert TM, Ziegler T (1999) J Phys Chem a 103:7535–7543. https://doi.org/10.1021/jp992202r
139. Truflandier LA, Sutter K, Autschbach J (2011) Inorg Chem 50:1723–1732. https://doi.org/10.1021/ic102174b
140. Le Guennic B, Autschbach J (2011) Can J Chem 89:814–821. https://doi.org/10.1139/v11-054
141. Tsipis AC, Karapetsas IN (2014) Dalt Trans 43:5409–5426. https://doi.org/10.1039/C3DT53594K
142. Rochon FD, Morneau A (1991) Magn Reson Chem 29:120–126. https://doi.org/10.1002/mrc.1260290205
143. Kerrison SJS, Sadler PJ (1985) Inorg Chim Acta 104:197–201. https://doi.org/10.1016/S0020-1693(00)86771-7
144. Rochon FD, Doyon M, Butler IS (1993) Inorg Chem 32:2717–2723. https://doi.org/10.1021/ic00064a023
145. Marenich AV, Cramer CJ, Truhlar DG (2009) J Phys Chem B 113:6378–6396. https://doi.org/10.1021/jp810292n
146. Cheeseman JR, Trucks GW, Keith TA, Frisch MJ (1996) J Chem Phys 104:5497–5509. https://doi.org/10.1063/1.471789

147. Wolinski K, Hinton JF, Pulay P (1990) J Am Chem Soc 112:8251–8260. https://doi.org/10.1021/ja00179a005
148. Hanwell MD, Curtis DE, Lonie DC, Vandermeersch T, Zurek E, Hutchison GR (2012) J Cheminform 4:17. https://doi.org/10.1186/1758-2946-4-17
149. Avogadro: an open-source molecular builder and visualization tool. Version 1.2.0. https://avogadro.cc/.
150. Chemcraft - graphical software for visualization of quantum chemistry computations. https://www.chemcraftprog.com.
151. BIOVIA, Dassault Systèmes, BIOVIA Draw, 19.1, San Diego: Dassault Systèmes, 2020.
152. Cherinka B, Andrews BH, Sánchez-Gallego J, Brownstein J, Argudo-Fernández M, Blanton M, Bundy K, Jones A, Masters K, Law DR, Rowlands K, Weijmans A-M, Westfall K, Yan R (2019) Astron J 158:74. https://doi.org/10.3847/1538-3881/ab2634
153. Laidler KJ, King MC (1983) J Phys Chem 87:2657–2664. https://doi.org/10.1021/j100238a002

Chapter 8
Adaptable Gaussian Bases for Quantum Dynamics of the Nuclei

Sophya Garashchuk

8.1 Introduction

Classical mechanics often gives an adequate representation of the nuclei in molecular dynamics simulations. Yet, the nuclear quantum-mechanical effects (**NQE**s) may play a significant role in chemical and physical processes in a wide range of molecular environments from the hydrogen storage within the metal-organic frameworks [1] to photovoltaic or spin-responsive materials [2–4], to enzyme activity [5]. The NQEs are typically the largest for light nuclei at low temperatures and energies, or more specifically when the characteristic energy of a process is comparable to the separation between the vibrational energy levels of chemical bonds.

The most general description of NQE comes from a solution to the time-dependent Schrödinger equation (**TDSE**), possibly, with the time-dependent potential,

$$\hat{H}\psi(\mathbf{x},t) = \imath\hbar\frac{\partial}{\partial t}\psi(\mathbf{x},t). \tag{8.1}$$

The Hamiltonian operator is a sum of the kinetic and potential energy operators, $\hat{K}$ and $\hat{V}$, the latter being a function of coordinates and time:

$$\hat{H} = \hat{K} + V(\mathbf{x},t), \quad \hat{K} = -\frac{\hbar^2}{2}\nabla^T \boldsymbol{M}^{-1}\nabla = -\frac{\hbar^2}{2}\sum_{n=1}^{d}\frac{1}{m_n}\frac{\partial^2}{\partial x_n^2}. \tag{8.2}$$

Throughout this chapter, we consider SE only for the nuclei, evolving on a single electronic potential energy surface (**PES**). For simplicity, we will describe the nuclei in Cartesian coordinates (unless indicated otherwise) with the diagonal kinetic energy operator. In other words, $\boldsymbol{M}$ is the diagonal matrix of the particle masses, $M_{nn} = m_n$,

S. Garashchuk (✉)
University of South Carolina, Columbia, SC 29208, USA
e-mail: garashchuk@sc.edu

E. Perlt (ed.), *Basis Sets in Computational Chemistry*, Lecture Notes in Chemistry 107,
https://doi.org/10.1007/978-3-030-67262-1_8

where index n enumerates the degrees of freedom (**DOF**s). Each particle is described by three coordinates in Cartesian space listed in a single vector $\mathbf{x}$. The total number of DOFs (and the size of the vectors and matrices) is $d = 3\times$ (number of particles). The atomic unit of the Planck's constant, $\hbar = 1$, is used henceforth. We will drop time as the argument in the external potential, $V \equiv V(\mathbf{x})$, with the understanding that all methodologies described below, except those based on the diagonalization of the Hamiltonian matrix of Sect. 8.3, are applicable to time-dependent potentials.

Computational efforts of describing a general quantum system, fully coupled by anharmonic interactions, scale exponentially with the system size. Therefore, the efficient basis representation of wavefunctions is essential for the studies of high-dimensional molecular systems. While the representation of wavefunctions in terms of standard bases is well established in the electronic structure theory, it remains an outstanding challenge in quantum mechanics of nuclei. This situation may be attributed to several factors.

(i) The classical description of nuclei is adequate in many situations (molecular dynamics is very useful). Therefore, quantum molecular dynamics has not received as much attention from theorists as the electronic structure.

(ii) The NQE are very sensitive to the quality of the electronic PES on which the dynamics unfolds. Thus, the development of practical methods (including theory, hardware, and software) of computing globally accurate PES was necessary before the NQE could be rigorously included.

(iii) Finally, the forces acting between the quantum nuclei are complicated many-body interactions and the systems undergo large-amplitude motion (reaction dynamics, isomerization, diffuse vibrational states, and highly excited vibrational states). Consequently, most multidimensional quantum dynamics approaches are fairly system-specific.

While there is a number of standard approaches to solve the time-dependent Schrodinger equation for the nuclei through explicit time-evolution of wavefunctions, such as the split-operator/Fast Fourier Transform and Chebyshev expansion of the Hamiltonian, or the Hamiltonian diagonalization (iterative schemes) [6–10], there is no standard approach of generating a wavefunction representation (beyond a few dimensions), which would remain accurate and practical in the course of dynamics.

The term *quantum dynamics* is often used to indicate that one is dealing with the Schrödinger equation for the *nuclei* as opposed to electrons, regardless of whether a time-dependent or time-independent wavefunction is required to answer the question at hand. Formally, for time-independent interactions, the time-dependent and time-independent solutions to the SE contain equivalent information, and are related through the Fourier transform. In other words, if a time-dependent solution $\psi(x, t)$ to the TDSE (8.1) is available at all times, one can extract *all* energy eigenstates $\chi(x, E)$ as the Fourier transform of $\psi(\mathbf{x}, t)$ into the energy domain,

$$\chi(x, E) = \mathcal{N}_E \int_{-\infty}^{\infty} \psi(x, t) e^{\imath Et} dt, \tag{8.3}$$

where $\mathcal{N}_E$ is the normalization constant. Vice versa, knowing all the eigenstates,

$$\hat{H}\chi(x, E) = E\chi(x, E), \tag{8.4}$$

one can reconstruct the time-evolution of *any* initial wavefunction $\psi(x, 0)$:

$$\psi(x, t) = \int_0^\infty dE\langle\chi(x, E)|\psi(x, 0)\rangle e^{-\imath Et}. \tag{8.5}$$

'All eigenstates' above means the ones that overlap with $\psi(x, 0)$; we omit the discussion of degenerate eigenstates for simplicity. The integral in Eq. (8.5) implies the integration over the states of the continuum spectrum and summation over the discrete eigenstates as appropriate for the problem.

The standard ways to represent a wavefunction (as a linear combination of fixed in time and space functions) are the finite basis representation (**FBR**) and the discrete variable representation (**DVR**). The eigenvalues of the Hamiltonian operator evaluated in FBR, traced to the variational minimization of the energy functional, give the upper bound on energy levels, a useful feature. The downside to FBR is the high cost of computing the matrix elements of the potential energy operator, especially if the basis functions are delocalized, and the dense character of the resulting Hamiltonian matrices. The DVR approach, introduced by Light in the 1980s [11–13] and reviewed in Ref. [14], is an elegant way of addressing both deficiencies of FBR. In DVR, which is equivalent to the evaluation of integrals by quadrature over the related to it finite basis, the potential energy matrix is diagonal and the high-energy regions of the coordinate space can be naturally excluded from consideration. Thus, the number of the PES evaluations is minimized, while the kinetic energy matrix remains fairly sparse. These two advantages in conjunction with the development of iterative diagonalization techniques, such as short iterative Lanczos [15–17], made DVR a method of choice when doing exact quantum dynamics for both vibrational and scattering calculations. Even to date the state-of-the-art vibrational calculations of spectroscopic accuracy (of a few wavenumbers) use DVR, sometimes in combination with FBR for selected coordinates. One of the most sophisticated calculations, performed by Viel and Leforestier [18] for HFCO, employed six-dimensional DVR of over 10^7 points, truncated by potential energy to about half a million points, to obtain about 150 accurate eigenstates. Even with all the advantages of DVR and a modest 10-point DVR per dimension in average, the sheer size of the basis, which scales exponentially with the number of dimensions d, makes an exact full-dimensional description of systems of more than five atoms ($d \geq 9$) impractical.

The unfavorable scaling of the basis size with d motivates the development of the quantum dynamics approaches based on the correlated, in other words not on the direct-product type, bases. While in general the scaling of the wavefunction complexity is at least exponential with d, the hope is that practical methods of generating efficient bases might be developed if the questions addressed by calculations are narrowed in some sense. For example, instead of the accurate calculation of a full wavefunction, one may target convergence of certain expectation values, of the energy levels within the limited range of energy, or of the correlation functions yielding spectra of medium resolution.

For high-dimensional problems, the central idea behind a manageable-size basis is to make it adaptable to the time-evolution of the target quantities. The most accomplished exact quantum method, used in many high-dimensional applications [19], is the multiconfiguration time-dependent Hartree method (**MCTDH**) [20–23], where multidimensional wavefunctions are built as products of single-particle functions, contracted from a general basis. The MCTDH-like method, closely related to the scope of this chapter, is the variational multiconfiguration Gaussian (vMCG) approach [24–26], based on Gaussian basis functions, whose parameters are formally defined by the evolving wavefunction through the time-dependent variational principle.

As in electronic structure theory, one reason to use Gaussian bases in quantum dynamics is their mathematical properties, such as 'the product of two Gaussians is a Gaussian', localized functions, analytic integrals, Gaussian quadrature, and Hermite polynomials. Another reason is that, unlike in the electronic structure (**ES**) theory, a Gaussian wavefunction (also a polynomial$\times$ a Gaussian) solves the TDSE for parabolic, possibly time-dependent, potentials. The harmonic oscillator model is the foundation for the analysis of molecular vibrations (the normal mode analysis of ES codes), while a Gaussian function is a standard description of a localized in space particle, such as a nucleus, moving close to the classical regime. Finally, the ever-growing efficiency of the ES calculations, enabling molecular dynamics with on-the-fly or on-the-grid *ab initio* ES evaluations and advances in the PES construction (such as the product representation [27, 28] and fitting/interpolation methods [29, 30]) made the evaluation of the potential matrix elements over a localized coordinate-space basis practical.

In this chapter, we discuss Gaussian basis methods of solving the time-independent and time-dependent SE for the nuclei. The same ideas could be used to solve the SE for electrons, or for both nuclei and electrons. An extension to electrons may be desirable, for example, in the presence of the time-dependent electric field of a laser. We limit ourselves to exact quantum methods, though we note that there are numerous semiclassical methods (thawed [31], frozen [32], cellular [33] dynamics, the Herman-Kluk propagator [34], and linearized semiclassical initial value representation [35]) based on the representation of a wavefunction as a superposition of Gaussians. Unlike exact QM methods, the semiclassical methods do not yield exact solutions of the SE in the limit of infinite basis. The remainder of this chapter is organized as follows. First, we review the Gaussian wavepacket (**GWP**) solution to the TDSE and describe a recent application to spectroscopy of NH_3 (Sect. 8.2). Then, we describe the *time-independent* Gaussian bases tailored to PES (Sect. 8.3) and the *time-dependent* Gaussian bases tailored to quantum dynamics of a wavefunction (Sect. 8.4). The concepts are illustrated by examples from our research, i.e. the quasi-random distributed Gaussian bases (**QDGB**) for the TISE and the quantum trajectory-guided Gaussian bases (**QTGB**) for the TDSE. Finally, we survey the 'intermediate' approaches employing the time-independent bases and wavefunction reexpansions at finite time intervals, adapted to the temporal changes of the wavefunction (Sect. 8.5). Finally, Sect. 8.6 concludes.

8.2 Evolution of a Gaussian Wavepacket

To set the stage, first we review an analytic solution to the TDSE with the parabolic potential, i.e. the Gaussian wavepacket, in d dimensions, as it is a useful model for molecular vibrations, and an inspiration for a multitude of Gaussian-based exact and semiclassical methods of quantum dynamics. The solution (see, for example, Ref. [36]) is given in a compact matrix form for a d-dimensional system described in Cartesian coordinates,

$$\psi(\mathbf{x},t) = \mathcal{N} \exp\left(-(\mathbf{x}-\mathbf{q}_t)^T \boldsymbol{A}_t (\mathbf{x}-\mathbf{q}_t) + \imath \mathbf{p}_t^T (\mathbf{x}-\mathbf{q}_t) + \imath s_t + \gamma_t\right), \tag{8.6}$$

where $\mathcal{N}$ is the initial normalization constant, so that $\gamma_0 = 0$,

$$\mathcal{N} = \left(\frac{2^N \det \boldsymbol{A}_{\Re}|_{t=0}}{\pi^N}\right)^{1/4}. \tag{8.7}$$

The wavefunction evolves according to the Hamiltonian,

$$\hat{H} = -\frac{1}{2}\nabla^T \boldsymbol{M}^{-1}\nabla + V, \quad V = \frac{1}{2}\mathbf{x}^T\, \boldsymbol{V}_2\, \mathbf{x}. \tag{8.8}$$

The matrix $\boldsymbol{M}$ is a diagonal matrix of particle masses, while $\boldsymbol{V}_2$ is a real symmetric matrix defining a quadratic potential, whose minimum is at the origin of the coordinate system and is equal to zero. In Eq. (8.6), the parameters $\mathbf{q}_t$, $\mathbf{p}_t$ are real d-dimensional vectors, and s_t and γ_t are real scalar functions of time, indicated as the subscript. The wavefunction is defined in terms of a *complex symmetric* matrix $\boldsymbol{A}_t$,

$$\boldsymbol{A}_t = \boldsymbol{A}_{\Re} + \imath \boldsymbol{A}_{\Im}.$$

The subscript t is omitted for clarity when unambiguous. The vector ∇ is the gradient with respect to the particle coordinates, $\mathbf{x}$.

Substituting Eq. (8.6) into the TDSE with the Hamiltonian (8.8), dividing the result by $\psi(x,t)$ and setting imaginary and real coefficients, multiplying powers of x, to zeroes, one obtains the following equations determining the time-evolution of the parameters:

$$\frac{d\mathbf{q}}{dt} = \boldsymbol{M}^{-1}\mathbf{p}, \quad \frac{d\mathbf{p}}{dt} = -\nabla V(\mathbf{q}), \tag{8.9}$$

$$\frac{ds}{dt} = \frac{\mathbf{p}^T \boldsymbol{M}^{-1}\mathbf{p}}{2} - V(\mathbf{q}) - \mathrm{Tr}(\boldsymbol{A}_{\Re}\boldsymbol{M}^{-1}), \tag{8.10}$$

$$\imath\frac{d\boldsymbol{A}}{dt} = 2\boldsymbol{A}\boldsymbol{M}^{-1}\boldsymbol{A} - \frac{1}{2}\boldsymbol{V}_2, \quad \frac{d\gamma}{dt} = \mathrm{Tr}(\boldsymbol{A}_{\Im}\boldsymbol{M}^{-1}). \tag{8.11}$$

Note, that Eqs. (8.9) are simply Newton's equations of motion for the center of the Gaussian wavepacket, $\mathbf{q}_t = \langle\psi|x|\psi\rangle_t$. Thus, $(\mathbf{q}_t, \mathbf{p}_t)$ are the coordinates and momenta of a classical trajectory. Equation (8.10) defines the evolution of the classical action function s_t for the trajectory $(\mathbf{q}_t, \mathbf{p}_t)$, except for the last term on the RHS,

$$U_0 \equiv \mathrm{Tr}(\boldsymbol{A}_{\Re}\boldsymbol{M}^{-1}).$$

This term does not affect the expectation values, but it does affect time-correlation functions and extracted from them spectra. U_0 can be interpreted as the zeroth order quantum effect, a time-dependent version of the ZPE. In a special case of $\psi(x,t)$ being the ground state of the harmonic oscillator, U_0 is equal to the ground state energy E_0, defining the trivial time-dependent phase of the ground state solution, $\psi(x,t) = \chi(x, E_0)\exp(-\imath E_0 t)$.

Finally, the time-dependence of $\boldsymbol{A}$ defines what is referred to as the breathing mode of the Gaussian, i.e. the change in the localization of $|\psi(\mathbf{x},t)|$, with accompanying it quadratic phase. The changes in the real 'width' of a Gaussian define the time-dependence of the function γ_t, which ensures the constant in the time wavefunction norm, $\langle|\psi(\mathbf{x},t)|^2\rangle = 1$. Because of the 'breathing' motion of the wavepacket, superimposed on the classical motion of the Gaussian center, this solution to the TDSE is also known as the *thawed* Gaussian wavepacket (**TGWP**). The breathing mode is the next order quantum effect, specific to non-coherent Gaussian wavepackets. For a special choice of the initial wavefunction, the width parameter $\boldsymbol{A}$ remains constant in time, while the wavepacket center executes classical motion within the parabolic well. (This is the so-called *coherent* wavepacket: in the normal modes coordinate $\boldsymbol{A}$ is diagonal, $A_{nn} = m_n\omega_n/2$, where ω_n and m_n are the frequency and mass of the nth normal mode.) An analytic solution to $\boldsymbol{A}$ is known for the harmonic oscillator [36], but in general, the time-dependent V solutions of Eq. (8.11), determining the complex Gaussian width $\boldsymbol{A}$, are solved numerically using sophisticated propagators, such as the Poisson propagator [37], to ensure stability. For the wavefunction to be normalizable, the eigenvalues of $\boldsymbol{A}_{\Re}$ have to be positive.

The total energy of ψ,

$$E = \langle\psi|\hat{H}|\psi\rangle = \frac{1}{2}\mathbf{p}^T\boldsymbol{M}^{-1}\mathbf{p} + V(\mathbf{q}) + \frac{1}{2^{d+1}}\mathrm{Tr}(\boldsymbol{A}_{\Re}^{-1}\boldsymbol{V}_2) + \mathscr{U}, \tag{8.12}$$

consists of the classical energy of the wavepacket center (first two RHS terms) and, in addition, of the potential energy term due to the wavepacket delocalization (the third term in RHS of Eq. (8.12)). The last term, $\mathscr{U}$, is the kinetic energy associated with the derivatives of the wavefunction amplitude and phase, respectively,

$$\mathscr{U} = \frac{1}{2}\left(\mathrm{Tr}(\boldsymbol{A}_{\Re}\boldsymbol{M}^{-1}) + \mathrm{Tr}(\boldsymbol{A}_{\Im}\boldsymbol{A}_{\Re}^{-1}\boldsymbol{A}_{\Im}\boldsymbol{M}^{-1})\right). \tag{8.13}$$

The wavefunction ansatz of Eq. (8.6), approximating a time-dependent solution to SE for *anharmonic* potentials, i.e. TGWP [38, 39], found applications in spectroscopy. TGWP evolves on the potential, expanded up to the second order in $\mathbf{x}$ around the wavepacket center, which simply means dynamics in a quadratic potential with time-dependent parameters. Obvious limitations of this approach are the inability to describe the wavefunction bifurcation and interference effects within a single Gaussian ansatz, and inaccuracy of the quadratic expansion of V, when TGWP spreads in coordinate space. A more rigorous version of TGWP is the variational GWP [40]: the evolution equations for the wavepacket parameters are derived from the McLachlan variational principle [41]. The resulting equations for the GWP parameters involve the potential and its first and second derivatives averaged over the wavefunction, rather than their values at the center of the GWP of the TGWP. Both the variational and thawed GWP dynamics are valid either at short times or for nearly harmonic potentials, while the Gaussian function is accurate over the space relevant to the computed quantity.

The variational GWP has conceptual appeal, but beyond model applications, the thawed GWP (the local harmonic approximation to V) has a big advantage that the PES information is needed only along a single trajectory. The PES can be rather straightforwardly computed on-the-fly, as demonstrated in recent applications to spectroscopy of floppy molecules from the Vanicek group [42, 43]. The authors have used the TGWP dynamics, while computing the PES, its gradient, and the Hessian on-the-fly (needed to solve Eqs. (8.9), (8.10), and (8.11)) to analyze the emission spectroscopy of oligothiophenes and absorption and photoelectron spectra of ammonia. What is remarkable in these applications is that the ES has to be solved for along a single trajectory, which enabled the efficient modeling of the oligothiophenes consisting of 2, 3, 4, and 5 rings. The latter system involves dynamics of 105 DOFs, carried out up to 0.2 ps. The TGWP emission spectra agree with the experimental ones quite well as shown in Fig. 8.1: the peak structure is reproduced, although there is an overall shift in peak positions. This shift may be attributed to the theory—local harmonic approximation of the PES, quality of the electronically excited PES computed using the Density Functional Theory, specifically CAM-B3LYP/6-31+G(d,p), or to the experimental conditions, i.e. interaction with the solvent.

For ammonia, the TGWP dynamics on the ground electronic PES yielded medium resolution photoelectron and absorption spectra in good agreement with experiments, despite the limitations of a Gaussian wavefunction form. This is surprising because the inversion of NH_3 (the 'umbrella' mode) is characterized by the double-well potential, typical for floppy molecules, and the local harmonic representation of the PES is a big approximation in this case. The TGWP description was reasonably accurate for up to 0.2 ps, yielding the medium resolution spectra, though the accuracy of dynamics was insufficient for longer times, when the wavepacket bifurcates, to obtain the higher spectral resolution. The ab initio on-the-fly TGWP setup also enabled useful mode decomposition analysis. For example, for the two thiophene ring system (system T2 in Fig. 8.1) 8 effective modes, comprised of 42 DOFs in full dimension, were identified as contributing to the emission spectrum. This type of analysis gives

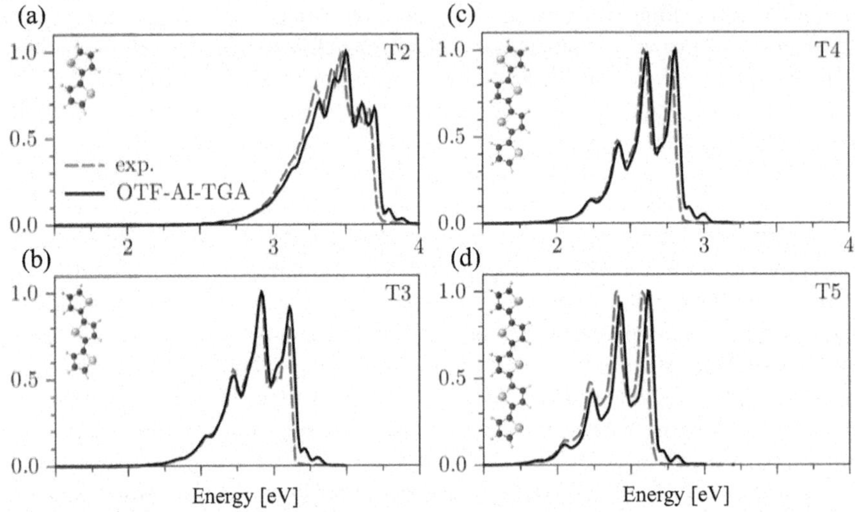

Fig. 8.1 Emission spectra of the oligothiophene chains Tn for n = {2, 3, 4, 5} shown in panels (**a–d**), respectively. The experimental emission spectra (exp., dashed green line) is compared with the full-dimensional TGWP calculations, which includes all normal modes (solid black line). The excited electronic energy, its gradient, and Hessian are computed on-the-fly ab initio (OTI-AI) using CAM-B3LYP density theory functional. Adapted with permission from Ref. [42]. Copyright 2014 American Institute of Physics

insight into the mode coupling and also paves the way for more accurate quantum dynamics studies in reduced dimensionality.

All in all, a single complex TGWP gives a very efficient description of 'mild' quantum effects (before the wavefunction bifurcates and quantum interference becomes important) and is useful in certain spectroscopic applications. In more challenging applications, the TGWPs have been used to represent passive (spectator or bath) modes characterized by the mild quantum effects in combination with more accurate basis representations for the active (reactive or subsystem) modes, such as in the multilayer Gaussian-MCTHD method [44, 45]. The simplicity and elegance of TGWP inspired numerous approximate and semiclassical dynamics methods, based on an expansion of a wavefunction or a propagator in terms of Gaussian functions. The common feature of these methods is that they are based on classical trajectory motion with the quantum effects incorporated from additional information, such as the trajectory action function and monodromy or stability matrix elements related to the width matrix $\boldsymbol{A}$ of TGWP. Some of the most popular semiclassical methods are based on the representation of a wavefunction or a propagator in terms of integrals (over the initial trajectory positions, momenta, or both) of Gaussians [33–35, 46–50]. An interesting recent methodology aims to reduce the semiclassical propagation error by 'slicing' the total propagation into short segments over which the Gaussians are evolved analytically, and reexpanding the wavefunction in a new set of Gaussians [51]. Another impact of the TGWP approach on quantum dynamics as a field is in the use of classical mechanics arguments, i.e. energy and phase space analyses, to construct compact efficient basis representations for exact QM dynamics approaches, including time-independent, time-dependent, and 'intermediate' methodologies, discussed and illustrated below.

8.3 Time-Independent Gaussian Bases Adapted to PES

In this section, we review the construction of efficient basis representation of the vibrational states employing Gaussian functions whose parameters are tuned to the features of the PES. Most of Sect. 8.3 is adapted with permission from [S. Garashchuk and J. C. Light, Quasirandom distributed Gaussian bases for bound problems, *J. Chem. Phys.*, 114:3929–3939, 2001]. Copyright 2001 American Institute of Physics.

The construction of a compact (in a sense of the number of functions) correlated basis for solving the time-independent SE is especially important for studies of highly excited vibrational states of polyatomic molecules or molecular clusters, corresponding to the large amplitude intermolecular motion. To construct a basis for such problems, Davis and Heller [52] considered complex Gaussian basis sets distributed in phase space through the Wigner distribution, though later it has been demonstrated that real distributed Gaussian basis sets (**DGB**s) [53] performed as well or better. In the latter approach, the distance between neighboring Gaussians was made proportional to the local de Broglie wavelength, and the exponents adjusted to give a desired average overlap of the (non-orthogonal) basis functions. The DGB representation, usually for the radial DOFs in conjunction with the DVRs in angles, was found efficient for a number of triatomic systems [54–59]. A later one-dimensional (**1D**) study of the *fully optimized* variational Gaussian bases [60] has shown that higher accuracy of the Hamiltonian eigenvalues was observed for very small reciprocal condition number (**RCN**) η. Small η, defined as the ratio of the smallest to largest eigenvalues of the overlap matrix $\boldsymbol{S}$, typically signals the linear dependence in the basis, but the strongly overlapping functions (responsible for small values of η) give better description of the wavefunction derivatives. The full basis optimization of Ref. [60] is a nonlinear problem, which scales as the number of basis functions N_b to the *fifth* power, making it impractical for multidimensional systems. In this section, we review the subsequent development of the quasi-random distributed Gaussian bases (QTGB) [61], which are not optimized, but incorporate features of the optimal bases known from the 1D studies, including the small RCN. The QTGB performance is illustrated on calculations of the rovibrational energy levels of H_2O.

8.3.1 *Optimized Distributed Gaussian Bases*

Our goal here is to construct an efficient basis comprised of real normalized Gaussians, $\{g_i\}$, $i \in [1, N_b]$,

$$g_i = \left(\frac{2\alpha_i}{\pi}\right)^{1/4} \exp\left(-\alpha_i(x - q_i)^2\right), \tag{8.14}$$

which can accurately represent the vibrational states whose energies are below a certain cutoff value, $E_{\rm cut}$. For clarity of presentation, we will consider a Cartesian-space Hamiltonian with separable kinetic energy given by Eq. (8.2), and outline the procedure in one dimension, x. The Gaussian basis is not orthogonal, thus, to find the eigenstates of $\hat{H}$ we need to compute the overlap matrix $\boldsymbol{S}$ with the elements

$$S_{ij} = \langle g_i | g_j \rangle, \tag{8.15}$$

the Hamiltonian matrix $\boldsymbol{H}$ with the elements,

$$H_{ij} = \langle g_i | \hat{H} | g_j \rangle, \tag{8.16}$$

and to solve the generalized eigenvalue problem for the matrix pair $(\boldsymbol{H}, \boldsymbol{S})$:

$$\boldsymbol{HB} = \boldsymbol{SBE}. \tag{8.17}$$

The diagonal matrix $\boldsymbol{E}$ contains the energy eigenvalues, while the columns of the matrix $\boldsymbol{B}$ are the respective eigenvectors. The basis function overlaps and the matrix elements of the kinetic energy operator are evaluated analytically. The potential matrix elements,

$$V_{ij} = \langle g_i | V | g_j \rangle, \tag{8.18}$$

can be efficiently computed using low-order Gauss–Hermite quadrature or local quadratic approximation to $V(x)$, taking advantage of the product property of Gaussians and provided that the Gaussian basis functions are sufficiently localized.

The kinetic energy associated with each g_i is proportional to its width parameter, $K_{ii} = \alpha_i/(2m)$. Thus, given the energy cutoff $E_{\rm cut}$, a reasonable expectation is that the optimal value of α_i is related to the 'residual' kinetic energy at the basis function center,

$$\alpha_i \sim m\left(E_{\rm cut} - V(q_i)\right). \tag{8.19}$$

As shown in Ref. [53], according to the semiclassical arguments, the density of the Gaussian centers should be proportional to the particle momentum, while according to numerical tests, use of the kinetic energy instead,

$$\rho(q_i) = \frac{1}{2(q_{i+1} - q_{i-1})} \sim (E_{\rm cut} - V(q_i)) \tag{8.20}$$

gives more accurate results. Both options generate the basis functions that are narrower in the regions of low V (or higher kinetic energy), thus the distribution of $\{q_i\}$ can be made denser in those regions; in the regions of high V (low kinetic energy), the basis functions can be broader and placed further apart.

To develop a cheap procedure of specifying the positions $\{q_i\}$ and width $\{\alpha_i\}$, as an alternative to the full variational optimization [60], we have analyzed optimal solutions to the following functional, $\tilde{F}$,

$$\tilde{F} = \mathrm{Tr}\,(\boldsymbol{H}) - \lambda \sum_{ij,i\neq j} \frac{S_{ij}}{1 - S_{ij}}. \tag{8.21}$$

The functional is minimized with respect to all q_i and α_i (without assumptions on their functional forms). The functional includes the energy term as the trace of $\boldsymbol{H}$ and the basis function 'repulsion' term controlled by the parameter λ. If the basis is orthogonal, the second term is equal to zero and $\tilde{F}$ reduces to the trace of the Hamiltonian matrix. For a non-orthogonal basis, minimization of the $\mathrm{Tr}(\boldsymbol{H})$ term alone would yield N_b copies of the ground state, $\mathrm{Tr}(\boldsymbol{H}) = N_b E_0$. The role of the repulsion term is to prevent large off-diagonal overlaps S_{ij} leading to the linear dependence in a basis, i.e. the degeneracy of $\boldsymbol{S}$. During the minimization of $\tilde{F}$, the parameter λ was set to a fixed value, but we found that the accuracy of the eigenstates could be further improved by the *uniform* scaling of α_i, which lowers the Hamiltonian eigenvalues. Thus, the procedure may be viewed as a two-step minimization of $\tilde{F}$: (i) optimize the Gaussian parameters for fixed λ; (ii) 'tune' the basis by uniform rescaling of $\{\alpha_i\}$ to lower the energy term, $\mathrm{Tr}(\boldsymbol{H})$; the rescaling factor effectively controls the Gaussian repulsion strength, which is the role of the parameter λ in Eq. (8.21).

The results of the simplified basis optimization above have been analyzed for several 1D models with the goal of identifying any trends usable in realistic multidimensional problems. In particular, the two-step minimization has been used to compute all bound energy levels of a 1D Morse oscillator for a particle of mass $m = 1$,

$$V(x) = D\,(\exp(-wx) - 1)^2\,, \tag{8.22}$$

which is 'potential B' from Ref. [53]. The parameter values are $D = 12.0$ E$_h$ and $w = 0.2041241$ a$_0^{-1}$; there are 24 energy levels below D. On the far right, the potential is set to $V(x) = 25.0$ E$_h$, $x > 70$ a$_0$. The nonlinear minimization procedure converged for the range of repulsion strength $\lambda = [0.5, 1.5]$, yielding the expected behavior of the optimal Gaussian density and width, shown in Fig. 8.2. For $\lambda = 1$, the obtained RCN of the overlap matrix was $\eta \approx 10^{-5}$. Thus, we could fully explore the effect of broadening the basis functions. Scaling of all α_i by a constant, toward $\eta = [10^{-9}, 10^{-12}]$, *increased* the accuracy of the energy levels by about three orders of magnitude. For yet smaller values of η the eigenvalues of $\boldsymbol{H}$ increased. Similar relation of the accuracy on RCN has been seen in other model systems (see Ref. [61] for more details), leading to a conclusion that there is a sizable optimal range of η spanning 2–4 orders of magnitude yielding low eigenenergies.

Figure 8.3 compares the highest bound eigenfunction computed in a basis to the analytic eigenfunction. The agreement is excellent and, given the diffuse nature of this eigenstate, the basis is highly efficient: all energy levels were obtained within the relative accuracy of 10^{-5}, using only two basis functions per eigenstate, $N_b = 48$. The best accuracy has been achieved by scaling the optimized widths $\{\alpha_i\}$ by 0.13,

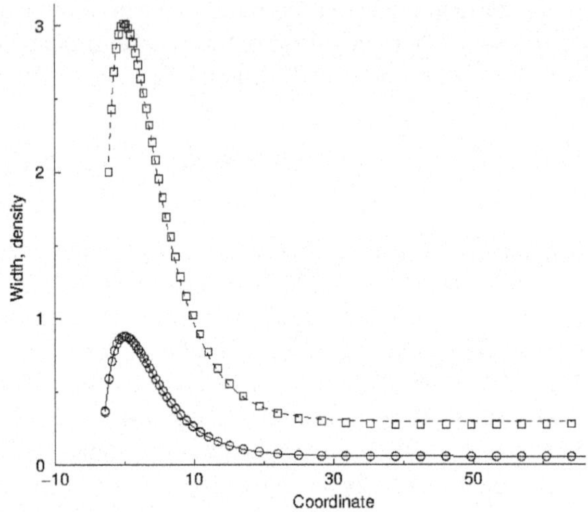

Fig. 8.2 The widths, $\{\alpha_i\}$, and the density $\{\rho_i\}$ of Gaussians as a function of their centers, $\{q_i\}$, for the Morse oscillator: circles mark α_i found from the minimization of the functional $\tilde{F}$, Eq. (8.21); solid line is the fit of α_i with the linear function of the potential; squares mark the optimized density and the dashed line is its linear fit. Adapted with permission from Ref. [61]. Copyright 2001 American Institute of Physics

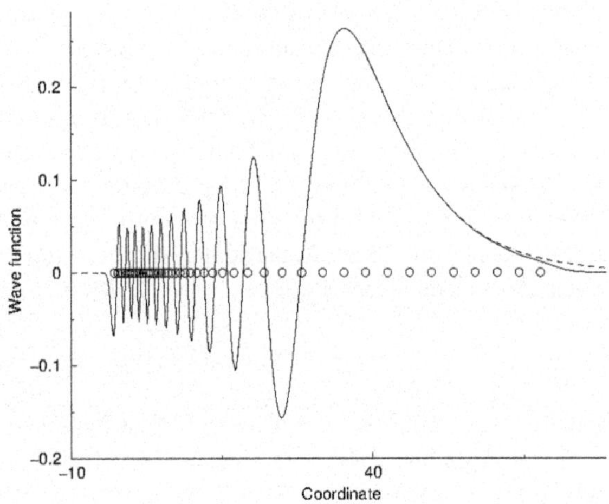

Fig. 8.3 The eigenfunction of the highest, $n = 23$, energy level for the Morse potential in coordinate space: dashed line shows the analytic result and the solid line shows the numerical eigenfunction. The centers of basis functions (circles) are also shown. Adapted with permission from Ref. [61]. Copyright 2001 American Institute of Physics

yielding RCN of $\eta = 1.47 \times 10^{-13}$. Another important observation is that the optimized widths and density could be accurately represented as linear functions of $V(x)$, as seen in Fig. 8.2.

Overall, according to the model analysis of the nonlinear minimization of $\tilde{F}$, we conclude that while the full optimization of the basis parameters is impractical in high dimensionality, efficient correlated basis can be simply constructed by introducing the linear dependence of the basis function widths on the kinetic energy, $\alpha_i = c(E_{\text{cut}} - V(q_i))$, followed by the uniform scaling of α_i to shift the RCN of the overlap matrix toward the range $\eta \sim 10^{-8} - 10^{-10}$. Then, the only two features of the basis, left to be tested directly, are the width scaling factor c and the basis size.

8.3.2 *Quasi-random Distributed Gaussians*

For high-dimensional problems, the full nonlinear optimization of Sect. 8.3.1 becomes expensive, if feasible at all. Thus, to generate a PES-adapted basis, we simply use the linear in V functional form of the basis function parameters to accomplish two tasks.

The first task is to place the Gaussian centers achieving the desired distribution of their centers. We use a quasi-random sequence to generate the Gaussian basis functions with potential-dependent density and widths within the energy contour $V(\mathbf{x}_i) < E_{\text{cut}}$. A general real Gaussian in d-dimensional coordinate space is

$$g_i(\mathbf{x}) = \mathcal{N}_i \exp\left(-(\mathbf{x}-\mathbf{q}_i)^T \boldsymbol{A}_i(\mathbf{x}-\mathbf{q}_i)\right), \qquad (8.23)$$

where the normalization constant $\mathcal{N}_i$ is given by Eq. (8.7) for $\boldsymbol{A} = \boldsymbol{A}_i$. The width parameters are arranged as a real symmetric matrix $\boldsymbol{A}_i$, in general, with nonzero off-diagonal elements and positive eigenvalues. Here, we take $\boldsymbol{A}_i$ as a diagonal matrix $A_{inn} = \alpha_{in}$, where n enumerates dimensions and i enumerates the basis functions.

The second task is to choose the Gaussian width parameters according to Eq. (8.19),

$$\alpha_{in} = c m_n \left(E_{\text{cut}} - V(\mathbf{q}_{in}) + \Delta\right), \quad n = 1 \ldots d \qquad (8.24)$$

with the same value of c for all basis functions and all dimensions. Guided by the results of full optimization in 1D of Sect. 8.3.1, the scaling parameter c is adjusted to have the reciprocal condition number in the range $\eta = [10^{-6}, 10^{-14}]$. The density of centers is also a linear function of the potential

$$\rho(\mathbf{q}_i) \sim E_{\text{cut}} - V(\mathbf{q}_i) + \Delta. \qquad (8.25)$$

In Eqs. (8.19) and (8.25), we have introduced a parameter Δ, affecting the ratio of the largest to smallest width parameter. Thus, Δ controls the sensitivity of the basis function to the PES. In the limit of large Δ, $\Delta \gg E_{\text{cut}}$, the Gaussians will have nearly equal width. Small values of Δ, $\Delta < E_{\text{cut}}$, will generate a basis whose center density and width are sensitive to the potential and mimic the optimal basis of Sect. 8.3.1. The scaling parameter c is chosen to minimize the sum of the energy eigenvalues. The value of c can be estimated from the lowest eigenvalue as

$$c \sim \frac{2E_0}{E_{\text{cut}} \sum_n m_n},$$

or better yet, from the normal modes vibrational analysis at the minimum of $V(\mathbf{x})$. Another option is to adjust both parameters, c and Δ, by computing a few eigenstates for a low E_{cut} in a small-sized QDGB, and then use those values for the target energy range (large E_{cut}). Overall, for a predetermined energy cutoff E_{cut}, the procedure has two adjustable parameters, c and Δ, and generates an efficient, correlated multidimensional basis in the coordinate space.

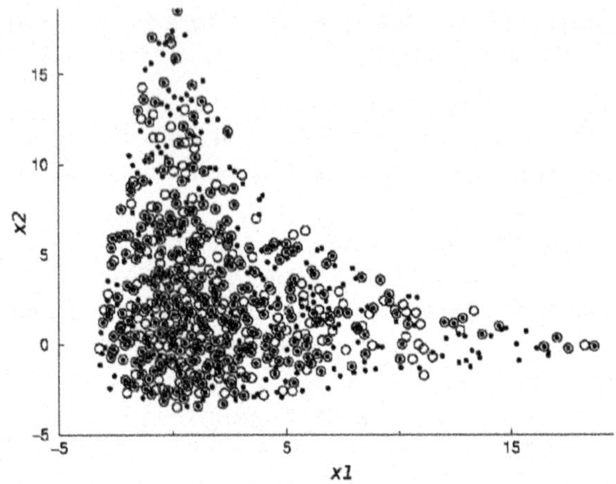

Fig. 8.4 The centers of the non-uniform QDGB 482 functions for the 2D Morse oscillator. The two sets, motivated by the 1D optimization, $\rho \sim E_{\rm cut} - V(\mathbf{q}) + \Delta$, and by the 1D semiclassical de Broglie wavelength, $\rho \sim (E_{\rm cut} - V(\mathbf{q}) + \Delta)^{1/2}$, are shown with open and filled circles, respectively. Adapted with permission from Ref. [61]. Copyright 2001 American Institute of Physics

Placing Gaussian basis functions according to the desired density of their centers $\rho(\mathbf{x})$ is accomplished by accepting randomly generated positions with the probability proportional to $\rho(\mathbf{x})$. In addition, to avoid excessive linear dependence in the basis, the new function is rejected if its maximal overlap with previously accepted functions exceeds a certain threshold. In fact, quasi-random or sub-random sequences of numbers are more efficient for our purposes. The quasi-random sequence has the advantage that the points $\mathbf{q}_i$ are generated as far apart as possible given the previous points in the sequence. For a relatively sparse distribution of points, this feature significantly reduces the probability that two points are so close to each other that one of them is rejected according to the Gaussian overlap criterion. The convergence of the energy eigenvalues with respect to the number of basis functions N_b can be monitored while the sequence of Gaussian centers $\mathbf{q}_i$ is being generated.

We use the Sobol sequence to generate the quasi-random points, and their density is modified according to Eq. (8.25) with the *rejection* method [62]. The point $\mathbf{q}_i$ is accepted if

$$\left[\frac{E_{\rm cut} - V(\mathbf{q}_i) + \Delta}{E_{\rm cut} + \Delta}\right] > b_i, \tag{8.26}$$

where numbers $b_i = [0, 1]$ belong to an independent sequence of random (not quasi-random) numbers. This basis is referred to as the non-uniform, i.e. width and density are adapted to PES, quasi-random distributed Gaussian basis or QDGB. For illustration, the positions of Gaussian centers adapted to a two-dimensional (2D) Morse oscillator potential are plotted in Fig. 8.4. All 122 energy eigenstates, with the eigenvalues below 95% of the dissociation energy, were computed using 482 functions, within the relative error of less than 10^{-4}.

8.3.3 Calculation of the Energy Levels for Triatomic Molecules

8.3.3.1 Water Molecule

As a chemically relevant demonstration of the QDGB generation scheme and performance, we have calculated the vibrational energy levels of water for the total angular momentum $J = 0$ using the potential energy surface of Ref. [63]. The target convergence of the energy levels below 25118 cm^{-1} (252 states in all) was the 'spectroscopic' accuracy of 0.1 cm^{-1}.

We have used the triatomic vibrational Hamiltonian in Radau coordinates, which allows the analytic evaluation of the kinetic energy operator,

$$\hat{H} = K^{2D} + K^{\theta} + V(R_1, R_2, \theta). \tag{8.27}$$

The 2D kinetic energy for the distance variables is

$$K^{2D} = -\frac{\hbar^2}{2m_1R_1^2}\frac{\partial}{\partial R_1}\left(R_1^2\frac{\partial}{\partial R}\right) - \frac{\hbar^2}{2m_2R_2^2}\frac{\partial}{\partial R_2}\left(R_2^2\frac{\partial}{\partial R_2}\right). \tag{8.28}$$

The angular part of the Hamiltonian is

$$K^{\theta} = -\frac{\hbar^2}{2}\left(\frac{1}{m_1R_1^2} + \frac{1}{m_2R_2^2}\right)\boldsymbol{j}^2, \tag{8.29}$$

where

$$\boldsymbol{j}^2 = \frac{1}{\sin\theta}\frac{\partial}{\partial\theta}\left(\sin\theta\frac{\partial}{\partial\theta}\right). \tag{8.30}$$

The volume element is $R_1^2R_2^2\sin\theta \mathrm{d}R_1\mathrm{d}R_2\mathrm{d}\theta$ [56, 64]. The QDGB is used to describe the distance variables, R_1 and R_2; the Legendre DVR is used to represent the angle θ. In DVR, the potential matrix is diagonal with respect to the discretized angle variable θ. Thus, we construct angle-specific 2D Gaussian bases $\{g^{(\mu)}\}$ for the two-dimensional Hamiltonians for each value, θ_μ [54, 64],

$$H_\mu^{2D} = K^{2D} + V(R_1, R_2, \theta_\mu). \tag{8.31}$$

The 2D Hamiltonians for different θ_μ are coupled via K^{θ} terms with the DVR expression for the $\boldsymbol{j}^2$ operator [65, 66]. The kinetic energy matrix elements are integrated on the range of the distance variables, i.e. $R_{1(2)} \in [0, \infty]$. The Gauss–Hermite quadrature of low order (4–7 points) is employed to compute the potential matrix elements. The 2D bases consist of Gaussians of variable width distributed quasi-randomly within the energy contour E_{cut} as described in Sect. 8.3.2. The width parameter of ith Gaussian centered at (R_{i1}, R_{i2}) is scaled by the masses m_1 and m_2,

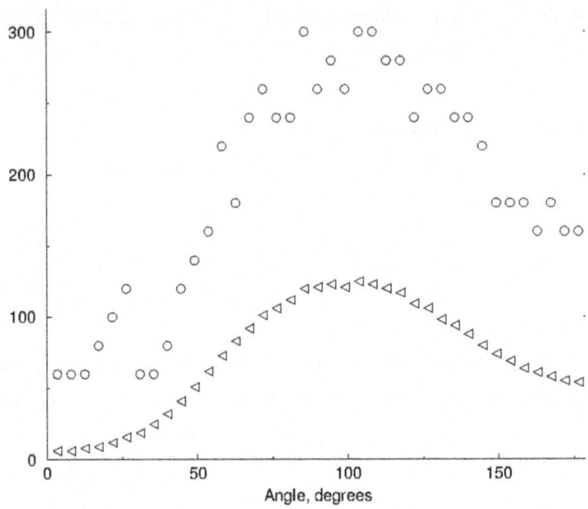

Fig. 8.5 Water molecule eigenstate calculation: the size of the 2D basis before and after truncation procedure as a function of the DVR angle. Adapted with permission from Ref. [61]. Copyright 2001 American Institute of Physics

$$\alpha_{i1(2)} = cm_{1(2)} \left(E_{\text{cut}} - V(R_{i1}, R_{i2}) + \Delta\right). \tag{8.32}$$

The optimal value of c is found from a scan of the trace of H^{2D} over the values of c. The convergence of the energy levels below E_{cut} is monitored during the construction of the basis. For each angle θ_μ, the size of the 2D Hamiltonian matrix is reduced by the number of eigenvalues above the truncation energy, E_{tr}. Then, the appropriately transformed angular kinetic energy K^θ is added to the diagonalized H^{2D}_μ; the resulting matrix, truncated according to the energy cutoff E_{tr}, forms the μth block of the full Hamiltonian matrix.

The numerical parameters for ten different QDGB/DVR calculations, performed to test the convergence and relative importance of various details of the setup, are listed in Table 8.1. The parameter Δ of Eq. (8.32) equals 0.01 E_h ($\approx$ 1800 cm^{-1}) for all calculations. The 2D QDGB sets were constructed by adding 20 functions at a time and their widths scaled. The smallest tolerated RCN was $\eta = 10^{-13}$. The basis size was fixed once either the target accuracy of the eigenvalues was met (in 2D) or until the number of functions exceeded 300. The number of Gaussian basis functions and the matrix size after the truncation are shown in Fig. 8.5 for the basis IV. On average, the truncation procedure reduces the basis size by 60%. For several angles, the target convergence of the 2D eigenvalues (better than 1 cm^{-1}) is not met. However, all levels below 32000 cm^{-1} are converged within 1.3 cm^{-1}, and those below 30000 cm^{-1} are converged within < 0.55 cm^{-1}. Since there are few 2D eigenvalues below E_{tr} at small angles, E_{cut} was increased for $\theta < 39°$, so that at least 40 QDGB functions are generated. The total matrix size of the full 3D calculations varied from 1574 to 3551. The symmetry of the molecule was not taken into account.

The dependence of the energy levels on (i) the initial seed of the random sequence $\{b_i\}$ of Eq. (8.25), (ii) on the number of the quadrature points for R_1 and R_2, and (iii) on the eigenvalue truncation energy parameter is illustrated in

Table 8.1 Convergence of the energy levels for the H_2O molecule for various QDGBs. The listed basis parameters are as follows: N_θ is the number of angular DVR points; N_q is the number of quadrature points per dimension for radial integrals; E_{tr} (E_h) is the truncation energy for the 2D eigenvalues; $E_{\rm cut}$ (E_h) is the cutoff energy for the placement of Gaussians; ϵ (cm^{-1}) is the convergence criterion for energy levels, used to construct 2D QDGB; N_{max} is the total size of the truncated matrix. The largest deviation for energy levels of water in cm^{-1} (taking those of the largest QDGB I as a reference) is listed for the lowest N_l eigenvalues. Adapted with permission from Ref. [61]. Copyright 2001 American Institute of Physics

Basis parameters								Maximal deviation (cm^{-1})				
Basis	N_θ	N_q	E_{tr}	$E_{\rm cut}$	ϵ^{2D}	$N_{\rm max}$	N_l	400	350	300	250	200
I	43	5	0.23	0.165	1.0	3551		–	–	–	–	–
II	39	5	0.23	0.165	1.0	3232		0.43	0.35	0.17	0.08	0.03
III	43	5	0.22	0.16	1.0	3129		1.02	0.33	0.18	0.07	0.02
IV	39	5	0.22	0.16	1.0	2834		0.62	0.42	0.09	0.05	0.03
V	39	5	0.23	0.16	1.0	3198		0.60	0.41	0.09	0.05	0.03
VI	39	7	0.22	0.16	1.0	2838		0.64	0.43	0.11	0.07	0.03
VII[a]	39	5	0.22	0.16	1.0	2777		8.38	3.76	1.90	0.50	0.09
VIII[b]	39	5	0.22	0.16	1.0	2838		0.73	0.34	0.10	0.06	0.03
IX	39	5	0.20	0.15	2.0	2138		6.02	3.80	1.26	0.31	0.12
X	39	4	0.18	0.145	2.0	1574		12.29	4.70	2.50	0.87	0.24

[a]small angles are excluded;
[b]different seed for the random sequence $\{b_i\}$ in Eq. (8.26) is used

Table 8.2 The number of levels converged to a specified accuracy in calculations with bases II, III, and IV. The bases are described in Table 8.1. The convergence is defined with respect to the basis I calculation. Adapted with permission from Ref. [61]. Copyright 2001 American Institute of Physics

Accuracy (cm^{-1})	1.02	0.50	0.20	0.10	0.05
Number of levels	413	378	321	292	211
Highest energy level (cm^{-1})	29785	28908	27357	26489	23552

Table 8.1. The table shows the maximal deviation among N_l lowest levels from the energy levels obtained from our largest, most accurate calculation employing basis I. The number of levels, $N_l = \{400, 350, 300, 250, 200\}$, correspond to the energies of $\{29520, 28183, 26714, 25069, 23153\}$ cm^{-1}, respectively. The number of levels obtained with the bases II, III, and IV that differ from the levels of the largest calculation by less than 1.0, 0.5, 0.2, 0.1, and 0.05 cm^{-1} is shown in Table 8.2: for the most efficient basis, 7 basis per energy level were required to achieve convergence below one wavenumber.

To briefly summarize this section, the non-uniform QDGB are found to be accurate and efficient, with good convergence properties. The obtained energy levels agree with those of Ref. [63] to sub-wavenumber accuracy; the highest energy reported in Ref. [63] ($n = 252$) is *higher* than its counterpart in the QDGB calculation by

$0.6\ \text{cm}^{-1}$. The non-uniform QDGB, being fully adaptable to the PES, including the functions' density and width, is more efficient than the equal-width Gaussian bases used in Ref. [56]. More challenging applications of the QDGB to the neon and argon trimers [61] support the overall conclusion on the QDGB performance.

8.4 Time-Dependent Gaussian Bases Adapted to the Wavefunction Dynamics

We start this section with a review of the basic features of solving the TDSE within a time-dependent basis representation of a wavefunction, as well as some of the established dynamics approaches. Then, the concept of the quantum trajectory dynamics is outlined and combined with the semiclassical idea of frozen Gaussians [32], to yield a formally exact dynamics employing the *quantum trajectory*-guided Gaussian bases (**QTGB**). Model applications and discussion conclude the section. Most of Sect. 8.4 is adapted with permission from [B. Gu and S. Garashchuk, Quantum Dynamics with Gaussian Bases Defined by the Quantum Trajectories, *J. Phys. Chem. A*, 19:3023–3031, 2016]. Copyright 2016 American Chemical Society.

8.4.1 *The Formalism*

A general approach to solving the TDSE (8.1) is to represent a wavefunction in a, possibly, non-orthogonal basis of N_b functions, $\{g_i(\mathbf{x}, t)\}$, $i \in [1, N_b]$. At time t, a wavefunction is expressed in terms of these basis function,

$$\psi(\mathbf{x}, t) = \sum_{i=1}^{N_b} c_i(t) g_i(\mathbf{x}, t), \tag{8.33}$$

where $\{c_i(t)\}$ are the expansion coefficients. The positions of all particles are specified by the vector $\mathbf{x}$ of length d (the number of DOFs). Let us assume that the ith basis function depends on time only through the time-dependent parameters enumerated by the index $\mu = 1 \ldots N_p$,

$$\mathbf{z}_i = \left(z_{i1}(t), \ldots, z_{iN_p}(t)\right), \tag{8.34}$$

where N_p is the number of the parameters specifying each basis function,

$$g_i(\mathbf{x}, t) := g_i(\mathbf{x}, z_{i1}(t), \ldots, z_{iN_p}(t)). \tag{8.35}$$

For simplicity, we assume here that N_b does not change in time, and that N_p is the same for all basis functions. The time-derivative of such a basis function is

$$\frac{\mathrm{d}g_i}{\mathrm{d}t} = \sum_{\mu=1}^{N_p} \dot{z}_{i\mu} \frac{\partial g_i}{\partial z_{i\mu}}, \quad \dot{z} := \frac{\mathrm{d}z}{\mathrm{d}t}. \tag{8.36}$$

Substitution of Eqs. (8.33) and (8.36) into the TDSE (8.1) and the integration of the resulting expression multiplied by g_j over the coordinates space gives the following matrix equation:

$$\imath\hbar \boldsymbol{S} \frac{\mathrm{d}\mathbf{c}}{\mathrm{d}t} = (\boldsymbol{H} - \imath \boldsymbol{D})\mathbf{c}. \tag{8.37}$$

As in Sect. 8.3, $\boldsymbol{S}$ is the overlap matrix, $S_{ij} = \langle g_i | g_j \rangle$; $\boldsymbol{H}$ denotes the Hamiltonian matrix, $H_{ij} = K_{ij} + V_{ij}$,

$$K_{ij} = -\frac{1}{2}\left\langle g_i \middle| \sum_{n=1}^{f} \frac{1}{m_n} \frac{\partial^2}{\partial x_n^2} \middle| g_j \right\rangle, \quad V_{ij} = \langle g_i | V(\mathbf{x}) | g_j \rangle. \tag{8.38}$$

The new matrix $\boldsymbol{D}$ is the non-Hermitian matrix, accounting for the time-dependence of the basis functions,

$$D_{ij} = \left\langle g_i \middle| \sum_{\mu=1}^{N_p} \frac{\mathrm{d}}{\mathrm{d}t} z_{j\mu} \middle| \frac{\partial g_j}{\partial z_{j\mu}} \right\rangle. \tag{8.39}$$

Equation (8.37) determines the evolution of the expansion coefficients $\mathbf{c}$, defining $\psi(\mathbf{x}, t)$ in a basis influenced by an external, possibly time-dependent potential, V.

The choice of the time-dependence of the basis functions, i.e. of the parameters $\mathbf{z}_i(t)$ (Eq. (8.34)), determines the accuracy and conservation properties of the dynamics. As shown, for example, in Ref. [67], the normalization of the wavefunction determined by Eq. (8.37) is conserved regardless of the quality of the basis representation or of the basis time-dependence. The total energy of a system,

$$E = \langle \psi(\mathbf{x}, t) | \hat{H} | \psi(\mathbf{x}, t) \rangle, \tag{8.40}$$

is conserved in three cases: (i) for *any* time-independent basis; (ii) for a time-dependent basis whose parameters are determined variationally, e.g. by applying the Dirac–Frenkel variational principle [68]; (iii) for a time-dependent basis which is *complete* in a sense of being sufficient to represent the wavefunction for a specific problem.

We have discussed the time-independent Gaussian bases, i.e. case (i), in Sect. 8.3. Out of the variational time-dependent basis methods, i.e. case (ii), the most relevant representative is the Gaussian-based vMCG method [69–72]. It has been noted, however, that the variational equations on the Gaussian parameters are ill-conditioned, and for general problems, the solutions become physically non-intuitive with time and are challenging to converge numerically [24, 73]. Limiting the Gaussian basis to the bath DOFs while using a more conventional description for the active ‘system' DOFs has been shown a much more practical approach enabling challenging high-dimensional applications [25, 26, 74].

All in all, time-dependent bases whose parameters are not variational (case (iii)), but instead come from classical or semiclassical theories of motion, have been actively explored. As mentioned above, with this type of bases, the energy is not formally conserved, but this feature may be used as the indicator of the basis completeness during the dynamics. The parameters of the non-variational Gaussian basis methods, many of which are developed for non-adiabatic dynamics on multiple coupled PESs, are defined by the positions and momenta of classical or Ehrenfest-type trajectories, often sampling the phase space of an initial wavefunction. One of such methods is the ab initio multiple spawning method, involving a GWP basis 'driven' by classical molecular dynamics, while additional GWPs are spawned during nonadiabatic events [75–78]. This method is incorporated into MOLPRO [78] and large applications include chromophores in complex molecular environments [79]. The multiconfigurational Ehrenfest method developed by Shalashilin and co-workers on the basis of coupled coherent states expansions [80–82] utilizes the bases simultaneously evolving on multiple electronic states according to the Ehrenfest dynamics.

The advantage of the methods with predefined time-dependence of the Gaussian basis functions is that the remaining dynamics equations (8.37) for the expansion coefficients are much easier to implement numerically than the complete set of equations for the fully variational basis. The usual concern is that defining the basis functions through classical dynamics may miss regions of space, inaccessible to classical trajectories, yet involved in exact quantum dynamics. Therefore, the *quantum* trajectories (**QT**s) representing a time-dependent wavefunction as a correlated ensemble provides have been considered as 'guides' for the Gaussian basis functions.

8.4.1.1 The Quantum Trajectory Dynamics

The Madelung-de Broglie-Bohm, also referred to as the hydrodynamic or the QT, formulation of the TDSE [83–85] is based on the polar representation of a complex wavefunction, expressed in terms of real amplitude $\mathscr{A}(\mathbf{x},t)$ and phase $\mathscr{S}(\mathbf{x},t)$,

$$\psi(\mathbf{x},t) = \mathscr{A}(\mathbf{x},t)\exp\left(\frac{\imath}{\hbar}\mathscr{S}(\mathbf{x},t)\right). \tag{8.41}$$

Substitution of the ansatz (8.41) into TDSE (8.1) leads to the following time-dependence of the wavefunction phase $\mathscr{S}$ and probability density ρ,

$$\rho(\mathbf{x},t) = |\psi(\mathbf{x},t)|^2 = \mathscr{A}^2, \tag{8.42}$$

$$\frac{\partial\mathscr{S}}{\partial t} = -\frac{1}{2}\nabla^T\mathscr{S}\boldsymbol{M}^{-1}\nabla\mathscr{S} - V - U, \tag{8.43}$$

$$\frac{\partial \rho}{\mathrm{d}t} = -\nabla^T \mathscr{S} \boldsymbol{M}^{-1} \nabla \rho - \rho \nabla^T \boldsymbol{M}^{-1} \nabla \mathscr{S}. \tag{8.44}$$

The time-dependent function $U \equiv U(\mathbf{x}, t)$ denotes the *quantum* potential, as opposed to the *classical* potential V,

$$U = -\frac{\hbar^2}{2\mathscr{A}} \nabla \boldsymbol{M}^{-1} \nabla \mathscr{A}. \tag{8.45}$$

To clarify the meaning of Eqs. (8.43) and (8.44), let us switch to the language of the trajectory dynamics. The gradient of the wavefunction phase,

$$\mathbf{p}(\mathbf{x}, t) = \nabla \mathscr{S}(\mathbf{x}, t), \tag{8.46}$$

at the position of the *quantum trajectory* $\mathbf{x}_t$ defines its momentum,

$$\mathbf{p}_t = \nabla \mathscr{S}|_{\mathbf{x}=\mathbf{x}_t}, \quad \frac{\mathrm{d}\mathbf{x}_t}{\mathrm{d}t} = \boldsymbol{M}^{-1} \mathbf{p}_t. \tag{8.47}$$

Then, in the Lagrangian frame-of-reference,

$$\frac{\mathrm{d}}{\mathrm{d}t} = \frac{\partial}{\partial t} + \boldsymbol{M}^{-1} \mathbf{p}_t^T \nabla, \tag{8.48}$$

one obtains the quantum Hamilton–Jacobi equation for the wavefunction phase $\mathscr{S}_t$, and the continuity equation for the probability density ρ_t,

$$\frac{\mathrm{d}\mathscr{S}_t}{\mathrm{d}t} = \frac{1}{2} \mathbf{p}_t^T \boldsymbol{M}^{-1} \mathbf{p}_t - (V + U)|_{\mathbf{x}=\mathbf{x}_t}, \tag{8.49}$$

$$\frac{\mathrm{d}\rho_t}{\mathrm{d}t} = -\rho_t \nabla^T \boldsymbol{M}^{-1} \mathbf{p}(\mathbf{x}, t)\big|_{\mathbf{x}=\mathbf{x}_t}. \tag{8.50}$$

The equation of motion for $\mathbf{p}_t$ is obtained by transforming the gradient of Eq. (8.43) into the moving frame-of-reference defined by Eq. (8.48),

$$\frac{\mathrm{d}p(\mathbf{x}_t)}{\mathrm{d}t} = -\nabla(V + U)|_{\mathbf{x}=\mathbf{x}_t}.$$

A comparison of the quantum Hamilton–Jacobi equation (8.49) to its classical counterpart shows that all ‘quantumness’ in the time-evolution of $\psi(\mathbf{x}, t)$ is expressed through the potential-like function $U = U(\mathbf{x}, t)$ in Eq. (8.45), known as the quantum potential [86]. The exponential scaling of quantum mechanics is traced to this non-local time-dependent function which, being added to the external classical potential V, generates the quantum features in dynamics. Certain simplifications, however, might be expected for heavy particles, such as nuclei: being inversely proportional to the particle mass, the quantum potential becomes negligible in the classical limit.

Therefore, nuclear motion may be treated as classical dynamics with quantum corrections due to the quantum potential [87, 88].

The efficiency of the QT description of a wavefunction follows from Eq. (8.50), according to which ρ within the volume element $\delta\mathbf{x}$ of each trajectory is conserved [87],

$$\rho(\mathbf{x}, t)\delta\mathbf{x}_t = \rho(\mathbf{x}, 0)\delta\mathbf{x}_0. \tag{8.51}$$

Thus, a single QT ensemble, accurately representing $\psi(\mathbf{x}, 0)$, will remain adequate to represent $\psi(\mathbf{x}, t)$ at all times. The exact QTs may be interpreted as a time-dependent grid optimized for a specific $\psi(\mathbf{x}, t)$. The catch is that the numerical implementation of Eq. (8.49) is, in general, impractical: U becomes singular as $\psi(\mathbf{x}, t) \to 0$. However, the QT dynamics may be used to define a nearly optimal time-dependent Gaussian basis, without solving the dynamics equations for QTs, which would require the computation of the quantum potential. Instead, we define the QT momentum, $\mathbf{p}$ from the wavefunction represented in a basis via Eq. (8.33), and use it to update the trajectory positions, which serve as the centers of the Gaussian basis functions.

8.4.1.2 Quantum Trajectory-Guided Basis of Real Frozen Gaussians

To implement the QT-guided Gaussian bases, we construct an adaptive basis out of time-dependent multidimensional Gaussians (d is the number of DOFs),

$$g_i = \sqrt[4]{\frac{\det \boldsymbol{A}}{\pi^d}} \exp\left(-\frac{1}{2}\sum_{n,n'=1}^{d} (x_n - q_{in}(t))A_{nn'}(x_{n'} - q_{in'}(t))\right). \tag{8.52}$$

The basis function g_i is centered at the position of the ith quantum trajectory $\mathbf{q}_i$, their number defining the size of the basis N_b. For simplicity, let us assume that the matrices $\boldsymbol{A}$ and $\boldsymbol{M}$ are diagonal, and their non-zero elements are $A_{nn} = \alpha_n$ are $M_{nn} = m_n$, respectively.

Note, that the Gaussian function of Eq. (8.52) above is different from the Gaussians of the Frozen Gaussian or Herman–Kluk semiclassical propagators [32, 34], which include linear in $\mathbf{x}$ phases, dependent on the momentum of a classical trajectory of the Gaussian center,

$$\mathcal{S}^{lin} = \mathbf{p}_i(t)\,(\mathbf{x} - \mathbf{q}_i(t)) + \gamma_i(t).$$

In the real basis formulation, the time-dependence of the expansions coefficients $c_i(t)$ incorporates the effect of $\gamma_i(t)$ associated with the classical action function at $\mathbf{q}_i$. Restriction for the basis functions in Eq. (8.52) to be real simplifies the formalism and makes implementation more robust: there is no term $d\mathbf{p}/dt$ in Eq. (8.36), thus unstable computation of ∇U is not needed. Omitting time t in the argument of functions for clarity, the matrix elements of Eqs. (8.39) and (8.38) are

$$S_{ij} = \prod_{n=1}^{d} \exp\left(-\frac{\alpha_n}{4}(q_{i_n} - q_{j_n})^2\right), \tag{8.53}$$

$$D_{ij} = \left\langle g_i \left| -\imath\frac{\partial}{\partial t}\right| g_j \right\rangle = \frac{\imath}{2}\sum_{n=1}^{d} \frac{\alpha_n}{m_n} p_{j_n}(q_{i_n} - q_{j_n}) S_{ij}. \tag{8.54}$$

The matrix elements for the kinetic energy operator are

$$K_{ij} = \sum_{n=1}^{d} \frac{\alpha_n}{4m_n}\left(1 - \frac{\alpha_n}{2}(q_{i_n} - q_{j_n})^2\right) S_{ij}. \tag{8.55}$$

The potential energy matrix elements in the examples below are evaluated within the local harmonic approximation to V at the midpoint $\bar{\mathbf{q}}$ of the Gaussian center $\mathbf{q}_i$ or $\mathbf{q}_j$:

$$V_{ij} = \left(V(\bar{\mathbf{q}}) + \sum_{n=1}^{d} \frac{1}{4\alpha_n}\frac{\partial^2 V(\bar{\mathbf{q}})}{\partial x_n^2}\right) S_{ij}, \tag{8.56}$$

where $\bar{\mathbf{q}} = (\mathbf{q}_i + \mathbf{q}_j)/2$. Some other PES approximations for the evaluation of V_{ij}, balancing the accuracy and numerical cost are a symmetric combination of the local harmonic approximations, a cheaper alternative for the *ab initio* PES evaluated on-the-fly; the linear local expansion of $V(\bar{\mathbf{q}})$ (no second derivative of V) as in the coupled coherent Gaussians method [80]; the product-basis fit of the PES developed by the MCTDH community [27, 28, 89]; evaluation by quadrature. The QT position, i.e. the Gaussian center of the ith function, is incremented at each time-step according to the momentum determined from $\psi(\mathbf{x}, t)$:

$$\frac{\mathrm{d}}{\mathrm{d}t} q_{i_n} = \frac{p_{i_n}}{m_n}, \quad \mathbf{p}_i = \Im\left(\frac{\nabla\psi}{\psi}\right)\bigg|_{\mathbf{x}=\mathbf{q}_i}. \tag{8.57}$$

In the *frozen* Gaussian implementation, outlined here, a QTGB simulation is started with the expansion of the initial wavefunction in terms of the Gaussian functions of fixed-width α. Their centers are the initial positions of the QTs. The number of Gaussians, N_b, depends on the target accuracy. For high-dimensional systems, the sampling of the initial QT positions is random or quasi-random. The initial momenta of QTs are defined by the initial wavefunction according to Eq. (8.57). The choice of α will affect the basis size and the accuracy of V_{ij} of Eq. (8.56) through the validity of the local harmonic (or other) approximation. Analogous to the QTGB approach of Sect. 8.3, the value of α and the density of the Gaussian centers should be such as to yield a reasonable condition number of the overlap matrix. Thus, it is necessary to try several values of α and basis sizes to assess convergence and the accuracy of the desired output quantities. The dynamics itself is accomplished by solving Eq. (8.37).

8.4.2 Implementation and Model Problems

In this subsection, we discuss certain aspects of the implementation of the QTGB dynamics, which are then applied to several models: the Morse oscillator representing the vibration of H_2, scattering on the Eckart barrier, and the 1D and 2D double-well potentials. The exact quantum-mechanical results come from the split-operator propagation implemented on a grid [90, 91].

8.4.2.1 Implementation

As discussed in Sect. 8.3, Gaussian bases of strongly overlapping functions may lead to ill-conditioned overlap matrix $\boldsymbol{S}$, which in the case of QTGB has to be inverted to solve Eq. (8.37). In dynamics, this situation may emerge in the course of time-evolution [52]. Formally, the non-crossing property of the QTs mitigates this problem for QTGBs: exact QTs never cross due to the strongly repulsive quantum potential developing as the trajectories approach each other. In practice, however, this property makes the QT dynamics inherently unstable and, moreover, for the Gaussians of fixed width, may become so close as to generate an ill-conditioned overlap matrix $\boldsymbol{S}$, indicating redundancy in a basis.

To deal with this problem, following Refs. [92–94], we occasionally 'restart' the basis, by reexpanding $\psi(\mathbf{x}, t)$ in a new set of Gaussian basis functions, i.e. $\psi(\mathbf{x}, t)$ is re-sampled in terms of the new QT ensemble. The reexpansion procedure prevents Gaussians from colliding, removes basis functions in the regions of negligible probability density, and adds more functions in the regions of increasing probability density. The exact QTs would track such changes in the probability density through the continuity Eq. (8.51). However, real frozen Gaussians do not have the flexibility to account for such changes, captured in QTGB by the expansion coefficients. Therefore, instantaneous adjustments of the basis increase the accuracy and stability of the dynamics.

In the examples below, we have used a simple reexpansion algorithm to generate the equidistant distribution of Gaussian centers of the new basis. We identify the leftmost outlying point $\mathbf{q}_1$. If the probability density at $\mathbf{q}_1$ exceeds a predefined threshold ϵ,

$$|\psi(\mathbf{q}_1, t)|^2 = |\sum_i c_i g_i(\mathbf{q}_1)|^2 > \epsilon,$$

then $\mathbf{q}_1$ is taken as a center of a Gaussian included in the new basis (or the next point $\mathbf{q}_2$ is tested and so on). Moving to the right in position in increments of Δq, until the region of negligible density on the right is reached, generates additional basis functions. In high dimensionality, a new set $\{\mathbf{q}'_i\}$ can be generated through the random importance sampling, similar to the construction of QDGBs of Sect. 8.3. The size of the resulting new basis, N'_b, is generally different from the old N_b, which adapts the basis to the wavefunction localization. The new expansions coefficients,

$\{c_i'\}$, are determined by minimization of the expansion error in the two bases,

$$I = ||\sum_i c_i' g_i' - \psi(\mathbf{x}, t)||^2. \tag{8.58}$$

The resulting matrix equation on $\{c_i'\}$ is

$$\mathbf{S}'\mathbf{c}' = \mathbf{b}', \tag{8.59}$$

where

$$S_{ij}' = \langle g_i' | g_j' \rangle, \quad b_i' = \langle g_i' | \psi \rangle, \tag{8.60}$$

and $\psi(\mathbf{x}, t)$ is expanded in the old basis, Eq. (8.33). The expansion coefficients are updated according to the second-order time differencing scheme [90],

$$\mathbf{c}(t + \Delta t) = \mathbf{c}(t - \Delta t) + 2\Delta t \frac{\mathrm{d}\mathbf{c}}{\mathrm{d}t}, \tag{8.61}$$

with $d\mathbf{c}/dt$ defined by Eq. (8.37).

To start the dynamics in the new basis at time T, we have to assign momenta to the QTs of a new basis. According to Eq. (8.57), the QT momentum is the gradient of the phase of $\psi(\mathbf{x}, T)$ evaluated at $\mathbf{x} = \mathbf{q}_i'$. To ensure smoothness of the momentum as a function of position for a *sparse* set of QTs, we use a convoluted wavefunction $\widetilde{\psi}_\beta(x, t)$ as given below in one dimension:

$$\widetilde{\psi}_\beta(x, t) = \left(\frac{\beta}{2\pi}\right)^{1/2} \int_{-\infty}^{\infty} e^{-\frac{\beta}{2}(x-y)^2} \psi(y, t) \mathrm{d}y. \tag{8.62}$$

Using $\widetilde{\psi}_\beta$ of Eq. (8.62) in Eq. (8.57), the momentum of the Gaussian center is

$$p = -\left(\frac{\beta}{\pi}\right)^{1/4} \Im \left(\frac{\beta}{\psi(x, t)} \int_{-\infty}^{\infty} (x - y) e^{-\frac{\beta}{2}(x-y)^2} \psi(y, t) \mathrm{d}y \right). \tag{8.63}$$

The integrals above are analytic. As $\beta \to \infty$, $\widetilde{\psi}_\beta \to \psi$, and the convolution has no effect on p. If $\beta \to 0$, then the momenta of all trajectories approach the average value of p and deviate significantly from the QT momenta. To generate smooth $p(q_i)$, we select such values of β, so that several Gaussians make significant contribution to $\widetilde{\psi}_\beta$ at each q_i. Note, that since the trajectory momenta p_i are not part of the basis function definition, the convolution procedure or other choices of the momenta do not affect the accuracy of the reexpansion. They will affect, however, the quality, i.e. completeness, of the basis at later times: having momenta closer to the QT values yields accurate basis representation of the wavefunction for longer times, while assigning all momenta equal values will, generally, necessitate frequent reexpansions to attain the desired accuracy.

8.4.2.2 Numerical Examples

The QTGB method is illustrated here for the double-well potentials in one and two dimensions. Additional model systems are described in Ref. [95].

The double-well potential is a prototypical model of reactions in condensed phase. It is an important test for any dynamics methods, as it presents a major challenge for the trajectory-based semiclassical dynamics of light particles due to the population transfer between the wells via tunneling at low energies: the classical trajectories with energies below the barrier top cannot overcome the barrier. Some of the quantum trajectories, however, gain energy from the QT ensemble and cross the barrier even if their initial energies are below the barrier top. The total energy of the QT ensemble, which fully describes $\psi(x, t)$, is conserved, but the energy of individual QTs is not and can be exchanged through the time-dependent quantum potential. Thus, ideally, in a double-well potential, the QT-guided Gaussian basis can be initially localized in the reactant well, yet describe the wavefunction density in the product well at a later time as the guiding QTs migrate to the product well. First we consider a 1D symmetric double-well potential modeling an electron transfer [92]:

$$V(x) = \frac{1}{16\zeta}x^4 - \frac{1}{2}x^2, \tag{8.64}$$

where $\zeta = 1.3544$ a_0^4/E_h. The barrier height is $V^\dagger = 1.3544$ E_h. The initial wavefunction is a Gaussian wavepacket,

$$\psi(x, 0) = \sqrt[4]{\frac{2\alpha_0}{\pi}} \exp\left(-\alpha_0(x - q_0)^2 + \imath p_0(x - q_0)\right), \tag{8.65}$$

whose parameters are $\{\alpha_0 = 0.5,\ q_0 = -2.5,\ p_0 = 0\}$ in atomic units. The particle mass is $m = 1$, and the time-step is 0.001 atomic units. The Gaussian width parameters are $\{\alpha = 8,\ \beta = 2\}$ a_0^{-2}. The wavefunction is localized in the left well. The wavefunction energy is about 2/3 of the barrier height. The barrier is wide: the distance between the two minima is around 4.7 a_0. Thus, tunneling is essential for population transfer, which presents a challenge for semiclassical methods [92]. The QTGB dynamics begins with $N_b = 16$ basis functions. The reexpansion is performed every 1500 time-steps resulting in increase in the basis size with time to $N_b = 23$. Figure 8.6 shows the wavefunction amplitude at $t = \{0, 1.5, 3.0, 4.5, 6.0\}$ a.u. in the process of population transfer with tunneling. The Gaussian centers $\{\mathbf{q}_i\}$, i.e. the QT positions, are plotted in the figure along the x-axis as functions of time up to $t = 6.0$ a.u. Thanks to the continuity Eq. (8.51), their behavior along illustrates the adaptation of the basis and the effect of reexpansions. Around $t = 5.4$ a.u., one of the QTs goes around a node, i.e. $\psi(x, t) = 0$, associated with the interference

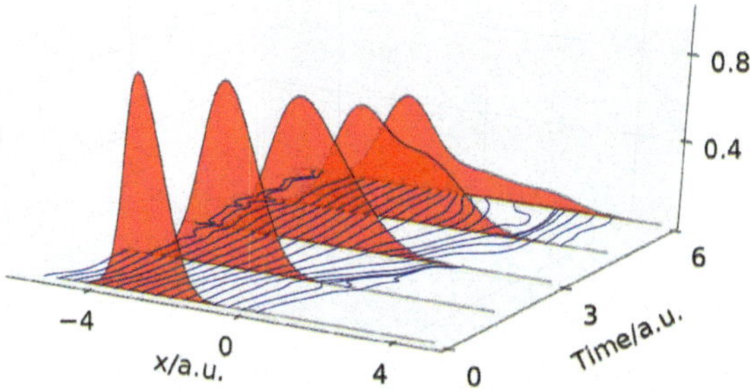

Fig. 8.6 QTGB dynamics in the one-dimensional double well. The blue lines mark the positions of the basis function centers $\{q\}_t$. The vertical axis represents the wavefunction amplitude, whose profiles are shown in red. The wavefunction has been reexpanded at $t = \{1.5,\ 3.0,\ 4.5\}$ a.u. The nodal pattern of the quantum trajectories going around the node at $x \approx 1.5$ a_0 is seen for $t \approx 5.4$ a.u. Adapted with permission from Ref. [95]. Copyright 2016 American Chemical Society

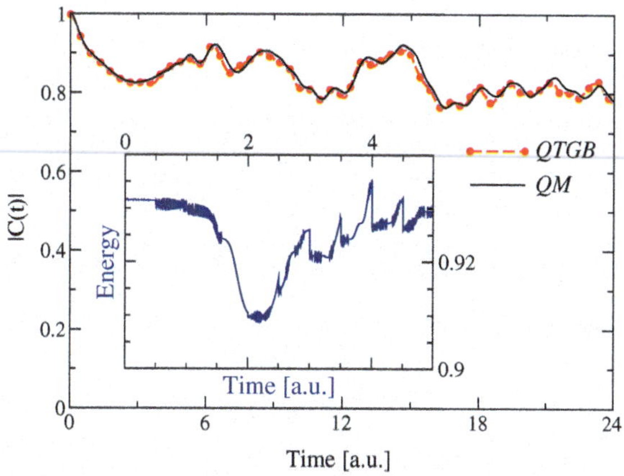

Fig. 8.7 QTGB dynamics in the one-dimensional double well: the absolute value of the autocorrelation function, $C(t) = \langle \psi^2(x, t/2) \rangle$, is compared to the exact QM result. Inset: the total energy is conserved within 2% in the course of the QTGB dynamics. Adapted with permission from Ref. [95]. Copyright 2016 American Chemical Society

effect. At the node near $x = 1.5$ a_0 the wavefunction changes sign, leading to singularities in the quantum potential and a breakdown of the QT dynamics. In QTGB dynamics, however, the interference pattern is reproduced through the superposition of the basis functions. To assess the accuracy once again the autocorrelation function, a phase-sensitive quantity, is calculated for $t = [0, 24]$ a.u. $C(t)$ is computed using Eq. (8.67), thus, $\psi(x, t)$ was propagated up to $t = 12$ a.u. $|C(t)|$ is shown in Fig. 8.7. The reexpansion procedure specified by Eq. (8.58) has been carried out until the error dropped below 5×10^{-5}. The time-dependence of the total energy is given in Fig. 8.7(inset). As discussed at the beginning of this section, in dynamics with the time-dependent bases, the energy conservation correlates not only the accuracy of the numerical solution to Eq. (8.37) but, importantly, with the basis completeness. In this example, with the energy minimum of $V(x)$ shifted to zero, the wavefunction energy remains constant within $1 - 2\%$ of its initial value.

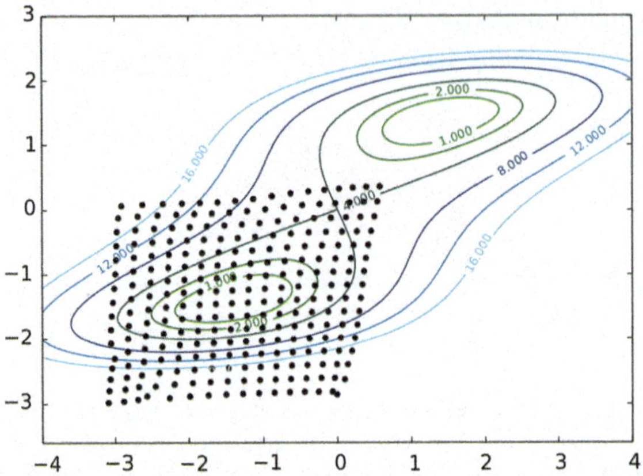

Fig. 8.8 QTGB dynamics in the two-dimensional double-well potential: positions $\{q_1, q_2\}$ of the basis function centers, shown as circles at $t = 3.0$ a.u., are superimposed on the contour plot of $V(x_1, x_2)$ given by Eq. (8.66). The horizontal and vertical axes are the x_1 and x_2 coordinates/center positions in atomic units, respectively. Adapted with permission from Ref. [95]. Copyright 2016 American Chemical Society

Table 8.3 The two-dimensional double-well model. The initial wavefunction and propagation parameters for three QTGB calculations, and the eigenfrequencies of the ground and first excited states are given in atomic units. The parameters α_0, q_0, and p_0 are defined in Eq. (8.65). N_b and α are the number and width of the basis functions given in Eq. (8.52)

Wavefunction and propagation parameters				
α_0	q_0	p_0	m	Δt
0.5	−1.4	0.0	1	0.001
	QT-guided Gaussian Bases			Exact QM
α	16	16	32	
N_b	10 × 10	12 × 12	16 × 16	128 × 128
Frequencies [a.u.]				
ν_g	0.4827	0.4822	0.4830	0.4829
ν_e	0.7110	0.7180	0.7209	0.7163

Next, we examine the QTGB performance for a 2D potential consisting of the double well linearly coupled to the harmonic oscillator of Ref. [96],

$$V(x_1, x_2) = x_2^2(\xi_1 x_2^2 - \xi_2) + \frac{1}{2}\xi_3(x_1 - x_2)^2 + \frac{\xi_2^2}{4\xi_1}. \tag{8.66}$$

The contour plot in Fig. 8.8 corresponds to the parameter values of $\xi_1 = 1$, $\xi_2 = 4$, and $\xi_3 = 4$. The minima of V are located at $(-\sqrt{2}, -\sqrt{2})$ and $(\sqrt{2}, \sqrt{2})$ a.u. The barrier top is $V^\dagger = 4$ a.u.

The wavefunction is initialized as the direct product of two Gaussians specified by Eq. (8.65), centered at the left-hand-side potential minimum. The parameters are listed in Table 8.3. In this model, we focus on the two lowest energy eigenvalues, extracted from the exact QM and QDGB dynamics from the spectrum of autocorrelation functions generated on the interval $t = [0, 6.0]$ a.u. Since the initial wavefunction is real, the correlation function is computed as

$$C(t) = \langle \psi_0 | e^{-\imath \hat{H} t} | \psi_0 \rangle = \int_{-\infty}^{\infty} \psi^2(x, t/2)\, \mathrm{d}x, \tag{8.67}$$

and transformed into the energy domain using harmonic inversion to enhance the resolution of the spectral features [97, 98]. The eigenvalues presented in Table 8.3 have been obtained from dynamics performed with 10×10, 12×12, and 16×16 basis functions. The positions of the Gaussian centers at $t = 3.0$ a.u. (see Fig. 8.8) illustrate the adaptation of the basis functions, initially centered on a square grid of positions $\{\mathbf{q}_i\}$. The frequencies of the symmetric ground and the first excited states are listed in Table 8.3,

$$\nu_{0(1)} \equiv \frac{E_{0(1)}}{2\pi},$$

and compared to the results from the conventional time-evolution on the spatial grid of 512×512 equidistant points. Calculations are performed with several sets of parameters listed in Table 8.3. The frequencies, which are very sensitive to the quality of the correlation function, agree with the QM results quite well.

In the 2D double-well example the wavefunction reexpansion was not necessary, since the eigenfrequencies were obtained from short-time dynamics. In general, however, this procedure gives a practical way of reducing the basis size and of controlling accuracy of dynamics. The wavefunction reexpansions enable the adaptation of the stationary Gaussian bases, which can be viewed as 'intermediate' between the time-dependent and time-independent representations. Several promising methods of this type are outlined in Sect. 8.5.

8.5 Time-Sliced Dynamics With Stationary Gaussian Bases

In quantum dynamics, the algorithms for expansion and reexpansion of wavefunctions in terms of Gaussian basis functions are essential for the practical treatment of delocalized wavefunctions, as it balances the basis size, accuracy, and stability of the time-evolution. For several Gaussian basis methods discussed in this section, the wavefunction reexpansion procedure is central to performing the dynamics. Two earlier methods, i.e. Matching-Pursuit Split-operator Fourier Transform [92, 99] and Quantum Wavepacket Ab Initio Molecular Dynamics [100, 101], combine exact short-time quantum evolution of the Gaussian basis functions with occasional reexpansions of a wavefunction in a new, presumably more efficient and accurate, Gaussian basis. The matching pursuit strategy is based on growing a 'new' basis by adding one basis function at a time. The functions are chosen, according to a certain algorithm, to minimize the residual difference between the wavefunction representations in the 'old' and the 'new' bases under construction, until the desired criterion is fulfilled. Thus, the basis is adjusted according to the time-evolution of a wavefunction encoded in the time-dependent expansion coefficients and in the parameters of the Gaussian basis functions.

More recent adaptable Gaussian-based methods, i.e. the basis expansion leaping multiconfigurational Gaussians (BEL-MCG) of Frankcombe [93] and and the trajectory-guided time-independent Gaussian basis of Saller and Habershon [94], forego the time-dependence of the basis functions altogether. In BEL-MCG just the basis reexpansion, or 'leaping', is left. The basis is stationary between the leaps, and is adjusted at certain intervals of time by constructing a more compact basis to express the time-evolved wavefunction. The advantage of the approach is that the Hamiltonian and overlap matrices are evaluated only once per each time-interval between the leaps. In the work by Habershon and co-workers, the choice of the Gaussian parameters comes from the classically evolved trajectories, sampling the Wigner transform of the initial wavefunction. The classical trajectory dynamics allows one to anticipate where the basis functions will be needed in the course of quantum dynamics of the wavefunction, and thus to construct an adequate basis of stationary Gaussians, tuned to the upcoming dynamics, though certainly not a 'minimal' basis for problems undergoing large amplitude motion. The Hamiltonian and overlap matrices are computed just once, which is a very appealing feature if the PES evaluations are expensive. The downside is that classical trajectories may not cover certain regions of space, such as those accessed through quantum tunneling. A very recent development of the trajectory-guided time-independent basis is to apply this idea to short segments of time to generate a compact adaptive basis [102]. The 'new' basis is constructed based on short-time classical trajectories, so the earlier deficiency of classical versus quantum spaces is mitigated. In addition, the basis functions of the 'old' basis with small expansions coefficients are removed at the reexpansion step, accomplished using the matching pursuit algorithm, to reduce the basis size. Figure 8.9 illustrates the method.

This time-sliced propagation scheme with basis adjustments allowed to reduce the basis size by an order of magnitude (compared to the original method of Ref. [94]) for challenging benchmark applications – for a 4-dimensional model of photoexcited pyrazine (shown in Fig. 8.10) and for a system tunneling in a double well coupled to 2–20 harmonic DOFs representing the environment. In the case of pyrazine, the time-slices are 10 fs, while the total propagation time is 150 fs long.

Another promising combination of short time-evolution and basis reduction is Time-Sliced Thawed Gaussian Propagation Method of Batista and co-workers [51]. In this method, the basis functions evolve as TGWPs, and the segments are combined employing the Husimi transform in the limit of highly overlapping Gaussians, at which step functions with negligible contribution to the wavefunction are removed.

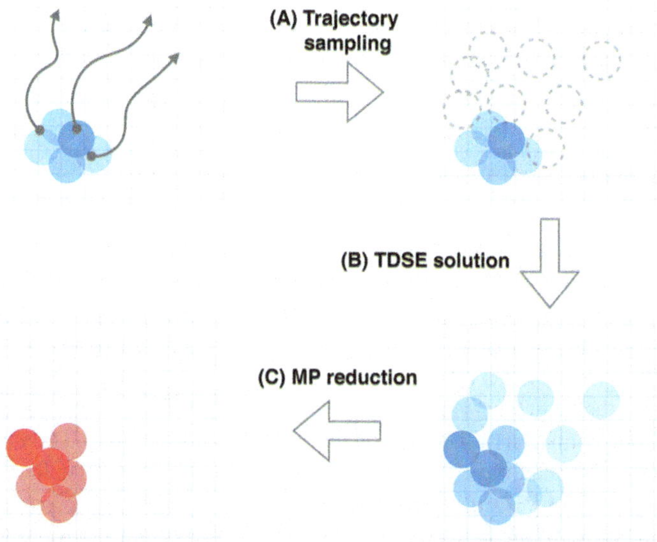

Fig. 8.9 The trajectory guiding algorithm. **a** An initial wavefunction is expanded in a basis of Gaussians (blue circles). Additional Gaussians (dashed circles) come from a sampling of short-time classical trajectories (gray lines). **b** Solution of the TDSE within the full basis is followed by **c** the basis reduction deploying the matching pursuit method. The cycle closes when a time-evolved wavefunction is expressed in a new, compact basis of Gaussian functions (red circles). Adapted with permission from Ref. [102]. Copyright 2017 American Chemical Society

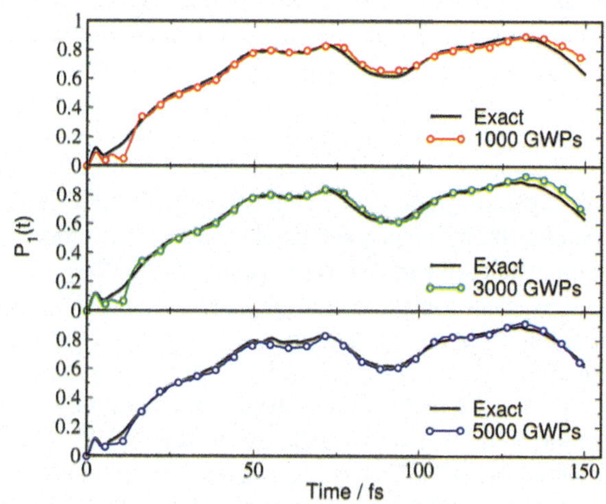

Fig. 8.10 Population of the lower diabatic S1 state, P1(t), as a function of time, calculated using adaptive basis sets with varying size, for the 4D pyrazine Hamiltonian. The basis set sizes given are approximately the number of GWPs which form the wavefunction during each short 10 fs propagation period. Adapted with permission from Ref. [102]. Copyright 2017 American Chemical Society

The emerging overall conclusion on the Gaussian basis representation of time-dependent wavefunctions is as follows: the basis reexpansion is a useful tool for improving the stability and accuracy of dynamics with the time-independent or time-dependent Gaussian bases. The basis reexpansion also helps to keep non-variational bases compact. The dynamics information, either from the time-dependent wavefunction itself, or from certain relevant exploratory trajectories, is beneficial for constructing physically meaningful, small yet accurate bases.

8.6 Summary and Outlook

In this chapter, we discussed Gaussian basis methods of solving the time-independent and time-dependent SE for the nuclei. The same ideas could be used to solve the SE for electrons, or for both nuclei and electrons. An extension to electrons may be desirable, for example, in the presence of the time-dependent electric field of a laser. We started by reviewing a time-dependent solution to SE for parabolic potentials, known as the thawed Gaussian wavepacket (TGWP) [31], when used with the local harmonic approximation to an arbitrary potential. A single complex TGWP is capable of describing mild quantum effects, associated with the wavepacket delocalization. In this regime, the TGWP is sufficiently accurate and highly efficient as demonstrated by recent applications to the spectroscopy of oligothiophenes and ammonium inversion [42, 43]. It is particularly appealing, when combined with on-the-fly ab initio electronic structure calculations, as information on the PES and its gradients is needed along a *single* classical trajectory. In more challenging applications, the TGWPs have been successfully used to represent environmental DOFs, while more accurate basis representations are employed for the reactive modes [44, 45]. More generally, the idea of using classical trajectory information, such as energy and phase space analyses, to construct compact efficient basis representations has been incorporated into many exact QM dynamics methods, including time-independent, time-dependent, and in between approaches.

Out of the time-independent methods, i.e. those based on Hamiltonian matrix diagonalization, we have reviewed a quasi-random distributed Gaussian bases (QDGBs). The advantage of QDGB is that it is correlated and adapted to a given PES. Checking the convergence of eigenvalues, while the basis is constructed, is an advantage for high-dimensional studies whose feasibility is determined by the basis size. Some other notable developments of the Gaussian basis methodology in time-independent context include phase space Gaussians on the von Neumann lattice with periodic boundary conditions [103, 104], and the wavelets representation obtained through canonical orthogonalization of the coherent state Gaussians [105]. Subsequent advances of the wavelets methodology (the truncation scheme and the momentum-symmetrized Gaussian basis) [106, 107] culminate in applications to acetonitrile (CH_3CN) and benzene, yielding thousands of eigenenergies [108, 109]. In the study of the benzene molecule, a 30-DOF problem, 500,000 eigenenergies below 6500 cm^{-1} were converged within 15 cm^{-1}. The approach is implemented in a massively parallel code SwitchBLADE, which is available for general use.[1]

[1]Interested researchers should send their request to Bill Poirier at Bill.Poirier@ttu.edu.

Next, we described the construction of compact time-dependent Gaussian bases (GB), guided by the quantum-like trajectories (QT). From the properties of QTs, it follows that the basis functions track the flow of the probability density, providing in a sense the optimal in coordinates space time-dependent grid, and adapt to the dynamics of a specific wavefunction. Therefore, for problems characterized by large amplitude motion in selected DOFs, the scaling properties of QTGB with the system size should be superior to those of conventional time-independent bases. The time-evolution of QTs in the exact limit of strongly quantum dynamics is, generally, unstable, due to singular features in the quantum potential, responsible for the formal exponential scaling of complexity with the system size [86]. Yet the trajectory framework is useful for quantum dynamics close to the classical regime, appropriate for the nuclei of polyatomic systems. In the QTGB approach, which is developed for this regime, the QTs move according to the momentum reconstructed from the wavefunction expanded in a Gaussian basis. Moreover, for real Gaussian basis functions, one can use modified QT momenta without making approximations, to improve the stability of propagation. The ensuing loss of efficiency is compensated by occasional reexpansions, performed to maintain the completeness of the basis in time. The total wavefunction energy, which is rigorously conserved in the limit of a complete time-dependent basis, serves as a convenient measure of the basis completeness and a criterion for the wavefunction reexpansion. So far, the QTGB method has been tested on low-dimensional model problems. We plan to extend QTGB to real Gaussians with adjustable width, and to improve the reexpansion procedure using ideas of matching pursuit and exploratory trajectory dynamics of Refs. [92, 93, 102]. Though seemingly technical, the wavefunction reexpansion is at the core of the 'intermediate' between the time-dependent and time-independent Gaussian bases, i.e. the stationary Gaussian bases adapted to the evolving wavefunction at the reexpansion stage. All in all, non-variational dynamics-guided Gaussian bases, combined with recent advances in the on-the-fly electronic structure calculation, is a promising practical approach to perform quantum dynamics of large molecular systems.

Acknowledgements The author thanks John Light, Vitaly Rassolov, and Bing Gu for the collaborative research in the area of quantum trajectory and Gaussian basis dynamics. The author thanks Jiri Vanicek and Scott Habershon for permission to use some of their figures. This material is based upon work supported by National Science Foundation under Grant No CHE-1056188, CHE-1565985, and OIA-1655740, and by a grant from US Department of Energy, DE-FG02-87ER13679.

References

1. Sumida K, Stueck D, Mino L, Chai J-D, Bloch ED, Zavorotynska O, Murray LJ, Dinca M, Chavan S, Bordiga S, Head-Gordon M, Long JR (2013) Impact of metal and anion substitutions on the hydrogen storage properties of M-BTT metal-organic frameworks. J Am Chem Soc 135(3):1083–1091
2. Shao M, Keum J, Chen J, He Y, Chen W, Browning JF, Jakowski J, Sumpter BG, Ivanov IN, Ma Y-Z, Rouleau CM, Smith SC, Geohegan DB, Hong K, Xiao K (2014) The isotopic effects of deuteration on optoelectronic properties of conducting polymers. Nat Commun 5:3180

3. Cai W, Dunuwille M, He J, Taylor TV, Hinton JK, MacLean MC, Molaison JJ, dos Santos AM, Sinogeikin S, Deemyad S (2017) Deuterium isotope effects in polymerization of benzene under pressure. J Phys Chem Lett 8(8):1856–1864 PMID: 28395511
4. Nguyen TD, Basel TP, Pu YJ, Li X-G, Ehrenfreund E, Vardeny ZV (2012) Isotope effect in the spin response of aluminum tris(8-hydroxyquinoline) based devices. Phys Rev B 85(24)
5. Klinman JP, Offenbacher AR (2018) Understanding biological hydrogen transfer through the lens of temperature dependent kinetic isotope effects. Acc Chem Res 51(9):1966–1974
6. Kosloff D, Kosloff R (1983) A Fouirier method solution for the time-dependent Schrodinger equation as a tool in molecular dynamics. J Comp Phys 52(1):35–53
7. Leforestier C, Bisseling RH, Cerjan C, Feit MD, Friesner R, Guldberg A, Hammerich A, Jolicard G, Karrlein W, Meyer HD, Lipkin N, Roncero O, Kosloff R (1991) A Comparison of different popagation schemes for the time-dependent Schrodinger equation. J Comput Phys 94(1):59–80
8. Kosloff R (1992) The Fourier Method. In Cerjan, C, editor, Numerical gird methods and their applications to Schrodinger equation, vol 412 of NATO ADVANCED SCIENCE INSTITUTES SERIES, SERIES C, MATHEMATICAL AND PHYSICAL SCIENCES, pp 175–194. NATO, SCI COMM, 1993. NATO Advanced Research Workshop on Grid Methods in Atomic and Molecular Quantum Calculations, Corte, Fracne
9. Kosloff R (1994) Propagation methods fo quantum molecular dynamics. Ann Rev Phys Chem 45:145–178
10. Billeter SR, VanGunsteren WF (1995) A comparison of different numerical propagation schemes for solving the time-dependent Schrodinger equation in the position representation in one dimension for mixed quantum- and molecular dynamics simulations. Mol Simul 15(5):301–322
11. Heather RW, Light JC (1983) Discrete variable theory of triatomic photodissociation. J Chem Phys 79:147
12. Light JC, Hamilton IP, Lill JV (1985) Generalized discrete variable representation in quantum mechanics. J Chem Phys 82:1400
13. Lill JV, Parker GA, Light JC (1986) The discrete variable finite basis approach to quantum scattering. J Chem Phys 85:900
14. Light JC, Carrington T Jr (2000) Discrete variable representations and their utilization. Adv Chem Phys 114:263–310
15. Lanczos C (1950) An iteration method for the solution of the eigenvalue problem of linear differential and integral operators. J Res Natl Bureau Stand 45:255–282. Research Paper 2133
16. Park TJ, Light JC (1986) Unitary quantum time evolution by iterative lanczos reduction. J Chem Phys 85:5870
17. Roy PN, Carrington T Jr (1996) A direct-operation lanczos approach for calculating energy levels. Chem Phys Lett 257:98–104
18. Viel A, Leforestier C (2000) Six-dimensional calculation of the vibrational spectrum of the hfco molecule. J Chem Phys 112:1212–1220
19. Gatti F, Lasorne B, Meyer A, AndNauts H-D (2017) Applications of quantum dynamics in chemistry. Springer, Berlin
20. Meyer HD, Manthe U, Cederbaum LS (1990) The multi-configurational time-dependent hartree approach. Chem Phys Lett 165(1):73–78
21. Burghardt I, Meyer H-D, Cederbaum LS (1999) Approaches to the approximate treatment of complex molecular systems by the multiconfiguration time-dependent hartree method. J Chem Phys 111(7):2927–2939
22. Meyer HD, Worth GA (2003) Quantum molecular dynamics: propagating wavepackets and density operators using the multiconfiguration time-dependent Hartree method. Theor Chem Acc 109:251–267
23. Wang HB, Thoss M (2003) Multilayer formulation of the multiconfiguration time-dependent Hartree theory. J Chem Phys 119:1289–1299
24. Burghardt I, Nest M, Worth GA (2003) Multiconfigurational system-bath dynamics using gaussian wave packets: Energy relaxation and decoherence induced by a finite-dimensional bath. J Chem Phys 119(11):5364–5378

25. Roemer S, Ruckenbauer M, Burghardt I (2013) Gaussian-based multiconfiguration time-dependent Hartree: A two-layer approach. I. Theory. J Chem Phys 138(6)
26. Roemer S, Burghardt I (2013) Towards a variational formulation of mixed quantum-classical molecular dynamics. Mol Phys 111(22–23, SI):3618–3624
27. Jackle A, Meyer HD (1996) Product representation of potential energy surfaces. J Chem Phys 104(20):7974–7984
28. Jackle A, Meyer HD (1998) Product representation of potential energy surfaces. II. J Chem Phys 109(10):3772–3779
29. Dawes R, Passalacqua A, Wagner AF, Sewell TD, Minkoff M, Thompson DL (2009) Interpolating moving least-squares methods for fitting potential energy surfaces: Using classical trajectories to explore configuration space. J Chem Phys 130(14):144107
30. Richings GW, Habershon S, Direct quantum dynamics using grid-based wave function propagation and machine-learned potential energy surfaces. J Chem Theory Comput
31. Huber D, Heller EJ (1987) Generalized gaussian wave packet dynamics. J Chem Phys 87(9):5302–5311
32. Heller EJ (1981) Frozen Gaussians: A very simple semiclassical approximation. J Chem Phys 75:2923
33. Heller EJ (1991) Cellular dynamics—a new semiclassical approach to time-dependent quantum mechanics. J Chem Phys 94:2723–2729
34. Herman Michael F, Kluk Edward (1984) A semiclasical justification for the use of non-spreading wavepackets in dynamics calculations. Chem Phys 91(1):27–34
35. Skinner DE, Miller WH (1999) Application of the semiclassical initial value representation and its linearized approximation to inelastic scattering. Chem Phys Lett 399:20–26
36. Tannor DJ (2006) Introduction to quantum mechanics: a time-dependent perspective. University Science Books
37. Faou E, Gradinaru V, Lubich C (2009) Computing semiclassical quantum dynamics with hagedorn wavepacketsal quantum dynamics with hagedorn wavepackets. SIAM J Sci Comput 31:3027
38. Heller EJ (1975) Time-dependent approach to semiclassical dynamics. J Chem Phys 62:1544
39. Heller EJ (1976) Wigner phase space method: Analysis for semiclassical applications. J Chem Phys 65:1289–1298
40. Coalson RD, Karplus M (1990) Multidimensional variational Gaussian wave packet dynamics with application to photodissociation spectroscopy. J Chem Phys 93:3919–3930
41. McLachlan A (1964) A variational solution of the time-dependent schrodinger equation. Mol Phys 8:39
42. Wehrle M, Sulc M, Vanicek J (2014) On-the-fly ab initio semiclassical dynamics: Identifying degrees of freedom essential for emission spectra of oligothiophenes. J Chem Phys 140:244114
43. Wehrle M, Oberli S, Vanicek J (2015) On-the-Fly ab initio semiclassical dynamics of floppy molecules: absorption and photoelectron spectra of ammonia. J Phys Chem A 119:5685–5690
44. Romer S, Burghardt I (2013) Towards a variational formulation of mixed quantum-classical molecular dynamics. Mol Phys 111:3618
45. Romer S, Ruckenbauer M, Burghardt I (2013) Gaussian-based multiconfiguration time-dependent hartree: A two-layer approach. i. Theory J Chem Phys 138:064106
46. Brown RC, Heller EJ (1981) Classical trajectory approach to photodissociation: The Wigner method. J Chem Phys 75:186–188
47. Kay KG (1994) Integral expressions for the semiclassical time-dependent propagator. J Chem Phys 100:4377–4392
48. Kay KG (1994) Semiclassical propagation for multidimensional systems by an initial-value method. J Chem Phys 101:2250–2260
49. Makri N, Miller WH (2002) Coherent state semiclassical initial value representation for the Boltzmann operator in thermal correlation functions. J Chem Phys 116:9207–9212
50. Liu J, Miller WH (2007) Linearized semiclassical initial value time correlation functions using the thermal Gaussian approximation: Applications to condensed phase systems. J Chem Phys 127(11):114506

51. Kong X, Markmann A, Batista VS (2016) Time-sliced thawed gaussian propagation method for simulations of quantum dynamics. J Phys Chem A 120(19):3260–3269
52. Davis MJ, Heller EJ (1979) Semi-classical Gaussian-basis set method for molecular vibrational wave-functions. J Chem Phys 71(8):3383–3395
53. Hamilton IP, Light JC (1986) On distributed gaussian bases for simple model multidimensional vibrational problems. J Chem Phys 84:306
54. Bacic Z, Light JC (1986) Highly excited vibrational levels of floppy triatomic molecules: A discrete variable representation-distributed gaussian basis approach. J Chem Phys 85:4594
55. Bacic Z, Whitnell RM, Brown D, Light JC (1988) Localized representations for large amplitude molecular vibrations. Comput Phys Commun 51:35
56. Bacic Z, Watt D, Light JC (1988) A variational localized representation calculation of the vibrational levels of the water molecule up to 27 000 cm^1. J Chem Phys 89(2):947–955
57. Peet AC (1989) The use of distributed gaussian basis sets for calculating energy levels of weakly bound complexes. J Chem Phys 90(8):4363–4369
58. Bacic Z, Light JC (1989) Theoretical methods for rovibrational states of floppy molecules. Ann Rev Phys Chem 40:469
59. Mladenovic M, Schmatz S (1998) Theoretical study of the rovibrational energy spectrum and the numbers and densities of bound vibrational states for the system hco+/hoc+. J Chem Phys 109(11):4456–4470
60. Poirier B, Light JC (2000) Efficient distributed gaussian basis for rovibrational spectroscopy calculations. J Chem Phys 113(1):211–217
61. Garashchuk S, Light JC (2001) Quasirandom distributed Gaussian bases for bound problems. J Chem Phys 114:3929–3939
62. Press WH, Flannery BP, Teukolsky SA, Vetterling WT (1992) Numerical recipes: the art of scientific computing, 2nd edn. Cambridge University Press, Cambridge
63. Polansky OL, Jensen PJ, Tennyson J (1996) The potential energy surface of $H_2^1 6O$. J Chem Phys 105:6490–6497
64. Choi SE, Light JC (1992) Highly excited vibrational eigenstates of nonlinear triatomic molecules: Applications to H_2O. J Chem Phys 97:7031
65. Bacic Z, Light JC (1987) Accurate localized and delocalized vibrational states of hcn/hnc. J Chem Phys 86(6):3065–3077
66. Light JC, Bacic Z (1987) Adiabatic approximation and nonadiabatic corrections in the discrete variable representation: Highly excited vibrational states of triatomic molecules. J Chem Phys 87(7):4008–4019
67. Habershon S (2012) Linear dependence and energy conservation in Gaussian wavepacket basis sets. J Chem Phys 136(1)
68. Broeckhove J, Lathouwers L, Kesteloot E, Leuven PV (1988) On the equivalence of time-dependent variational principles. Chem Phys Lett 149:547
69. Burghardt I, Meyer H-D, Cederbaum LS (1999) Approaches to the approximate treatment of complex molecular systems by the multiconfiguration time-dependent hartree method. J Chem Phys 111:2927
70. Worth GA, Burghardt I (2003) Full quantum mechanical molecular dynamics using gaussian wavepackets. Chem Phys Lett 368:502
71. Worth GA, Robb MA, Burghardt I (2004) A novel algorithm for non-adiabatic direct dynamics using variational gaussian wavepackets. Faraday Discuss 127:307
72. Burghardt I, Giri K, Worth GA (2008) Multimode quantum dynamics using gaussian wavepackets: The gaussian-based multiconfiguration time-dependent hartree (g-mctdh) method applied to the absorption spectrum of pyrazine. J Chem Phys 129:174104
73. Shalashilin DV, Burghardt I (2008) Gaussian-based techniques for quantum propagation from the time-dependent variational principle: Formulation in terms of trajectories of coupled classical and quantum variables. J Chem Phys 129(8):084104
74. Richings GW, Polyak I, Spinlove KE, Worth GA, Burghardt I, Lasorne B (2015) Quantum dynamics simulations using gaussian wavepackets: the vmcg method. Int Rev Phys Chem 34(2):269–308

75. Ben-Nun M, Martínez TJ (2000) A multiple spawning approach to tunneling dynamics. J Chem Phys 112(14):6113–6121
76. Ben-Nun M, Quenneville J, Martinez TJ (2001) Ab initio multiple spawning: Photochemistry from first principles quantum molecular dynamics. J Chem Phys 104:5161–5175
77. Toniolo A, Ciminelli C, Persico M, Martínez TJ (2005) Simulation of the photodynamics of azobenzene on its first excited state: Comparison of full multiple spawning and surface hopping treatments. J Chem Phys 123(23):234308
78. Levine BG, Coe JD, Virshup AM, Martinez TJ (2008) Implementation of ab initio multiple spawning in the Molpro quantum chemistry package. Chem Phys 347(1–3):3–16
79. Olsen S, Lamothe K, Martinez TJ (2010) Protonic gating of excited-state twisting and charge localization in GFP chromophores: a mechanistic hypothesis for reversible photoswitching. J Am Chem Soc 132:1192+
80. Shalashilin DV, Child MS (2004) Real time quantum propagation on a Monte Carlo trajectory guided grids of coupled coherent states: 26D simulation of pyrazine absorption spectrum. J Chem Phys 121:3563–3568
81. Shalashilin DV (2010) Nonadiabatic dynamics with the help of multiconfigurational ehrenfest method: Improved theory and fully quantum 24d simulation of pyrazine. J Chem Phys 132:244111
82. Makhov DV, Glover WJ, Martinez TJ, Shalashilin DV (2014) Ab initio multiple cloning algorithm for quantum nonadiabatic molecular dynamics. J Chem Phys 141:054110
83. Madelung E (1927) Quantum theory in hydrodynamic form. Z Phys 40:322–326
84. de Broglie L (1930) An introduction to the study ot wave mechanics. E. P. Dutton and Company Inc, New York
85. Bohm D (1952) A suggested interpretation of the quantum theory in terms of "hidden" variables I and II. Phys Rev 85:166–193
86. Garashchuk S, Rassolov VA (2008) Stable long-time semiclassical description of zero-point energy in high-dimensional molecular systems. J Chem Phys 129:024109
87. Garashchuk S, Rassolov VA (2004) Energy conserving approximations to the quantum potential: Dynamics with linearized quantum force. J Chem Phys 120:1181–1190
88. Garashchuk S, Rassolov V, Prezhdo O (2011) Reviews in computational chemistry, vol 27, chapter Semiclassical Bohmian dynamics, pp 111–210. Wiley
89. Otto F (2014) Multi-layer Potfit: An accurate potential representation for efficient high-dimensional quantum dynamics. J Chem Phys 140(1)
90. Kosloff R (1988) Time-dependent quantum-mechanical methods for molecular dynamics. J Phys Chem 92:2087–2100
91. Feit MD, Fleck JA Jr, Steiger A (1982) Solution of the Schrödinger equation by a spectral method. J Comp Phys 47:412–433
92. Wu YH, Batista VS (2003) Matching-pursuit for simulations of quantum processes. J Chem Phys 118:6720–6724
93. Koch W, Frankcombe TJ (2013) Basis expansion leaping: a new method to solve the time-dependent schrodinger equation for molecular quantum dynamics. Phys Rev Lett 110(26)
94. Maximilian AC (2015) Saller and Scott Habershon. Basis set generation for quantum dynamics simulations using simple trajectory-based methods. J Chem Theory Comput 11(1):8–16
95. Bing G, Garashchuk S (2016) Quantum dynamics with gaussian bases defined by the quantum trajectories. J Phys Chem A 120(19):3023–3031
96. Garashchuk S, Dell'Angelo D, Rassolov VA (2014) Classical limit of quantum nuclear dynamics based on selective use of the quantum potential. J Chem Phys 141:234107
97. Mandelshtam VA, Taylor HS (1997) Harmonic inversion of time signals and its applications. J Chem Phys 107:6756–6769
98. Mandelshtam VA, Taylor HS (1998) Harmonic inversion of time signals and its applications (vol 107, pg 6756, 1997). J Chem Phys 109:4128
99. Chen X, Wu YH, Batista VS (2005) Matching-pursuit/split-operator-Fourier-transform computations of thermal correlation functions. J Chem Phys 122(6)

100. Iyengar SS, Jakowski J (2005) Quantum wave packet ab initio molecular dynamics: An approach to study quantum dynamics in large systems. J Chem Phys 122(11)
101. Jakowski J, Sumner I, Iyengar SS (2006) Computational improvements to quantum wave packet ab initio molecular dynamics using a potential-adapted, time-dependent deterministic sampling technique. J Chem Theory Comput 2(5):1203–1219
102. Maximilian AC (2017) Saller and Scott Habershon. Quantum dynamics with short-time trajectories and minimal adaptive basis sets. J Chem Theory Comput 13(7):3085–3096
103. Shimshovitz A, Tannor DJ (2012) Phase-Space Approach to Solving the Time-Independent Schrodinger Equation. Phys Rev Lett 109(7)
104. Machnes S, Assémat E, Larsson HR, Tannor DJ (2016) Quantum dynamics in phase space using projected von neumann bases. J Phys Chem A 120(19):3296–3308
105. Poirier B (2003) Using wavelets to extend quantum dynamics calculations to ten or more degrees of freedom. J Theor Comput Chem 2(1):65–72
106. Poirier B, Salam A (2004) Quantum dynamics calculations using symmetrized, orthogonal Weyl-Heisenberg wavelets with a phase space truncation scheme. II. Construction and optimization. J Chem Phys 121(4):1690–1703
107. Halverson T, Poirier B (2012) Accurate quantum dynamics calculations using symmetrized Gaussians on a doubly dense Von Neumann lattice. J Chem Phys 137(22)
108. Halversona T, Poirier B (2015) Large scale exact quantum dynamics calculations: Ten thousand quantum states of acetonitrile. Chem Phys Lett 624:37–42
109. Halverson T, Poirier B (2015) One million quantum states of benzene. J Phys Chem A 119(50):12417–12433

Index

E. Perlt (ed.), *Basis Sets in Computational Chemistry*, Lecture Notes in Chemistry 107,
https://doi.org/10.1007/978-3-030-67262-1

The manufacturer's authorised representative in the EU is Springer Nature Customer Service Centre GmbH, Europaplatz 3, 69115 Heidelberg, Germany. If you have any concerns regarding our products, please contact ProductSafety@springernature.com

Printed and bound by CPI Group (UK) Ltd, Croydon, CR0 4YY
28/09/2026
02224078-0002